Spring+Spring MVC+MyBatis 整合框架开发技术（微课视频版）

■ 李雷孝　云　静　邢红梅　翟　娜　德世洋　主编

清华大学出版社
北　京

内容简介

本书全面讲解使用最新流行轻量级框架 SSM(Spring＋Spring MVC＋MyBatis)进行 JavaEE Web 开发的技术，并以综合案例“学生选课管理系统”为主线贯穿全书，知识内容层层推进，将知识点有机地串联起来，便于读者掌握与理解。

本书共 16 章，分 4 部分。第 1 部分是 MyBatis 篇，包括第 1～5 章，涵盖 MyBatis 基础、MyBatis 的核心 XML 配置文件等内容；第 2 部分是 Spring 篇，包括第 6～10 章，涵盖 Spring 基础、使用 Spring 管理 Bean 等；第 3 部分是 Spring MVC 篇，包括第 11～15 章，涵盖 Spring MVC 基础、Spring MVC 常用注解等内容；第 4 部分是 MyBatis ＋ Spring ＋ Spring MVC 整合篇，包括第 16 章，涵盖 MyBatis＋Spring＋Spring MVC 整合思路、基础环境搭建等内容。本书附有知识点案例源代码、综合案例源代码、数据库文件、教学大纲、教学 PPT、课后习题参考答案等配套资源。

本书可作为普通高校计算机科学与技术、软件工程、网络工程、物联网工程、数据科学与大数据技术、人工智能等相关专业的本科生教材，也可作为 SSM 框架技术学习者的参考书。

图书在版编目（CIP）数据

Spring＋Spring MVC＋MyBatis 整合框架开发技术 ：微课视频版 / 李雷孝等主编. -- 北京 ：清华大学出版社，2024. 9. --（清华开发者学堂）. -- ISBN 978-7-302-67283-8

Ⅰ. TP312.8

中国国家版本馆 CIP 数据核字第 20246W0N97 号

责任编辑： 张　玥
封面设计： 吴　刚
责任校对： 王勤勤
责任印制： 宋　林

出版发行： 清华大学出版社

网　　址： https://www.tup.com.cn，https://www.wqxuetang.com
地　　址： 北京清华大学学研大厦 A 座　　**邮　　编：** 100084
社 总 机： 010-83470000　　**邮　　购：** 010-62786544
投稿与读者服务： 010-62776969，c-service@tup.tsinghua.edu.cn
质量反馈： 010-62772015，zhiliang@tup.tsinghua.edu.cn
课件下载： https://www.tup.com.cn，010-83470236

印 装 者： 三河市君旺印务有限公司
经　　销： 全国新华书店
开　　本： 185mm×260mm　　**印　　张：** 26.25　　**字　　数：** 639 千字
版　　次： 2024 年 9 月第 1 版　　**印　　次：** 2024 年 9 月第1 次印刷
定　　价： 85.00 元

产品编号：094694-01

前言

在当今 IT 行业，Java 拥有世界上数量最多的程序员和岗位需求，95%以上的服务器端均是由 Java 开发的。SSM 框架是一种经典的 JavaEE Web 开发框架，它由 Spring、Spring MVC 和 MyBatis 三个开源项目整合而成。SSM 框架各个组件之间相互配合，提供了一套完整的解决方案，用于开发企业级的 Java Web 应用程序。作为一种成熟稳定的 Java Web 开发框架，SSM 框架在企业级应用系统、电子商务平台、社交网络应用系统、内容管理系统等实际项目中得到了广泛应用，具有高效性、灵活性、易用性、可靠性等特点。未来，它可以与微服务结合，提供更加灵活、可扩展的解决方案；集成新技术和新特性，以适应新的需求；持续进行性能优化和改进，减少不必要的开销、提升框架的稳定性和安全性，更好地满足项目的需求。本书从 SSM 框架开发的基础技术知识点入手，辅以知识点案例，以实际工程项目为主线，重点讲解 SSM 框架开发技术在综合项目案例开发中的应用。

本书是一本既培养学生软件开发技术，又培养学生工程实践能力的教材。教材以 IT 企业对开发人员的技术能力要求为基础，以工程能力培养为目标，梳理了软件工程对计算机软件开发技术要求的知识点，并形成相应的知识单元；按照工程需求顺序组织课程内容，便于学生学习和掌握；提供一定量的案例，注重实践能力的培养。本书可以作为计算机类专业各层次学生的教材，也可以作为 JavaEE Web 应用开发者的参考用书。

全书分为 4 部分，共 16 章，具体如下。

第 1 部分：MyBatis 篇。

第 1 章：MyBatis 基础，包括 MyBatis 概述、SqlSessionFactoryBuilder、SqlSessionFactory、SqlSession 和 SQL Mapper。

第 2 章：XML 文件，包括 MyBatis 配置文件、映射文件等内容。

第 3 章：关联查询，包括 MyBatis 实现一对一、一对多、多对多的处理与查询等内容。

第 4 章：动态 SQL，包括 MyBatis 动态组装功能、动态 SQL 语句拼接语法、常用元素等内容。

第 5 章：缓存和存储过程，包括 MyBatis 缓存机制、一级缓存和二级缓存的使用与原理、MyBatis 存储过程的调用等内容。

第 2 部分：Spring 篇。

第 6 章：Spring 基础，包括 Spring 框架介绍、Spring 框架体系结构、Spring 开发环境构建、Spring 核心容器、依赖注入与控制反转思想原理等内容。

第 7 章：使用 Spring 管理 Bean，包括 Spring Bean 的配置、实例化、装配方式、作用域以及生命周期等内容。

第 8 章：面向切面编程，包括面向切面思想、AOP 概念和术语、动态代理以及 AOP 的实现和 AspectJ 的开发等内容。

第 9 章：Spring 框架的数据库编程，包括 Spring JDBC 简介、Spring JDBC 配置并实现对数据库的操作等内容。

第 10 章：Spring 框架的数据库事务管理，包括数据库事务介绍、Spring 支持事务管理的核心接口、声明事务的实现等内容。

第 3 部分：Spring MVC 篇。

第 11 章：Spring MVC 基础，包括 Spring MVC 概念与基本组件、Spring MVC 工作流程与原理、Spring MVC 开发环境构建等内容。

第 12 章：常用注解，包括 Spring MVC 常用注解的介绍以及使用的内容。

第 13 章：Spring MVC 数据处理，包括数据转换处理、数据格式化处理、数据校验处理、域对象共享、视图跳转方式等内容。

第 14 章：Spring MVC 拦截器和异常处理，包括拦截器介绍、拦截器执行原理、Spring MVC 框架异常处理等内容。

第 15 章：Spring MVC 其他功能，包括 Spring MVC 框架实现文件上传与下载、Spring MVC 国际化实现等内容。

第 4 部分：MyBatis + Spring + Spring MVC 整合篇。

第 16 章：MyBatis + Spring + Spring MVC 框架整合，包括整合思路、开发环境和依赖处理、框架逐步搭建等内容。

本书具有以下特点：

(1) 遵照教指委计算机科学与技术和软件工程专业及相关专业的培养目标和培养方案，合理安排 SSM 框架开发技术知识体系，结合 JavaEE Web 开发技术方向的先行课程和后续课程组织相关知识点与内容。

(2) 注重理论和实践的结合，教材融入面向对象软件开发过程和工程实践背景的综合项目案例，使学生在掌握理论知识的同时提高分析问题和解决问题的能力，提高创新意识。

(3) 每个知识点都包括基础案例，每章都有一个综合案例，知识内容层层推进，便于学生接受。每章综合案例以“学生选课管理系统”为基础，以开发过程为主线，将知识点有机地串联在一起，便于学生掌握与理解。

(4) 每章的习题提供一定数量的课外实践题目，采用课内外结合的方式，培养学生对软件开发的兴趣，以满足当前社会对软件开发人员的需求。

（5）教材提供配套的课件、例题案例、章节案例和综合案例的源码。

本书由李雷孝、云静、邢红梅、翟娜和德世洋 5 位作者共同编写。其中，李雷孝编写了第 11～15 章，云静编写了 6～9 章，邢红梅编写了第 5、10 章和综合案例，翟娜编写了第 1～4 章，德世洋编写了第 16 章、各章习题、综合案例、电子资源，并统稿全书。在编写过程中，本书参阅了甲骨文公司、青岛软件园、上海杰普软件等公司的教学科研成果，也吸取了国内外教材的精髓，对这些作者的贡献表示由衷的感谢。本书在出版过程中，得到了刘利民教授、马志强教授和张世娥老师的支持和帮助；还得到了清华大学出版社张玥编辑的大力支持，在此表示诚挚的感谢。本书受到全国高等学校计算机教育研究会高等学校计算机教材建设项目、内蒙古自治区软件开发技术系列课程教学创新团队等项目资助。

由于编者水平有限，书中难免有不妥和疏漏之处，恳请各位专家、同仁和读者不吝赐教和批评指正。

编者

2024 年 3 月

目录

第1部分 MyBatis篇

第 10 章 Spring 框架的数据库事务管理 /250

第 3 部分 Spring MVC 篇

第 11 章 Spring MVC 基础 /273

第 4 部分 MyBatis+Spring+Spring MVC 整合篇

第1部分

MyBatis 篇

第1章 MyBatis基础

本章主要讲解 MyBatis 概述、SqlSessionFactoryBuilder、SqlSessionFactory、SqlSession 和 SQL Mapper。

1.1 MyBatis 概述

1.1.1 MyBatis 简介

MyBatis 本是 Apache 的一个开源项目 iBatis，2010 年迁移到了 Google Code，并改名为 MyBatis。

MyBatis 是基于 Java 的持久层框架，包括 SQL Maps 和 Data Access Objects（DAO），消除了几乎所有 JDBC 代码和参数的手工设置以及结果集的检索。MyBatis 使用简单的 XML 或注解，用于配置和原始映射，将接口和 Java 的 POJO 映射成数据库中的记录。

目前，Java 的持久层框架产品很多，常见的有 Hibernate 和 MyBatis。Hibernate 是一个全表映射的框架，只需要提供 POJO 和映射关系即可；而 MyBatis 是一个半自动映射的框架，需要手动匹配 POJO、SQL 和映射关系。Hibernate 是一个强大、高效、简洁、全自动化的持久层框架；MyBatis 是一个小巧、高效、半自动化的持久层框架。这两个持久层框架各有优缺点，开发者应根据实际需要选择。

1.1.2 MyBatis 的工作流程

要掌握好一门技术，需要理解其设计思想，即明白其工作机制，MyBatis 的工作流程如图 1-1 所示。

MyBatis 的工作流程如下。

首先，使用配置信息创建 SqlSessionFactory，调用 SqlSessionFactoryBuilder 的 build 方

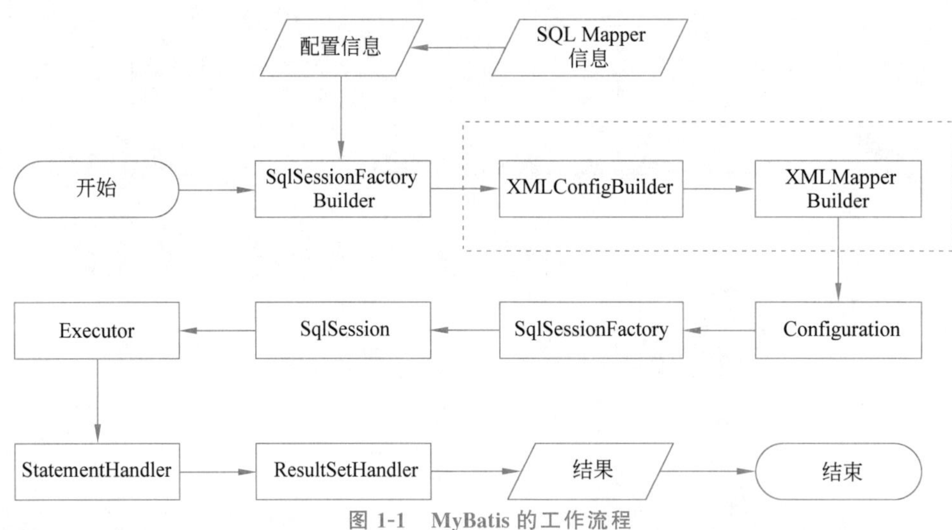

图 1-1　MyBatis 的工作流程

法，将配置信息封装为 Configuration 对象，传递给 build 方法。配置信息有两种形式：XML 文件或 Java 代码，对应的 build 方法参数也不同。如果是 XML 配置文件，则以输入流形式传递给 build 方法，并使用 XMLConfigBuilder 和 XMLMapperBuilder 来解析 XML 文件，并构建配置信息和 SQL Mapper 信息。

其次，使用 SqlSessionFactory 的 openSession 方法创建 SqlSession。在 openSession 方法中调用 openSessionFromDataSource 方法，在这个过程中创建 Executor（执行器）。然后，Executor 会找到适合的 StatementHandler 来创建 Statement 对象，并执行查询（类似 JDBC 操作）。最后，使用 ResultSetHandler 来处理结果集，并返回结果。

总体来说，MyBatis 的工作流程包括创建 SqlSessionFactory、创建 SqlSession、创建 Executor 执行器、创建 Statement 对象、执行查询和处理结果集。具体的实现类和接口会根据使用场景和配置信息的不同而变化。

1.2　SqlSessionFactoryBuilder

SqlSessionFactoryBuilder 是用于创建 SqlSessionFactory 的构造器。使用建造者（Builder）模式可以有效降低编程复杂度，让开发者专注于 SqlSessionFactory 的创建。一旦创建完 SqlSessionFactory，SqlSessionFactoryBuilder 就不再有用处，不宜一直保留，最好作为局部方法变量使用。下面介绍 SqlSessionFactoryBuilder 使用两种方式创建 SqlSessionFactory 的代码。

例 1-1：演示 XML 文件保存配置信息。共包含 3 个文件：mybatis-config.xml、db.properties（创建数据库连接的属性配置文件）和 CoursesMapper.xml。其中前两个文件位于 src 目录下。CoursesMapper.xml 文件内容详见 1.5.1 节。

db.properties 的代码如下。

```
jdbc.driver=com.mysql.cj.jdbc.Driver
```

```
jdbc.url=jdbc:mysql://localhost:3306/db_ssm?characterEncoding=
utf8&serverTimezone=GMT&useSSL=false&serverTimezone=Asia/Shanghai
jdbc.username=root
jdbc.password=123456
```

mybatis-config.xml 的代码如下。

```
<?xml version="1.0" encoding="UTF-8"?>
<!DOCTYPE configuration PUBLIC "-//mybatis.org//DTD Config 3.0//EN" "http://
mybatis.org/dtd/mybatis-3-config.dtd">
<configuration>
<properties resource="db.properties"/>
<typeAliases>
    <package name="com.imut.pojo"/>
</typeAliases>
<environments default="development">
    <environment id="development">
    <transactionManager type="JDBC" />
    <dataSource type="POOLED">
        <property name="driver" value="${jdbc.driver}" />
        <property name="url" value="${jdbc.url}" />
        <property name="username" value="${jdbc.username}" />
        <property name="password" value="${jdbc.password}" />
    </dataSource>
    </environment>
</environments>
<mappers>
    <mapper resource="com/imut/mapper/CoursesMapper.xml"/>
</mappers>
</configuration>
```

例 1-1 配置文件 mybatis-config.xml 包含了对其他属性文件的引用(通过<properties>元素)、默认别名的实体类包路径(通过<typeAliases>元素)、运行环境(通过<environments>元素)和映射文件路径(通过<mappers>元素)等信息。具体的属性和元素规范可以参考官方文档 https://mybatis.org/mybatis-3/zh/configuration.html。

在配置文件中引用外部属性文件时,需要使用 ${key}的形式来获取属性值,其中 key 与属性文件中的键一致。

通过以上配置,可以灵活地管理和配置 MyBatis 框架,满足不同的需求。

例 1-2:演示读取 XML 配置文件创建 SqlSessionFactory 的功能。程序为 MyBatisUtil.java,位于 src/com/imut/utils 目录下。

代码如下。

```
public SqlSessionFactory getSqlSessionFactory() throws Exception{
    String resource = "mybatis-config.xml";
    InputStream inputStream = Resources.getResourceAsStream(resource);
    SqlSessionFactory sqlSessionFactory = new SqlSessionFactoryBuilder().build
(inputStream);
```

```
    return sqlSessionFactory;
}
```

编写例 1-2 前,需要在项目中包含 MyBatis 的 jar 包,以便使用其提供的类或接口。

例 1-3:演示使用 Java 代码创建 SqlSessionFactory 的功能。程序为 MyBatisUtil1.java。Courses.java 的文件内容详见例 1-4,CoursesMapper.java 的文件内容详见 1.5.1 节。

代码如下。

```
public SqlSessionFactory getSqlSessionFactory()  throws Exception{
    PooledDataSource dataSource = new PooledDataSource();
    dataSource.setDriver("com.mysql.cj.jdbc.Driver");
    dataSource.setUrl("jdbc:mysql://localhost:3306/db_ssm?characterEncoding
=utf8&serverTimezone=GMT&useSSL=false&serverTimezone=Asia/Shanghai");
    dataSource.setUsername("root");
    dataSource.setPassword("123456");
    TransactionFactory transactionFactory = new JdbcTransactionFactory();
    Environment environment = new Environment("development", transactionFactory,
dataSource);
    Configuration configuration = new Configuration(environment);
    configuration.getTypeAliasRegistry().registerAlias("courses", Courses.class);
    configuration.addMapper(CoursesMapper.class);
    SqlSessionFactory ssf = new SqlSessionFactoryBuilder().build(configuration);
    return ssf;
}
```

本书以 XML 形式讲解为主,故后续均保留例 1-2 的方式。

1.3 SqlSessionFactory

SqlSessionFactory 接口是 MyBatis 的核心对象,可理解为数据库连接池,用于创建 SqlSession。采用工厂(Factory)模式能够灵活地支持创建不同的 SqlSession,多个 SqlSessionFactory 实例会导致资源浪费,甚至应用卡死。通常情况下只需要一个实例,可以使用单例模式将其作为一个共享的单例对象。具体的实现类可以参考图 1-2。

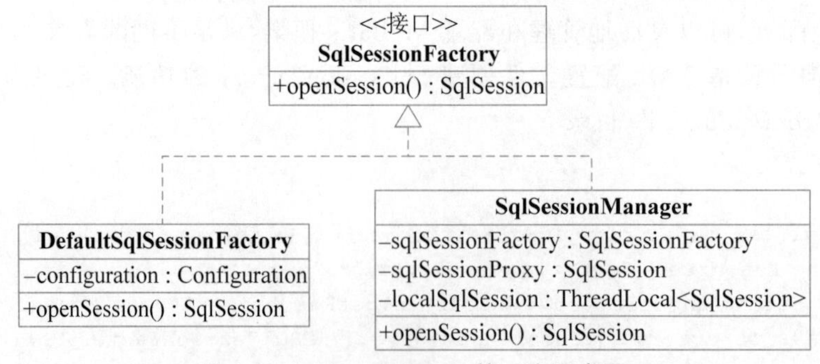

图 1-2 SqlSessionFactory 接口实现类图

在 1.1.2 节中介绍了通过 build 方法创建 SqlSessionFactory，其类型是 DefaultSqlSessionFactory。实际上，DefaultSqlSessionFactory 是 SqlSessionFactory 接口的一个实现类，也是大多数情况下常用的实现类。DefaultSqlSessionFactory 类中包含一个 Configuration 类型的属性，用于提供给 SqlSessionFactory 所需的配置信息。它重写了 openSession 方法，该方法会调用 openSessionFromDataSource 方法。在 openSessionFromDataSource 方法中，会根据配置信息获取 Environment 对象，然后再获取 TransactionFactory 对象。接下来，使用 TransactionFactory 对象创建 Transaction 对象，并根据传入的执行器类型参数创建 Executor 对象。最后，将 Configuration、Executor 对象和传入的自动提交标识符传递给 DefaultSqlSession 类的构造方法，进行成员变量的初始化，并将其作为 openSessionFromDataSource 方法的返回值，类型为 SqlSession。通过上述流程完成 SqlSession 的创建。

需要注意的是，SqlSession 的返回类型是 DefaultSqlSession，它们之间是一种关联关系。

1.4 节将会介绍 SqlSession 的相关内容。

1.4 SqlSession

SqlSession 接口用于执行持久化操作。它包含了对数据库的一系列操作，如获取连接、插入、删除、更新、查询、关闭连接、提交事务和回滚事务等。使用 Mapper 接口可以更加简洁、类型安全地进行增、删、改、查操作。SqlSession 负责与数据库交互，通过封装 JDBC 来发送 SQL，控制事务以及获取 Mapper 接口。每个连接都会占用连接池的资源，因此使用完毕后，应将其还给数据库连接池 SqlSessionFactory，以避免资源浪费。SqlSession 的生命周期通常与请求的生命周期相对应，即请求结束后其生命也结束。为了确保连接的关闭，可以使用 finally 块来完成关闭连接的任务。另外，SqlSession 实例不应该被共享，否则可能导致数据错误。SqlSession 有多个实现类，如图 1-3 所示。

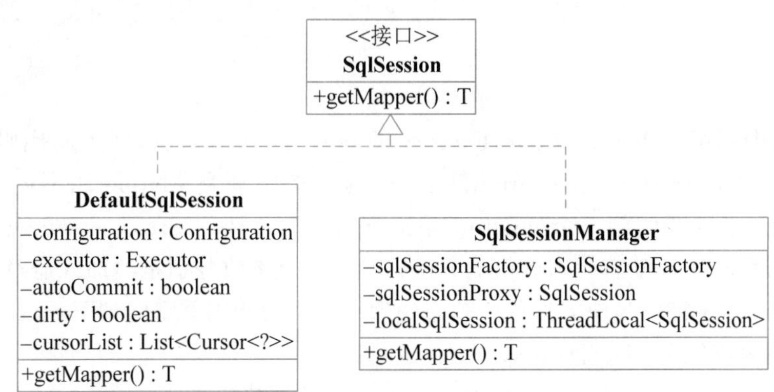

图 1-3　SqlSessionFactory 接口实现类图

DefaultSqlSession 是 SqlSession 的一个实现类，也是大多数场景下会使用的实现类。DefaultSqlSession 类包含多个属性，其中 Configuration 属性用于提供 SqlSession 所需的配置信息，Executor 属性负责执行实际的数据库操作，autoCommit 属性表示是否自动提交，

dirty 属性用于判断事务提交或回滚的状态。其中，重写的 getMapper 方法通过配置信息获取要使用的 Mapper，并通过 configuration 的 getMapper 方法调用 MapperRegistry 的 getMapper 方法来获取 Mapper 的实例。SqlSessionManager 使用了代理来确保线程安全的 SqlSession。

上述过程仅仅是找到了要使用的 Mapper 接口，要执行具体的 SQL 语句，还需要进行映射。1.5 节将介绍映射器的内容。

1.5 SQL Mapper

SQL Mapper 是 SQL 映射器，它使用基于 JDK 的动态代理来实现。1.4 节介绍了 MapperRegistry 的 getMapper 方法，该方法会获取并赋值给一个 MapperProxyFactory 类型的变量。从命名方式可以看出，MapperProxyFactory 是一个工厂类，它通过调用 newInstance 方法来生成传入 Mapper 接口的代理类。

MapperRegistry.java 中 getMapper 方法的源代码如下。

```
public <T> T getMapper(Class<T> type, SqlSession sqlSession) {
    MapperProxyFactory < T > mapperProxyFactory = (MapperProxyFactory) this.
knownMappers.get(type);
    if (mapperProxyFactory == null) {
        throw new BindingException ( " Type " + type + " is not known to the
MapperRegistry.");
    } else {
        try {
            return mapperProxyFactory.newInstance(sqlSession);
        } catch (Exception var5) {
            throw new BindingException("Error getting mapper instance. Cause: "
+ var5, var5);
        }
    }
}
```

以上代码中调用了 MapperProxyFactory 的 newInstance 方法，即下面代码中的第 2 个 newInstance 方法，用于创建 InvocationHandler 接口的实现类 MapperProxy。然后将该实现类对象作为参数传递给第一个 newInstance 方法。通过调用 Proxy.newProxyInstance 方法生成动态代理类(JDK 动态代理提供的方法)，其中 3 个参数分别是类加载器对象、接口实例和 InvocationHandler 接口实现类对象。动态代理将接口的调用重定向到 InvocationHandler 的 invoke 方法，此时指的是 MapperProxy 的 invoke 方法。

MapperProxyFactory.java 中 newInstance 方法的源代码如下。

```
public class MapperProxyFactory<T> {
    protected T newInstance(MapperProxy<T> mapperProxy) {
        return Proxy.newProxyInstance(this.mapperInterface.getClassLoader(),
new Class[]{this.mapperInterface}, mapperProxy);
```

```
    }

    public T newInstance(SqlSession sqlSession) {
        MapperProxy<T> mapperProxy = new MapperProxy(sqlSession, this.
mapperInterface, this.methodCache);
        return this.newInstance(mapperProxy);
    }
}
```

动态代理可以减少不必要的编码工作，只需编写 Mapper 接口就可以找到对应的 SQL 并执行(可以是 Mapper.xml 文件中定义的 SQL，也可以是注解中定义的 SQL)。下面介绍这两种映射方式。介绍这两种方式之前，需要先创建一个实体类 Courses，用于表示数据库中的课程信息表 courses，并存储相关数据。

例 1-4：演示 XML 映射器的使用方法，共包含 7 个文件：Courses.java(实体类)、CoursesMapper.java(Mapper 接口)、CoursesMapper.xml(XML 映射器)、mybatis-config.xml、db.properties、MyBatisUtil.java 和 TestMB(测试类)。创建实体类前，需要先创建数据库表 courses，并填充表中数据。

courses 表结构及数据如表 1-1 所示。

表 1-1　课程(courses)表结构及数据

cid(课程 ID，主键，非空)	cname(课程名称，默认为 NULL)	tid(授课教师 ID，默认为 NULL)
1	math	1
2	chinese	2
3	english	3
4	C++	2

Courses.java 代码如下，位于 src/com/imut/pojo 目录下。

```
package com.imut.pojo;
public class Courses {
private Integer cid;
private String cname;
private Integer tid;
//此处省略属性对应的 getter()、setter()和 toString()方法
}
```

1.5.1 XML

使用 XML 方式映射，一般需要满足以下 4 个条件：①Mapper 接口方法名与 Mapper.xml 中的 id 属性名称一致；②接口方法的返回类型与 Mapper.xml 中的返回类型一致；③接口方法的传入参数类型与 Mapper.xml 中传入的参数类型一致；④Mapper.xml 中<mapper>元素的 namespace(命名空间)属性的值与接口文件的全限定类名一致。

首先创建 Mapper 接口文件 CoursesMapper，CoursesMapper.java 的代码如下，位于

src/com/imut/mapper 目录下。

```
package com.imut.mapper;
import java.util.List;
import com.imut.pojo.Courses;
public interface CoursesMapper {
    int deleteByPrimaryKey(Integer cid);
    int insert(Courses record);
    Courses selectByPrimaryKey(Integer cid);
    List<Courses> selectAll();
    int updateByPrimaryKey(Courses record);
}
```

接着创建与 CoursesMapper.java 接口文件相对应的 XML 映射器 CoursesMapper.xml。CoursesMapper.xml 的代码如下，位于 src/com/imut/mapper 目录下。

```
<?xml version="1.0" encoding="UTF-8"?>
<!DOCTYPE mapper PUBLIC "-//mybatis.org//DTD Mapper 3.0//EN" "http://mybatis.
org/dtd/mybatis-3-mapper.dtd">
<mapper namespace="com.imut.mapper.CoursesMapper">
    <resultMap id="BaseResultMap" type="com.imut.pojo.Courses">
        <id column="cid" jdbcType="INTEGER" property="cid" />
        <result column="cname" jdbcType="VARCHAR" property="cname" />
        <result column="tid" jdbcType="INTEGER" property="tid" />
    </resultMap>
    <delete id="deleteByPrimaryKey" parameterType="java.lang.Integer">
        delete from courses where cid = #{cid,jdbcType=INTEGER}
    </delete>
    <insert id="insert" parameterType="com.imut.pojo.Courses">
        insert into courses (cid, cname, tid) values (#{cid,jdbcType=INTEGER},
#{cname,jdbcType=VARCHAR}, #{tid,jdbcType=INTEGER})
    </insert>
    <update id="updateByPrimaryKey" parameterType="com.imut.pojo.Courses">
        update courses set cname = #{cname, jdbcType=VARCHAR}, tid = #{tid,
jdbcType=INTEGER} where cid = #{cid,jdbcType=INTEGER}
    </update>
    <select id="selectByPrimaryKey" parameterType="java.lang.Integer"
resultMap="BaseResultMap">
        select cid, cname, tid from courses where cid = #{cid,jdbcType=INTEGER}
    </select>
    <select id="selectAll" resultMap="BaseResultMap">
        select cid, cname, tid from courses
    </select>
</mapper>
```

mybatis-config.xml 文件内容和 db.properties 文件内容详见例 1-1，MyBatisUtil.java 文件内容详见例 1-2。还需要一个测试类，用于测试 XML 映射器中 SQL 的执行情况。文件位于 src/com/imut/test 目录下，测试类使用 Junit 测试的方式，故在创建测试类前，需要先引入 Junit 相关的 jar 包。又因为需要连接 MySQL 数据库查询数据，还需要引入驱动 jar

包，以便使用它们提供的类或接口。TestMB.java 的代码如下，位于 src/com/imut/test 目录下。

```
package com.imut.test;
import com.imut.mapper.CoursesMapper;
import com.imut.pojo.Courses;
import com.imut.utils.MyBatisUtil;
import org.apache.ibatis.session.SqlSession;
import org.apache.ibatis.session.SqlSessionFactory;
import java.util.List;
import org.junit.Test;
public class TestMB {
    @Test
    public void testCoursesMapper() throws Exception {
        SqlSessionFactory ssf = new MyBatisUtil().getSqlSessionFactory();
        SqlSession session = ssf.openSession();
        try {
            CoursesMapper mapper = session.getMapper(CoursesMapper.class);
            Courses courses = mapper.selectByPrimaryKey(3);
            System.out.println("根据 cid(3)查询的课程信息:" + courses);
            courses = new Courses();
            courses.setCid(5);
            courses.setCname("C");
            courses.setTid(3);
            int rcode = mapper.insert(courses);
            System.out.println("插入影响的数据条数:" + rcode);
            List<Courses> list = mapper.selectAll();
            System.out.println("插入后,查询全部课程信息:" + list);
            courses.setCid(4);
            courses.setCname("C++");
            courses.setTid(1);
            rcode = mapper.updateByPrimaryKey(courses);
            System.out.println("更新影响的数据条数:" + rcode);
            list = mapper.selectAll();
            System.out.println("更新后,查询全部课程信息:" + list);
            mapper.deleteByPrimaryKey(2);
            list = mapper.selectAll();
            System.out.println("删除后,查询全部课程信息:" + list);
        } finally {
            session.close();
        }
    }
}
```

testCoursesMapper 是测试方法，测试了 CoursesMapper.java 中的所有方法。运行结果如图 1-4 所示。

从图 1-4 中可以看到测试结果为通过，并且正确地输出了每个方法的查询结果。

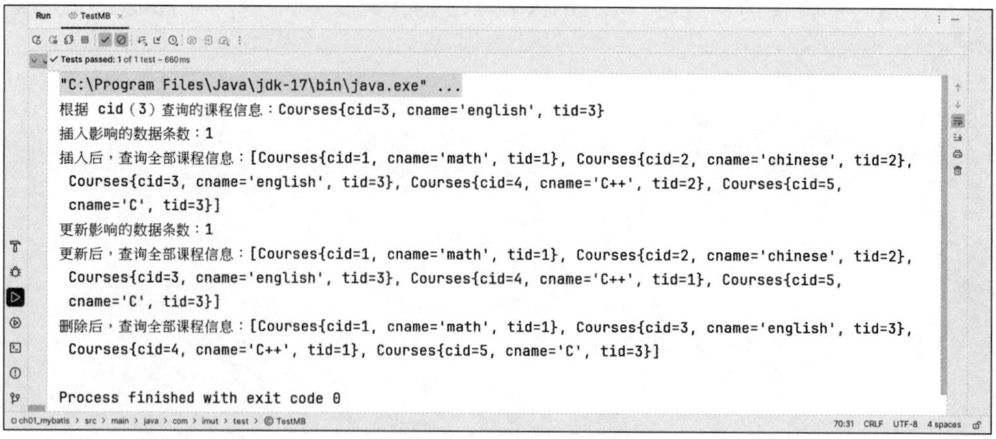

```
Run  TestMB ×
✓ Tests passed: 1 of 1 test – 660 ms
"C:\Program Files\Java\jdk-17\bin\java.exe" ...
根据 cid（3）查询的课程信息：Courses{cid=3, cname='english', tid=3}
插入影响的数据条数：1
插入后，查询全部课程信息：[Courses{cid=1, cname='math', tid=1}, Courses{cid=2, cname='chinese', tid=2},
 Courses{cid=3, cname='english', tid=3}, Courses{cid=4, cname='C++', tid=2}, Courses{cid=5,
 cname='C', tid=3}]
更新影响的数据条数：1
更新后，查询全部课程信息：[Courses{cid=1, cname='math', tid=1}, Courses{cid=2, cname='chinese', tid=2},
 Courses{cid=3, cname='english', tid=3}, Courses{cid=4, cname='C++', tid=1}, Courses{cid=5,
 cname='C', tid=3}]
删除后，查询全部课程信息：[Courses{cid=1, cname='math', tid=1}, Courses{cid=3, cname='english', tid=3},
 Courses{cid=4, cname='C++', tid=1}, Courses{cid=5, cname='C', tid=3}]

Process finished with exit code 0
ch01_mybatis > src > main > java > com > imut > test > TestMB        70:31  CRLF  UTF-8  4 spaces
```

图 1-4　XML 映射运行结果

1.5.2　注解

为了区分不同类型的接口，需要创建一个新的接口文件 CoursesMapper1.java。如果要使用注解方式，首先需要修改 mybatis-config.xml 文件中的<mapper>元素，这样映射器才能找到带有注解的 Mapper 接口。

例 1-5：演示使用注解映射器的方法。它包含 4 个文件：mybatis-config.xml、Courses.java（实体类，沿用例 1-4 中的类）、CoursesMapper1.java（Mapper 接口）、TestMB（测试类）。在 mybatis-config.xml 中，需要注册新的接口文件 CoursesMapper1。需要修改的 mybatis-config.xml 代码如下。

```
<mappers>
    <mapper class="com.imut.mapper.CoursesMapper1" />
</mappers>
```

接着编写 Mapper 接口文件。将例 1-4 的 CoursesMapper.java 复制并重命名为 CoursesMapper1.java，把注解添加到原有方法的上方，CoursesMapper1.java 的代码如下。

```
package com.imut.mapper;
import com.imut.pojo.Courses;
import org.apache.ibatis.annotations.Delete;
import org.apache.ibatis.annotations.Insert;
import org.apache.ibatis.annotations.Select;
import org.apache.ibatis.annotations.Update;
import java.util.List;
public interface CoursesMapper1 {
    @Delete("delete from courses where cid = #{cid,jdbcType=INTEGER}")
    int deleteByPrimaryKey(Integer cid);
    @Insert ("insert into courses (cid, cname, tid) values (#{cid,jdbcType=
INTEGER}, #{cname,jdbcType=VARCHAR}, #{tid,jdbcType=INTEGER})")
    int insert(Courses record);
    @Select("select cid, cname, tid from courses where cid = #{cid,jdbcType=
INTEGER}")
```

```
    Courses selectByPrimaryKey(Integer cid);
    @Select("select cid, cname, tid from courses")
    List<Courses> selectAll();
    @Update("update courses set cname = #{cname,jdbcType=VARCHAR}, tid = #{tid,
jdbcType=INTEGER} where cid = #{cid,jdbcType=INTEGER}")
    int updateByPrimaryKey(Courses record);
}
```

最后编写测试方法。将例 1-4 的 TestMB.java 中的 testCoursesMapper()方法复制并重命名为 testAnnotationCoursesMapper(),再将获取 CoursesMapper 接口那一行的代码修改为获取 CoursesMapper1 接口(该行已加粗)。TestMB.java 的代码如下。

```
package com.imut.test;
import com.imut.mapper.CoursesMapper;
import com.imut.mapper.CoursesMapper1;
import com.imut.pojo.Courses;
import com.imut.utils.MyBatisUtil;
import org.apache.ibatis.session.SqlSession;
import org.apache.ibatis.session.SqlSessionFactory;
import java.util.List;
import org.junit.Test;
public class TestMB {
    @Test
    public void testAnnotationCoursesMapper() throws Exception {
        SqlSessionFactory ssf = new MyBatisUtil().getSqlSessionFactory();
        SqlSession session = ssf.openSession();
        try {
            CoursesMapper1 mapper = session.getMapper(CoursesMapper1.class);
            Courses courses = mapper.selectByPrimaryKey(3);
            System.out.println("根据 cid(3)查询的课程信息:" + courses);
            courses = new Courses();
            courses.setCid(5);
            courses.setCname("C");
            courses.setTid(3);
            int rcode = mapper.insert(courses);
            System.out.println("插入影响的数据条数:" + rcode);
            List<Courses> list = mapper.selectAll();
            System.out.println("插入后,查询全部课程信息:" + list);
            courses.setCid(4);
            courses.setCname("C++");
            courses.setTid(1);
            rcode = mapper.updateByPrimaryKey(courses);
            System.out.println("更新影响的数据条数:" + rcode);
            list = mapper.selectAll();
            System.out.println("更新后,查询全部课程信息:" + list);
            mapper.deleteByPrimaryKey(2);
```

```
                list = mapper.selectAll();
                System.out.println("删除后,查询全部课程信息:" + list);
            } finally {
                session.close();
            }
        }
    }
```

testAnnotationCoursesMapper 是测试方法,测试了 CoursesMapper1.java 中的所有方法。运行结果如图 1-5 所示。

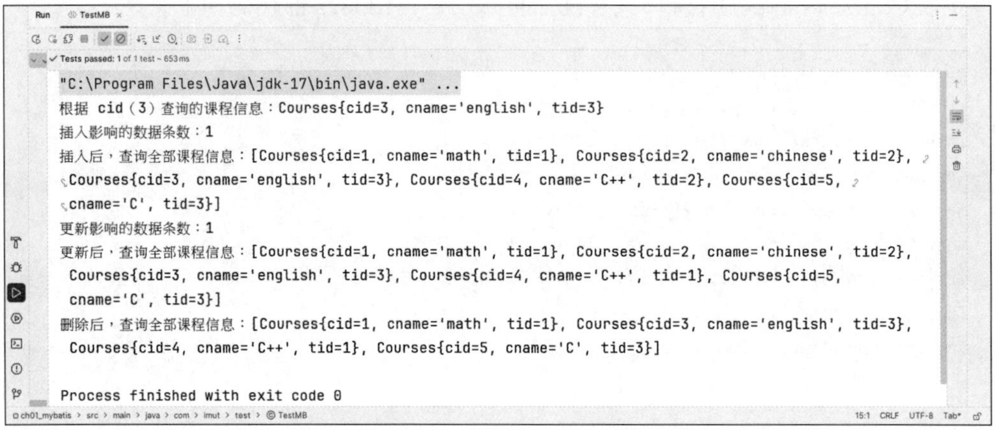

```
Run    TestMB
Tests passed: 1 of 1 test – 653 ms
"C:\Program Files\Java\jdk-17\bin\java.exe" ...
根据 cid（3）查询的课程信息：Courses{cid=3, cname='english', tid=3}
插入影响的数据条数：1
插入后，查询全部课程信息：[Courses{cid=1, cname='math', tid=1}, Courses{cid=2, cname='chinese', tid=2},
Courses{cid=3, cname='english', tid=3}, Courses{cid=4, cname='C++', tid=2}, Courses{cid=5,
cname='C', tid=3}]
更新影响的数据条数：1
更新后，查询全部课程信息：[Courses{cid=1, cname='math', tid=1}, Courses{cid=2, cname='chinese', tid=2},
 Courses{cid=3, cname='english', tid=3}, Courses{cid=4, cname='C++', tid=1}, Courses{cid=5,
 cname='C', tid=3}]
删除后，查询全部课程信息：[Courses{cid=1, cname='math', tid=1}, Courses{cid=3, cname='english', tid=3},
 Courses{cid=4, cname='C++', tid=1}, Courses{cid=5, cname='C', tid=3}]

Process finished with exit code 0
ch01_mybatis > src > main > java > com > imut > test > TestMB          15:1  CRLF  UTF-8  Tab*
```

图 1-5　注解映射运行结果

从图 1-5 中可以看到测试结果为通过,并且正确地输出了每个方法的查询结果。

从上述两种实现方式的代码可以看出,使用注解方式比 XML 方式的代码更少。尽管如此,MyBatis 官方更倾向使用 XML 方式,一是复杂的语句使用注解会使得 Java 文件中的代码看起来更混乱;二是注解会破坏 Java 代码的优雅性。因此,选择使用哪种映射方式取决于实际的应用场景,如果只涉及非常简单的 SQL 操作,可以考虑使用注解方式;如果涉及较复杂的 SQL 操作,官方建议使用 XML 方式。

1.6　综合案例

1.6.1　开发环境

本系统使用的开发环境如下。

技术要求:Spring 5.2.5＋Spring MVC＋MyBatis 3.5.6;

Java 版本:JDK 17;

服务器:Tomcat 8.5;

数据库:MySQL 8.0;

开发工具:IDEA2023。

1.6.2 案例设计

“学生选课管理系统”是全书的综合案例，包含各章节知识点和内容，并结合实际应用，将每章技术与系统的功能实现相结合，逐步实现整个系统。为了符合书中知识点的组织结构，我们对原系统进行了裁剪，每个章节实现一个功能或模块。设计综合案例的总体思路是：功能简单易懂，并尽可能涵盖大部分知识点。

1. 系统功能分析与设计

学生选课管理系统主要包括用户登录、学生信息管理、课程信息管理以及选课管理4大模块。功能模块如图1-6所示。

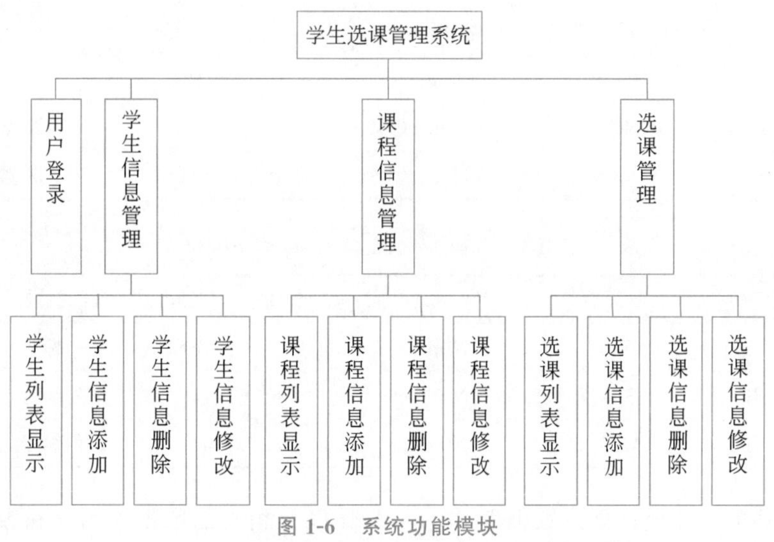

图1-6 系统功能模块

学生选课管理系统各模块功能具体描述如下。

学生信息管理包括列出所有学生信息，增加学生信息，修改学生信息以及删除学生信息。

课程信息管理包括列出所有课程信息，增加课程信息，修改课程信息以及删除课程信息。

选课管理包括列出所有选课信息，增加选课信息，修改选课信息以及删除选课信息。

2. 数据库系统分析与设计

通过分析系统功能模块，系统所要管理的数据实体有用户、学生、课程。根据系统功能模块分析，设计系统数据库表，如表1-2～表1-5所示。

表1-2 用户表(user)逻辑结构设计

序号	中文名称	字段名称	数据类型	外键	主键	说明
1	ID	ID	Number		PK	唯一标识
2	用户名	USERNAME	Varchar(20)			登录的用户名
3	密码	PASSWORD	Varchar(20)			登录的密码

表 1-3　学生表(student)逻辑结构设计

序号	中文名称	字段名称	数据类型	外键	主键	说明
1	SID	SID	Number		PK	唯一标识
2	学号	SNO	Varchar(15)			学生学号
3	姓名	SNAME	Varchar(30)			学生姓名
4	性别	SGENDER	Varchar(10)			学生性别
5	年龄	SAGE	Number			学生年龄

表 1-4　课程表(course)逻辑结构设计

序号	中文名称	字段名称	数据类型	外键	主键	说明
1	CID	CID	Number		PK	唯一标识
2	课程编码	CNO	Varchar(15)			课程编码
3	课程名称	CNAME	Varchar(30)			课程名称

表 1-5　选课表(ccourse)逻辑结构设计

序号	中文名称	字段名称	数据类型	外键	主键	说明
1	ID	ID	Number		PK	唯一标识
2	学生 ID	SID	Number	FK		学生编号,关联学生表
3	课程 ID	CID	Number	FK		课程编号,关联课程表

根据上述数据库设计创建数据库 db_student 以及相关的数据库用户和数据库表。创建数据库的 SQL 语句参见随书电子资源。

根据对系统的分析和数据库设计画出系统的类图,如图 1-7 所示。

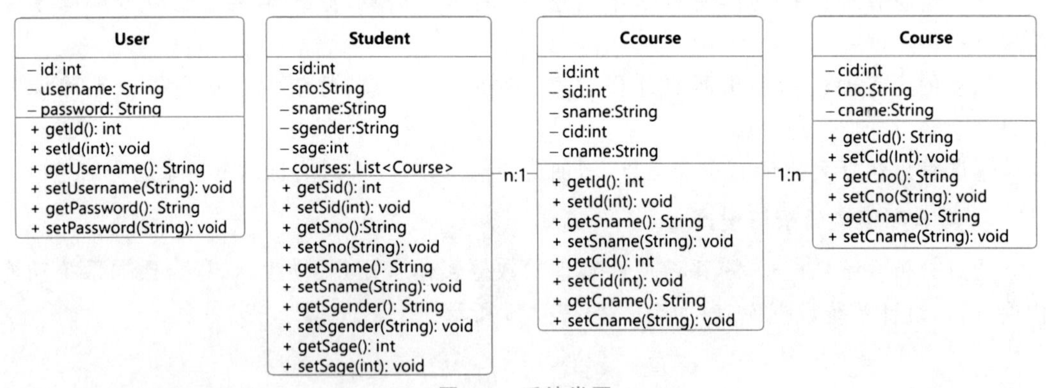

图 1-7　系统类图

下面描述本章案例的实现过程。

本章案例主要实现“学生信息管理”模块中根据学生 SID 查询学生信息的功能。案例使用的主要文件如表 1-6 所示。

表 1-6 本章案例使用的文件

文　　件	所在包/路径	功　　能
log4j.properties	src/	日志输出配置文件
db.properties	src/	数据库连接的属性文件
mybatis-config.xml	src/	MyBatis 配置文件
Student.java	com.imut.pojo	封装学生信息的类
StudentMapper.java	com.imut.mapper	定义操作学生表方法的接口
StudentMapper.xml	com.imut.mapper	Student 类的映射文件
StudentTest.java	com.imut.test	测试操作学生信息方法的类

1.6.3 案例演示

使用 Junit4 测试执行 StudentTest.java 中的 getStudentByIdTest()方法，控制台输出的结果如图 1-8 所示。

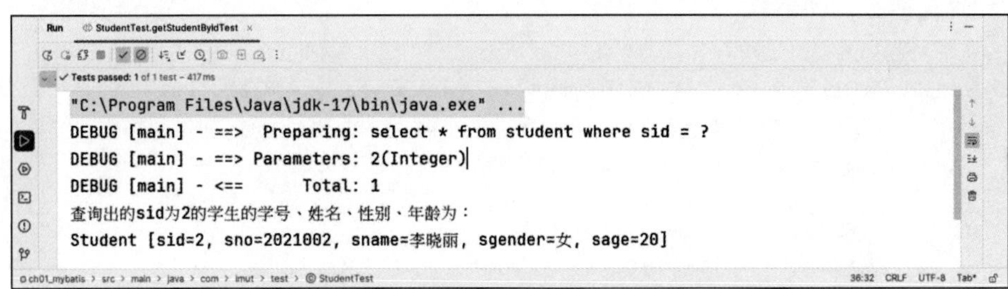

图 1-8 查询 sid 为 2 的学生信息的运行结果

1.6.4 代码实现

1. 创建数据库表

在 MySQL 数据库中创建名为 db_student 的数据库。在该数据库下创建 student 表，并插入数据。student 表的具体信息如表 1-3 所示。

2. 创建项目，添加依赖

在 IDEA 中创建名为 ch01_mybatis 的 Maven 工程。在 pom.xml 文件中添加 junit、log4j、mybatis、mysql-connector-java 4 个依赖，代码如下。

```
<dependencies>
    <dependency>
        <groupId>junit</groupId>
        <artifactId>junit</artifactId>
        <version>4.12</version>
    </dependency>
    <dependency>
        <groupId>log4j</groupId>
```

```
        <artifactId>log4j</artifactId>
        <version>1.2.17</version>
    </dependency>
    <dependency>
        <groupId>org.mybatis</groupId>
        <artifactId>mybatis</artifactId>
        <version>3.5.6</version>
    </dependency>
    <dependency>
        <groupId>mysql</groupId>
        <artifactId>mysql-connector-java</artifactId>
        <version>8.0.11</version>
    </dependency>
</dependencies>
```

在 pom.xml 文件添加设置编译打包/src/main/java 和/src/main/resource 下所有文件的方式。代码如下。

```
<build>
    <resources>
        <resource>
            <directory>src/main/java</directory><!--所在的目录-->
            <includes><!--包括目录下的.properties,.xml 文件都会扫描到-->
                <include>**/*.properties</include>
                <include>**/*.xml</include>
            </includes>
            <filtering>false</filtering>
        </resource>
        <resource>
            <directory>src/main/resources</directory><!--所在的目录-->
            <includes><!--包括目录下的.properties,.xml 文件都会扫描到-->
                <include>**/*.properties</include>
                <include>**/*.xml</include>
            </includes>
            <filtering>false</filtering>
        </resource>
    </resources>
</build>
```

3. 创建配置文件

在/src/main/resources 下创建数据库连接属性配置文件 db.properties，代码如下。

```
jdbc.djdbc.driver=com.mysql.cj.jdbc.Driver
jdbc.url=jdbc:mysql://localhost:3306/db_student?characterEncoding=
utf8&serverTimezone=GMT&useSSL=false
jdbc.username=root
jdbc.password=123456
```

在/src/main/resources 下创建日志输出配置文件 log4j.properties，代码如下。

```
# Global logging configuration
log4j.rootLogger=ERROR, stdout
# MyBatis logging configuration...
log4j.logger.com.imut=DEBUG
# Console output...
log4j.appender.stdout=org.apache.log4j.ConsoleAppender
log4j.appender.stdout.layout=org.apache.log4j.PatternLayout
log4j.appender.stdout.layout.ConversionPattern=%5p [%t] - %m%n
```

在/src/main/resources 下创建 mybatis 配置文件 mybatis-config.xml，代码如下。

```
<?xml version="1.0" encoding="UTF-8" ?>
<!DOCTYPE configuration PUBLIC "-//mybatis.org//DTD Config 3.0//EN"
"http://mybatis.org/dtd/mybatis-3-config.dtd">
<configuration>
    <!-- 引入数据源的配置文件 db.properties -->
    <properties resource="db.properties"/>
    <!-- 配置环境 -->
    <environments default="development">
        <environment id="development">
            <transactionManager type="JDBC" />
            <!-- 配置连接池 -->
            <dataSource type="POOLED">
                <!-- 使用${}取出配置文件 db.properties 中的值 -->
                <property name="driver" value="${jdbc.driver}" />
                <property name="url" value="${jdbc.url}" />
                <property name="username" value="${jdbc.username}" />
                <property name="password" value="${jdbc.password}" />
            </dataSource>
        </environment>
    </environments>
    <!-- 配置映射文件的位置 -->
    <mappers>
        <!-- resource:从类路径下查找资源 -->
        <mapper resource="com/imut/mapper/StudentMapper.xml" />
    </mappers>
</configuration>
```

4. 创建实体类

在/src/main/java 下创建包 com.imut.pojo，在 com.imut.pojo 包下新建与 student 表相对应的实体类 Student.java，Student.java 的代码如下。

```
package com.imut.pojo;
import java.io.Serializable;
public class Student implements Serializable{
    private int sid;
    private String sno;
    private String sname;
    private String sgender;
```

```
    private int sage;
    //此处省略 getXxx()、setXxx()和 toString()方法
}
```

5. 创建映射文件

在 com.imut 下新建 mapper 包，在 com.imut.mapper 创建接口文件 StudentMapper.java，StudentMapper.java 的代码如下。

```
package com.imut.mapper;
import com.imut.pojo.Student;
public interface StudentMapper {
    //根据学生 SID 查询学生信息
    public Student getStudentById(int sid);
}
```

在 com.imut.mapper 目录创建映射器文件 StudentMapper.xml，StudentMapper.xml 代码如下。

```
<?xml version="1.0" encoding="UTF-8" ?>
<!DOCTYPE mapper PUBLIC "-//mybatis.org//DTD Mapper 3.0//EN"
"http://mybatis.org/dtd/mybatis-3-mapper.dtd">
<mapper namespace="com.imut.mapper.StudentMapper">
    <!-- 根据学生 SID 查询学生记录-->
    <select id="getStudentById" resultType="com.imut.pojo.Student">
        select * from student where sid = #{sid}
    </select>
</mapper>
```

6. 创建测试类

在 com.imut 下新建 test 包，在 com.imut.test 下创建测试类 StudentTest.java，StudentTest.java 的代码如下。

```
package com.imut.test;
import java.io.InputStream;
import org.apache.ibatis.io.Resources;
import org.apache.ibatis.session.SqlSession;
import org.apache.ibatis.session.SqlSessionFactory;
import org.apache.ibatis.session.SqlSessionFactoryBuilder;
import org.junit.Test;
import com.imut.mapper.StudentMapper;
import com.imut.pojo.Student;
public class StudentTest {
    @Test
    public void getStudentByIdTest() {
        String resource = "mybatis-config.xml";
        SqlSession sqlSession=null;
        try {
```

```
            //1.读取配置文件 mybatis-config.xml
            InputStream inputStream = Resources.getResourceAsStream(resource);
            //2.根据配置文件 mybatis-config.xml 构建 SqlSessionFactory 对象
            SqlSessionFactory sqlSessionFactory = new SqlSessionFactoryBuilder
().build(inputStream);
            //3.通过 SqlSessionFactory 对象创建 SqlSession 对象
            sqlSession = sqlSessionFactory.openSession();
            //4.使用 SqlSession 操作数据库,先获取接口的实现类
            StudentMapper mapper = sqlSession.getMapper(StudentMapper.class);
            Student student = mapper.getStudentById(2);
            System.out.println("查询出的 SID 为 2 的学生的学号、姓名、性别、年龄为:");
            System.out.println(student);
        }catch(Exception e) {
            e.printStackTrace();
        }finally {
            sqlSession.close();
        }
    }
}
```

1.7 习题

1. 单选题

(1) 以下关于 MyBatis 核心对象的说法,错误的是(   )。

A. SqlSessionFactoryBuilder 实例的作用是创建 SqlSession

B. SqlSessionFactory 实例可以理解为数据库连接池

C. SqlSession 实例可以理解为数据库连接

D. SqlSessionFactory 实例的作用是创建 SqlSession

(2) 以下关于 MyBatis 核心对象生命周期的说法,错误的是(   )。

A. SqlSessionFactoryBuilder 实例的最佳作用域范围是方法,也就是定义为本地方法变量即可

B. SqlSessionFactory 实例的生命周期应该在整个应用的执行期间都存在

C. SqlSession 实例是线程不安全的,因此其生命周期范围应该是请求或方法

D. SqlSession 实例通常定义为一个类的静态变量

2. 填空题

(1) SQL 映射器的两种实现方式是________和________。

(2) XML 形式的 SQL 映射器由________和________组成。

3. 简答题

(1) 什么是 MyBatis?

(2) MyBatis 与 Hibernate 有哪些不同?

(3) MyBatis 的三大核心对象是什么,作用分别是什么?

第2章 XML文件

本章主要讲解两种XML文件：配置文件（mybatis-config.xml）和映射文件（SQL Mapper）的常用元素。

2.1 配置文件

在配置文件中，各元素有固定的位置顺序，如果元素的顺序不正确，将会导致错误。

2.1.1 properties元素

properties元素用于添加属性信息，有两种方式可以使用。第一种方式是直接在properties元素下使用property子元素进行属性的定义；第二种方式是引入外部属性文件。引入外部属性文件可以避免配置文件内容过多的问题，也可以在多个项目中共享属性文件，减少修改属性值的工作量，实现一处修改，处处生效。

例2-1：演示在XML文件properties元素下使用property子元素列出属性配置信息的方式。程序为mybatis-config.xml。

```
<?xml version="1.0" encoding="UTF-8"?>
<!DOCTYPE configuration PUBLIC "-//mybatis.org//DTD Config 3.0//EN" "http://
mybatis.org/dtd/mybatis-3-config.dtd">
<configuration>
    <properties>
        <property name="jdbc.driver" value="com.mysql.cj.jdbc.Driver">
</property>
        <property name="jdbc.url" value="jdbc:mysql://localhost:3306/db_ssm?
characterEncoding=utf8&serverTimezone=GMT&useSSL=false&serverTimezone=Asia/
Shanghai"></property>
```

```
        <property name="jdbc.username" value="root"></property>
        <property name="jdbc.password" value="123456"></property>
    </properties>
    <environments default="development">
        <environment id="development">
            <transactionManager type="JDBC" />
            <dataSource type="POOLED">
                <property name="driver" value="${jdbc.driver}" />
                <property name="url" value="${jdbc.url}" />
                <property name="username" value="${jdbc.username}" />
                <property name="password" value="${jdbc.password}" />
            </dataSource>
        </environment>
    </environments>
    <mappers>
        <mapper resource="com/imut/mapper/CoursesMapper.xml" />
    </mappers>
</configuration>
```

以上代码中的 driver、url、username、password 属性值将会被 properties 元素下对应的 property 子元素中设置的 values 属性值替代。

为了保持 XML 文件的简洁，可以将 property 子元素中的属性配置放在一个独立的文件中，并通过 properties 元素的 resource 属性引用。

例 2-2：演示在 XML 文件中使用 properties 元素的 resource 属性引用属性配置文件的方式。程序为 mybatis-config.xml。

1.2 节的例 1-1 已列出了 properties 元素在 mybatis-config.xml 中的一种代码引用方式，此处不再列举。

properties 元素除了 resource 属性，还有 url 属性，该属性值的书写标准参照统一资源定位符(Uniform Resource Locator，URL)。

例 2-3：演示在 XML 文件中使用 properties 元素的 url 属性引用配置文件的方式。程序为 mybatis-config.xml，请根据项目实际情况修改 url 路径。

```
<properties url="file:///E:/ch02/src/db.properties"/>
```

两种使用 properties 元素配置属性的方式都比较简洁，开发者可以根据自己的需求选择。但本书推荐使用 resource 属性的方式，因为对于文件系统的依赖性低，受文件系统影响小。

在实际应用中，当开发人员结合多种方式使用 properties 元素时，可能会引发同一个属性在不同地方配置且具有不同值时哪个值会生效的问题。规则如下：通过方法参数传递的属性具有最高优先级；其次是 properties 元素的 resource 或 url 属性指定的配置文件；最低优先级是 properties 元素下的 property 子元素指定的属性。

2.1.2 settings 元素

settings 元素用于配置应用程序运行时的行为。在 XML 配置文件中，它不是必需的元素，通常可以不写，使用默认值即可。settings 元素下只有一个子元素 setting，该元素具有两个属性：name 和 value，setting 子元素的常用配置项如表 2-1 所示。

表 2-1 setting 子元素属性配置信息表

设置名(name)	描述	有效值(value)	默认值
cacheEnabled	全局性地开启或关闭所有映射器配置文件中已配置的任何缓存	true | false	true
lazyLoadingEnabled	延迟加载的全局开关。开启时，所有关联对象都会延迟加载。特定关联关系中可通过设置 fetchType 属性来覆盖该项的开关状态	true | false	false
aggressiveLazyLoading	开启时，任一方法的调用都会加载该对象的所有延迟加载属性。否则，每个延迟加载属性会按需加载（参考 lazyLoadTriggerMethods）	true | false	false
useGeneratedKeys	允许 JDBC 支持自动生成主键，需要数据库驱动支持。如果设置为 true，将强制使用自动生成主键。尽管一些数据库驱动不支持此特性，但仍可正常工作（如 Derby）	true | false	false
autoMappingBehavior	指定 MyBatis 应如何自动映射列到字段或属性。NONE 表示关闭自动映射；PARTIAL 只会自动映射没有定义嵌套结果映射的字段。FULL 会自动映射任何复杂的结果集(无论是否嵌套)	NONE | PARTIAL | FULL	PARTIAL
defaultExecutorType	配置默认的执行器。SIMPLE 就是普通的执行器；REUSE 执行器会重用预处理语句（PreparedStatement）；BATCH 执行器不仅重用语句还会执行批量更新	SIMPLE | REUSE | BATCH	SIMPLE
mapUnderscoreToCamelCase	是否开启驼峰命名自动映射，即从经典数据库列名 A_COLUMN 映射到经典 Java 属性名 aColumn	true | false	false
logPrefix	指定 MyBatis 增加到日志名称的前缀	任何字符串	
logImpl	指定 MyBatis 所用日志的具体实现方式，未指定时将自动查找	SLF4J | LOG4J | LOG4J2 | JDK_LOGGING | COMMONS_LOGGING | STDOUT_LOGGING | NO_LOGGING	

表 2-1 所列配置项并非 MyBatis 的全部配置信息，随着版本的更新，配置项也会随之变化。

例 2-4：演示在 XML 文件中使用 settings 元素的子元素 setting 设置配置项的方式。

程序为 mybatis-config.xml。

```
<settings>
    <setting name="cacheEnabled" value="true"/>
    <setting name="lazyLoadingEnabled" value="true"/>
    <setting name="multipleResultSetsEnabled" value="true"/>
    <setting name="useColumnLabel" value="true"/>
    <setting name="useGeneratedKeys" value="false"/>
    <setting name="autoMappingBehavior" value="PARTIAL"/>
    <setting name="autoMappingUnknownColumnBehavior" value="WARNING"/>
    <setting name="defaultExecutorType" value="SIMPLE"/>
    <setting name="defaultStatementTimeout" value="25"/>
    <setting name="defaultFetchSize" value="100"/>
    <setting name="safeRowBoundsEnabled" value="false"/>
    <setting name="mapUnderscoreToCamelCase" value="false"/>
    <setting name="localCacheScope" value="SESSION"/>
    <setting name="jdbcTypeForNull" value="OTHER"/>
    <setting name="lazyLoadTriggerMethods" value="equals,clone,hashCode,
toString"/>
</settings>
```

上述示例中配置的 lazyLoadingEnabled、autoMappingUnknownColumnBehavior、defaultStatementTimeout 和 defaultFetchSize 是有意义的，而其他配置项可以选择性地配置，因为它们的 value 属性值与默认值相同。

2.1.3 typeAliases 元素

typeAliases 元素用于给类起别名，从而减少全限定类名（包名＋类名）导致的书写冗余。MyBatis 依旧提供 XML 和注解两种方式。且在使用 XML 提供的 typeAliases 元素的时候依然会有两种方式，其中一种是使用 typeAlias 子元素。

例 2-5：演示在 XML 文件中使用 typeAliases 元素的子元素 typeAlias 为单个类指定别名的方式。程序为 mybatis-config.xml。

```
<typeAliases>
    <typeAlias type="com.imut.pojo.Courses" alias="Courses"/>
</typeAliases>
```

在 typeAlias 的子元素中，type 属性值为类的全限定类名，alias 属性值为别名。

另外一种方式即在 typeAliases 元素下使用 package 元素。将类的包名赋值给 name 属性，则该包下的所有类都会以相同的方式获得别名。通过这种方式，可以一次性告知 MyBatis 类的位置，实现批量操作。

例 2-6：演示在 XML 文件中使用 typeAliases 元素的子元素 package 为多个类定义别名的方式。程序为 mybatis-config.xml。

```
<typeAliases>
    <package name="com.imut.pojo"/>
</typeAliases>
```

根据配置的包名(类的路径)扫描包下的所有类,并为类设置别名。

例 2-7:演示使用注解为类指定别名的方式。程序为 Courses.java。

```
package com.imut.pojo;
import org.apache.ibatis.type.Alias;
@Alias("Course")
public class Courses {
    …
}
```

别名也可以在 Mapper.xml 中使用,通过 resultType 属性指定返回类型的别名。

例 2-8:演示别名的使用方式,以注解名为 Course 为例。程序为 CoursesMapper.xml。

```
<select id="selectByPrimaryKey" parameterType="java.lang.Integer" resultType
="Course">
    select cid, cname, tid from courses where cid = #{cid,jdbcType=INTEGER}
</select>
```

2.1.4 environments 元素

environments 元素用于配置数据库环境,支持连接多种数据库,以满足不同的业务场景。在 environments 元素中可以定义多个 environment 子元素,每个子元素代表一个数据源,通过设置 default 属性可以指定项目使用的数据源。设置 environment 元素的 id 属性,即可指定数据源名称。

例 2-9:演示多环境下项目使用单一数据源的方式。程序为 mybatis-config.xml。

```
<environments default="development">
    <environment id="development">
        <transactionManager type="JDBC" />
        <dataSource type="POOLED">
            <property name="driver" value="${jdbc.driver}" />
            <property name="url" value="${jdbc.url}" />
            <property name="username" value="${jdbc.username}" />
            <property name="password" value="${jdbc.password}" />
        </dataSource>
    </environment>
    <environment id="test">
        <transactionManager type="MANAGED" />
        <dataSource type="UNPOOLED">
            …
        </dataSource>
    </environment>
    <environment id="production">
        <transactionManager type="JDBC" />
        <dataSource type="JNDI">
            …
        </dataSource>
    </environment>
</environments>
```

例 2-9 共定义了 3 个数据源，分别代表了开发(development)、测试(test)和生产(production) 3 种环境。其中，数据源选择的是开发环境。每个数据源的配置包含了事务管理器和数据源连接信息。事务管理器的类型可以是 JDBC 或 MANAGED，而数据源的连接形式可以是连接池(POOLED)、非连接池(UNPOOLED)或 JNDI。在 dataSource 元素中，可以设置多个 property 子元素，用于指定连接数据源所需的各种属性。

例 2-10：演示单一环境下项目配置数据源的方式。程序为 mybatis-config.xml。

```
<environments default="development">
    <environment id="development">
        <transactionManager type="JDBC" />
        <dataSource type="POOLED">
            <property name="driver" value="${jdbc.driver}" />
            <property name="url" value="${jdbc.url}" />
            <property name="username" value="${jdbc.username}" />
            <property name="password" value="${jdbc.password}" />
        </dataSource>
    </environment>
</environments>
```

2.1.5 mappers 元素

mappers 元素用于配置映射器，支持多种引入方式，以满足不同的业务需求。MyBatis 提供了 XML 和注解两种方式来定义映射器。

例 2-11：演示使用 mapper 子元素的 resource 属性引入映射器的方式，resource 属性值为 XML 映射器文件的类路径。程序为 mybatis-config.xml。

```
<mappers>
    <mapper resource="com/imut/mapper/CoursesMapper.xml"/>
    <mapper resource="com/imut/mapper/StudentMapper.xml"/>
</mappers>
```

Student.java、StudentMapper.java 和 StudentMapper.xml 文件内容详见 1.6.3 节。

例 2-12：演示使用 mapper 子元素的 class 属性引入映射器的方式，class 属性值为接口文件的全限定类名。程序为 mybatis-config.xml。

```
<mappers>
    <mapper class="com.imut.mapper.CoursesMapper" />
</mappers>
```

因为例 2-11 已经注册了 CoursesMapper，所以运行例 2-12 前需要注释例 2-11 中相应 Mapper 的注册代码。

例 2-13：演示使用 mapper 子元素的 url 属性引入映射器的方式，url 属性值为接口文件的全限定类名。程序为 mybatis-config.xml。

```
<mappers>
    <mapper url="file:///E:/ch02/src/com/imut/mapper/CoursesMapper.xml" />
</mappers>
```

MyBatis 提供另一种方式，即在 mappers 元素下使用 package 子元素。将映射器的包名赋值给 name 属性，则包下的所有映射器文件都会以相同的方式被注册。通过这种方式告知 MyBatis 映射器的位置，相当于一种批量操作。本书将注释其他方式，保留使用 package 元素的方式。

例 2-14：演示使用 package 子元素的 name 属性引入映射器的方式。程序为 mybatis-config.xml。

```
<mappers>
    <package name="com.imut.mapper"/>
</mappers>
```

根据映射器的位置扫描包下的所有文件并引入。要想实现上述功能，MyBatis 也提供两个注解：@Mapper 和@MapperScan。映射器数量少时，可以通过@Mapper 注解的方式一个一个地标识映射器接口。映射器数量较多时，可以使用注解@MapperScan 扫描包。通常@MapperScan 注解使用度更高。

2.2 映射文件

映射文件通常由接口文件和 XML 文件组成，用于定义数据库操作。接口文件可以直接使用注解来定义简单的 SQL 语句，而 XML 文件则更适用于复杂场景。本书主要讲解 XML 形式的映射文件。

XML 文件包含多个元素，比如用于不同操作的元素 select、insert、update、delete，还有用于处理结果集的元素 resultMap，用于定义 SQL 语句块的元素 sql 等。每个元素又包含多个属性，通过元素与元素、元素与属性的组合，可以实现强大的功能。

2.2.1 select 元素

select 元素主要用于查询数据，可以自动生成 SQL 语句，也可以使用自定义的 SQL 语句。常用的属性如表 2-2 所示。

表 2-2 select 元素常用的属性

属 性	描 述
id	在命名空间中唯一的标识符，可以被用来引用这条语句
parameterType	传入 SQL 语句的参数的全限定类名或别名。这个属性是可选的，因为 MyBatis 可以通过类型处理器(TypeHandler)推断出传入语句的参数，默认值为未设置(unset)
resultType	SQL 语句的返回结果的全限定类名或别名。注意：如果返回的是集合，那应该设置为集合包含的类型，而不是集合本身的类型。resultType 和 resultMap 只能使用一个

续表

属性	描述
resultMap	用于对外部定义好的 resultMap 的引用。结果映射是 MyBatis 最强大的特性，如果对其理解透彻，许多复杂的映射问题都能迎刃而解。resultType 和 resultMap 只能使用一个
useCache	用于标识是否使用缓存。将其设置为 true 后，会导致本条语句的结果被二级缓存缓存起来，默认值：select 元素为 true

例 2-15：演示使用 select 元素查询选课表中全部数据的方式。共包含 7 个文件：Ccourses.java（选课实体类）、CcoursesMapper.java（选课 Mapper 接口）、CcoursesMapper.xml（选课 XML 映射器）、mybatis-config.xml、db.properties、MyBatisUtil.java 和 TestMB.java。创建实体类前，需要先创建数据库表 ccourses 并填充表中数据。又因为 ccourses 表的 sno 字段是外键，与 students 表有关联，需要先创建表 students，表结构及数据详见例 3-1。ccourses 表结构及数据如表 2-3 所示。

表 2-3　选课（ccourses）表结构及数据

id（选课 ID，主键，自增，非空）	cid（课程 ID，默认为 NULL，外键）	sno（学号，默认为 NULL，外键）
1	2	3
2	2	2
3	2	3
⋮	⋮	⋮

cid 用于与课程信息关联，即与课程表中的课程 ID 关联；sno 用于与学生信息关联，即与学生表中的学号关联。

Ccourses.java 代码如下，位于 src/com/imut/pojo 目录下。

```
package com.imut.pojo;
public class Ccourses {
    private Integer id;
    private Integer cid;
    private Integer sno;

}
```

CcoursesMapper.java 代码如下，位于 src/com/imut/mapper 目录下。

```
package com.imut.mapper;
import com.imut.pojo.Ccourses;
import java.util.List;
public interface CcoursesMapper {
```

注意：① 方法名需要与 CcoursesMapper.xml 中对应 SQL 语句的 id 属性值一致。

② 返回类型需要和 CcoursesMapper.xml 中对应 SQL 语句的 resultType 属性一致。因为会返回多条数据，所以使用 List 接收。

```
    List<Ccourses> selectAll();
}
```

CcoursesMapper.xml 代码如下，位于 src/com/imut/mapper 目录下。

```
<?xml version="1.0" encoding="UTF-8"?>
<!DOCTYPE mapper PUBLIC "-//mybatis.org//DTD Mapper 3.0//EN" "http://mybatis.
org/dtd/mybatis-3-mapper.dtd">
<mapper namespace="com.imut.mapper.CcoursesMapper">
    <!--方式一:使用 resultMap 属性-->
    <select id="selectAll" resultMap="BaseResultMap">
        select id, cid, sno
        from ccourses
    </select>
    <!--方式二:使用 resultType 属性-->
    <select id="selectAll" resultType="ccourses">
        select id, cid, sno
        from ccourses
    </select>
</mapper>
```

例 2-15 中的方式一不需要传入参数，由 Mybatis 自动生成。但是需要配合 resultMap 元素使用，内容详见例 2-23。方式二的 resultType 属性值为别名。运行时需要注释其中一种方式。mybatis-config.xml 文件沿用例 2-14 后形成的内容，db.properties 文件内容详见例 2-2，MyBatisUtil.java 文件内容详见例 1-2。

TestMB.java 代码如下，位于 src/com/imut/test 目录下。

```
package com.imut.test;
import com.imut.mapper.CcoursesMapper;
import com.imut.mapper.CoursesMapper;
import com.imut.pojo.Ccourses;
import com.imut.pojo.Courses;
import com.imut.utils.MyBatisUtil;
import org.apache.ibatis.session.SqlSession;
import org.apache.ibatis.session.SqlSessionFactory;
import org.junit.Test;
import java.util.List;
public class TestMB {
    @Test
    public void testCcoursesMapper() throws Exception {
        SqlSessionFactory ssf = new MyBatisUtil().getSqlSessionFactory();
        SqlSession session = ssf.openSession();
        try {
            CcoursesMapper mapper = session.getMapper(CcoursesMapper.class);
            List<Ccourses> ccoursesList = mapper.selectAll();
            System.out.println("全部选课信息:" + ccoursesList);
```

```
        } finally {
            session.close();
        }
    }
}
```

例 2-15 的运行结果如图 2-1 所示。

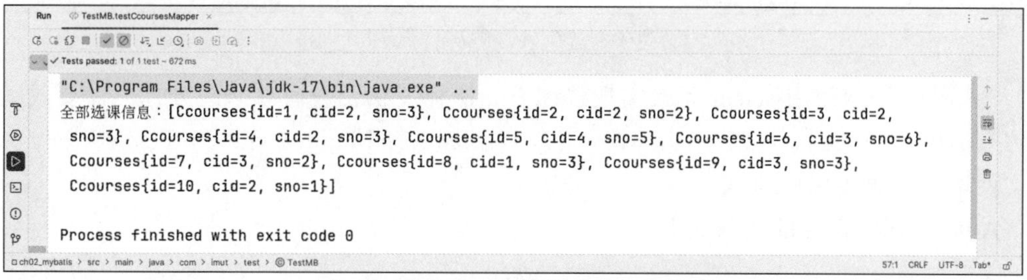

```
"C:\Program Files\Java\jdk-17\bin\java.exe" ...
全部选课信息：[Ccourses{id=1, cid=2, sno=3}, Ccourses{id=2, cid=2, sno=2}, Ccourses{id=3, cid=2,
 sno=3}, Ccourses{id=4, cid=2, sno=3}, Ccourses{id=5, cid=4, sno=5}, Ccourses{id=6, cid=3, sno=6},
 Ccourses{id=7, cid=3, sno=2}, Ccourses{id=8, cid=1, sno=3}, Ccourses{id=9, cid=3, sno=3},
 Ccourses{id=10, cid=2, sno=1}]

Process finished with exit code 0
```

图 2-1　查询全部选课记录的运行结果

例 2-16：演示使用 select 元素实现根据 id 查询选课表中数据的方式。需要改动的文件有 CcoursesMapper.java、CcoursesMapper.xml 和 TestMB.java。其余文件均沿用例 2-15 的即可。

CcoursesMapper.java 的新增代码如下。

```
Ccourses selectByPrimaryKey(Integer cid);
```

CcoursesMapper.xml 的新增代码如下。

```
<select id="selectByPrimaryKey" parameterType="java.lang.Integer" resultMap
="BaseResultMap">
    select id, cid, sno
    from ccourses
    where id = #{id,jdbcType=INTEGER}
</select>
```

例 2-16 中的 CcoursesMapper.xml 里新增的语句需要传入 Integer 类型的参数 id，使用了 parameterType 属性，其类型为 java.lang.Integer，由 Mybatis 自动生成。同样需要配合 resultMap 元素使用，内容详见例 2-23。SQL 语句接收传入参数值的形式为 #{参数名，数据库对应列的类型}。在多数情况下，{}内可以只写参数名，并且参数名可以与 Mapper 接口中传入的参数名不一致。但是，如果传入多个参数，则需要使用注解，且参数名需要与注解内的名称保持一致。

TestMB.java 的新增代码如下，位于 testCcoursesMapper 方法中。

```
Ccourses ccourses = mapper.selectByPrimaryKey(3);
System.out.println("根据 id(3)查询的选课信息：" + ccourses);
```

例 2-16 的运行结果如图 2-2 所示。

```
Run  TestMB
Tests passed: 1 of 1 test – 662 ms
"C:\Program Files\Java\jdk-17\bin\java.exe" ...
全部选课信息：[Ccourses{id=1, cid=2, sno=3}, Ccourses{id=2, cid=2, sno=2}, Ccourses{id=3, cid=2,
sno=3}, Ccourses{id=4, cid=2, sno=3}, Ccourses{id=5, cid=4, sno=5}, Ccourses{id=6, cid=3, sno=6},
Ccourses{id=7, cid=3, sno=2}, Ccourses{id=8, cid=1, sno=3}, Ccourses{id=9, cid=3, sno=3},
Ccourses{id=10, cid=2, sno=1}]
根据 id（3）查询的选课信息：Ccourses{id=3, cid=2, sno=3}
ch02_mybatis > src > main > java > com > imut > test > TestMB          50:49  CRLF  UTF-8  Tab*
```

图 2-2　例 2-16 的运行结果

例 2-17：演示使用 select 元素实现选课表和课程表联表查询数据的方式。需要改动的文件有 AllCcourses.java、CcoursesMapper.java、CcoursesMapper.xml 和 TestMB.java。其余文件均沿用例 2-16 的即可。

AllCcourses.java 的代码如下。

```
package com.imut.pojo;
public class AllCcourses {
    private Integer id;
    private Integer cid;
    private String cname;
    private Integer sno;
    //此处省略属性对应的 getter()、setter()和 toString()方法
}
```

CcoursesMapper.java 的新增代码如下。

```
List<AllCcourses> selectMulTables();
```

CcoursesMapper.xml 的新增代码如下。

```
<select id="selectMulTables" resultMap="JoinResultMap">
    select id, ccourses.cid, cname, sno
    from ccourses, courses where ccourses.cid = courses.cid
</select>
```

TestMB.java 的新增代码如下，位于 testCcoursesMapper 方法中。

```
List<AllCcourses> allCcoursesList = mapper.selectMulTables();
System.out.println("联表查询后的全部选课信息:" + allCcoursesList);
```

例 2-17 中的 CcoursesMapper.xml 里新增的语句不需要传入参数，由 Mybatis 自动生成。但是需要配合 resultMap 元素使用，内容详见例 2-24。

上述 3 个例子传入的参数个数小于或等于 1，但在实际场景中，还存在不少传入多个参数的情况。为了确保 Mapper.xml 文件可以正确地接收所有参数，需要使用@Param 注解标识每一个参数。

例 2-17 的运行结果如图 2-3 所示。

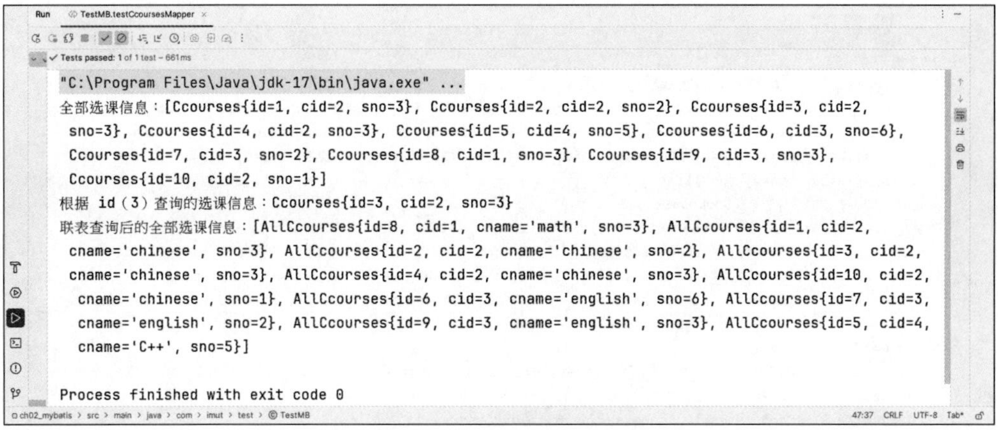

```
"C:\Program Files\Java\jdk-17\bin\java.exe" ...
全部选课信息：[Ccourses{id=1, cid=2, sno=3}, Ccourses{id=2, cid=2, sno=2}, Ccourses{id=3, cid=2,
 sno=3}, Ccourses{id=4, cid=2, sno=3}, Ccourses{id=5, cid=4, sno=5}, Ccourses{id=6, cid=3, sno=6},
 Ccourses{id=7, cid=3, sno=2}, Ccourses{id=8, cid=1, sno=3}, Ccourses{id=9, cid=3, sno=3},
 Ccourses{id=10, cid=2, sno=1}]
根据 id（3）查询的选课信息：Ccourses{id=3, cid=2, sno=3}
联表查询后的全部选课信息：[AllCcourses{id=8, cid=1, cname='math', sno=3}, AllCcourses{id=1, cid=2,
 cname='chinese', sno=3}, AllCcourses{id=2, cid=2, cname='chinese', sno=2}, AllCcourses{id=3, cid=2,
 cname='chinese', sno=3}, AllCcourses{id=4, cid=2, cname='chinese', sno=3}, AllCcourses{id=10, cid=2,
  cname='chinese', sno=1}, AllCcourses{id=6, cid=3, cname='english', sno=6}, AllCcourses{id=7, cid=3,
  cname='english', sno=2}, AllCcourses{id=9, cid=3, cname='english', sno=3}, AllCcourses{id=5, cid=4,
  cname='C++', sno=5}]

Process finished with exit code 0
```

图 2-3　例 2-17 的运行结果

例 2-18：演示使用 select 元素实现根据课程编号或学号查询选课表中数据的方式。需要改动的文件有 CcoursesMapper.java、CcoursesMapper.xml 和 TestMB.java。其余文件均沿用例 2-17 的即可。

CcoursesMapper.java 的新增代码如下。

```
List<Ccourses> selectByCidOrSno(@Param("cid") Integer id, @Param("sno")
Integer sno);
```

@Param 注解写在参数的前面，括号中的内容为真正传入 CcoursesMapper.xml 对应 SQL 语句的参数名，可以和 CcoursesMapper.java 中的参数名不一致。当有多个参数时，每个参数都需要有@Param 注解标识。

CcoursesMapper.xml 的新增代码如下。

```
<select id="selectByCidOrSno" parameterType="integer" resultType="ccourses">
    select id, cid, sno
    from ccourses
    where cid = #{cid} or sno = #{sno}
</select>
```

例 2-18 中的 CcoursesMapper.xml 里新增的语句传入的两个参数类型一致，故指定 parameterType 的属性值为 Integer，是 MyBatis 中 java.lang.Integer 的内置别名。SQL 语句接收传入参数值的形式省略了数据库对应列的类型。

TestMB.java 的新增代码如下，位于 testCcoursesMapper 方法中。

```
int cid = 2;
int sno = 3;
ccoursesList = mapper.selectByCidOrSno(cid, sno);
System.out.println("根据 cid(2)或 sno(3)查询的选课信息:" + ccoursesList);
```

例 2-18 的运行结果如图 2-4 所示。

select 元素还包含许多其他属性，需根据实际业务场景选择使用，通常情况下选择默认

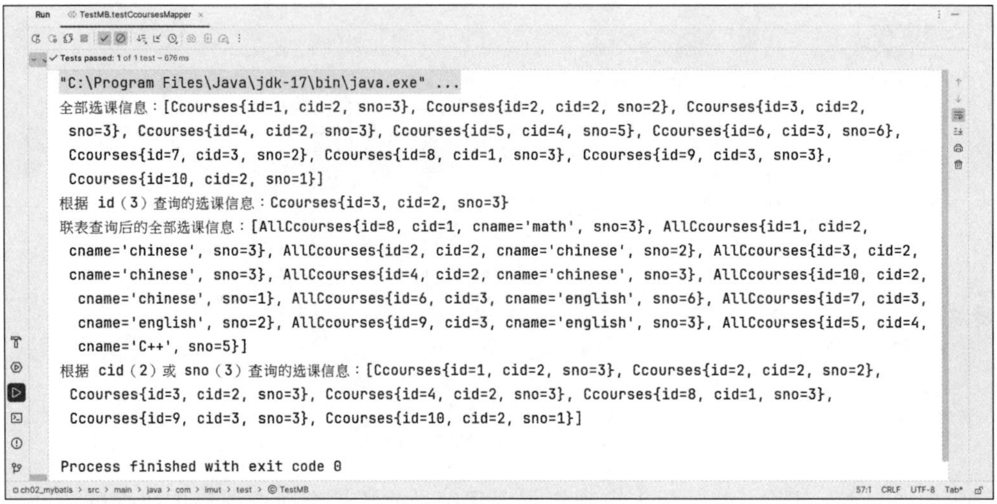

```
Run   TestMB.testCcoursesMapper
Tests passed: 1 of 1 test – 676 ms
"C:\Program Files\Java\jdk-17\bin\java.exe" ...
全部选课信息：[Ccourses{id=1, cid=2, sno=3}, Ccourses{id=2, cid=2, sno=2}, Ccourses{id=3, cid=2,
 sno=3}, Ccourses{id=4, cid=2, sno=3}, Ccourses{id=5, cid=4, sno=5}, Ccourses{id=6, cid=3, sno=6},
 Ccourses{id=7, cid=3, sno=2}, Ccourses{id=8, cid=1, sno=3}, Ccourses{id=9, cid=3, sno=3},
 Ccourses{id=10, cid=2, sno=1}]
根据 id（3）查询的选课信息：Ccourses{id=3, cid=2, sno=3}
联表查询后的全部选课信息：[AllCcourses{id=8, cid=1, cname='math', sno=3}, AllCcourses{id=1, cid=2,
 cname='chinese', sno=3}, AllCcourses{id=2, cid=2, cname='chinese', sno=2}, AllCcourses{id=3, cid=2,
 cname='chinese', sno=3}, AllCcourses{id=4, cid=2, cname='chinese', sno=3}, AllCcourses{id=10, cid=2,
  cname='chinese', sno=1}, AllCcourses{id=6, cid=3, cname='english', sno=6}, AllCcourses{id=7, cid=3,
  cname='english', sno=2}, AllCcourses{id=9, cid=3, cname='english', sno=3}, AllCcourses{id=5, cid=4,
  cname='C++', sno=5}]
根据 cid（2）或 sno（3）查询的选课信息：[Ccourses{id=1, cid=2, sno=3}, Ccourses{id=2, cid=2, sno=2},
 Ccourses{id=3, cid=2, sno=3}, Ccourses{id=4, cid=2, sno=3}, Ccourses{id=8, cid=1, sno=3},
 Ccourses{id=9, cid=3, sno=3}, Ccourses{id=10, cid=2, sno=1}]

Process finished with exit code 0
ch02_mybatis > src > main > java > com > imut > test > TestMB          57:1  CRLF  UTF-8  Tab*
```

图 2-4　例 2-18 的运行结果

值即可。

2.2.2　insert 元素

insert 元素用于插入数据，支持自动生成 SQL 语句、自定义 SQL 语句。其包含多种属性，如表 2-4 所示。

表 2-4　insert、update、delete 元素属性配置信息表

属　　性	描　　述
id	在命名空间中唯一的标识符，可以被用来引用这条语句
parameterType	传入 SQL 语句的参数的全限定类名或别名。这个属性是可选的，因为 MyBatis 可以通过类型处理器（TypeHandler）推断出具体传入语句的参数，默认值为未设置（unset）
statementType	可选值为 STATEMENT、PREPARED（默认值）或 CALLABLE。与可选值相对应的接口分别是 Statement、PreparedStatement 或 CallableStatement
useGeneratedKeys	（仅适用 insert 和 update 元素）MyBatis 使用 JDBC 的 getGeneratedKeys 方法来取出由数据库内部生成的主键（比如 MySQL 和 SQL Server 这样的关系型数据库管理系统的自动递增字段），默认值为 false。
keyProperty	（仅适用 insert 和 update 元素）指定能够唯一识别对象的属性，MyBatis 会使用 getGeneratedKeys 的返回值或 insert 语句的 selectKey 子元素设置它的值，默认值为未设置（unset）。如果生成列不止一个，可以用逗号分隔多个属性名称
keyColumn	（仅适用 insert 和 update 元素）设置生成键值在表中的列名，在某些数据库（像 PostgreSQL）中，当主键列不是表中第一列的时候，是必须设置的。如果生成列不止一个，可以用逗号分隔多个属性名称

例 2-19：演示使用 insert 元素实现向选课表插入新数据的方式。需要改动的文件为 CcoursesMapper.java、CcoursesMapper.xml 和 TestMB.java。其余文件均沿用例 2-18 的即可。

CcoursesMapper.java 的新增代码如下。

```
int insert(Ccourses ccourses);
```

CcoursesMapper.xml 的新增代码如下。

```
<insert id="insert" parameterType="com.imut.pojo.Ccourses">
    insert into ccourses (id, cid, sno) values (#{id,jdbcType=INTEGER}, #{cid,
jdbcType=INTEGER}, #{sno,jdbcType=INTEGER})
</insert>
```

例 2-19 中的 CcoursesMapper.xml 里新增的语句需要传入选课表类型参数，parameterType 属性使用的是全限定类名的方式，由 Mybatis 自动生成。因为返回类型只能是 int，并且只有一个返回值，所以没有 resultType 属性和 resultMap 属性。

TestMB.java 的新增代码如下，位于 testCcoursesMapper 方法中。

```
ccourses.setId(11);
ccourses.setCid(3);
ccourses.setSno(3);
int result = mapper.insert(ccourses);
if (result==1) {
    System.out.println("选课数据插入成功");
}else
    System.out.println("选课数据插入失败");
```

例 2-19 的运行结果如图 2-5 所示。

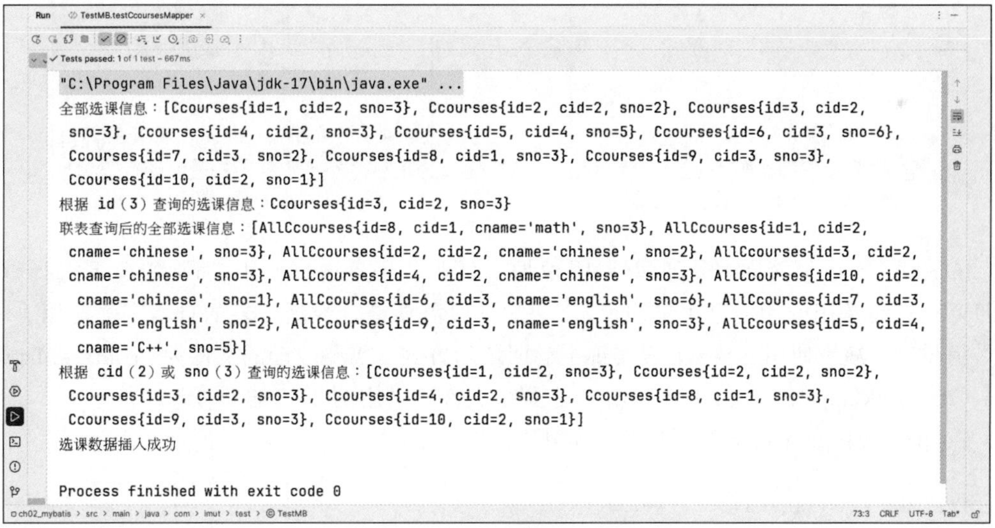

```
Run  TestMB.testCcoursesMapper
Tests passed: 1 of 1 test – 667ms
"C:\Program Files\Java\jdk-17\bin\java.exe" ...
全部选课信息：[Ccourses{id=1, cid=2, sno=3}, Ccourses{id=2, cid=2, sno=2}, Ccourses{id=3, cid=2,
 sno=3}, Ccourses{id=4, cid=2, sno=3}, Ccourses{id=5, cid=4, sno=5}, Ccourses{id=6, cid=3, sno=6},
 Ccourses{id=7, cid=3, sno=2}, Ccourses{id=8, cid=1, sno=3}, Ccourses{id=9, cid=3, sno=3},
 Ccourses{id=10, cid=2, sno=1}]
根据 id（3）查询的选课信息：Ccourses{id=3, cid=2, sno=3}
联表查询后的全部选课信息：[AllCcourses{id=8, cid=1, cname='math', sno=3}, AllCcourses{id=1, cid=2,
 cname='chinese', sno=3}, AllCcourses{id=2, cid=2, cname='chinese', sno=2}, AllCcourses{id=3, cid=2,
 cname='chinese', sno=3}, AllCcourses{id=4, cid=2, cname='chinese', sno=3}, AllCcourses{id=10, cid=2,
 cname='chinese', sno=1}, AllCcourses{id=6, cid=3, cname='english', sno=6}, AllCcourses{id=7, cid=3,
 cname='english', sno=2}, AllCcourses{id=9, cid=3, cname='english', sno=3}, AllCcourses{id=5, cid=4,
 cname='C++', sno=5}]
根据 cid（2）或 sno（3）查询的选课信息：[Ccourses{id=1, cid=2, sno=3}, Ccourses{id=2, cid=2, sno=2},
 Ccourses{id=3, cid=2, sno=3}, Ccourses{id=4, cid=2, sno=3}, Ccourses{id=8, cid=1, sno=3},
 Ccourses{id=9, cid=3, sno=3}, Ccourses{id=10, cid=2, sno=1}]
选课数据插入成功

Process finished with exit code 0
ch02_mybatis > src > main > java > com > imut > test > TestMB        73:3  CRLF  UTF-8  Tab*
```

图 2-5　例 2-19 的运行结果

例 2-19 运行后数据库 ccourse 表中的数据如图 2-6 所示。

按照例 2-19 操作完毕，会发现虽然控制台输出选课数据插入成功，但实际数据库表中并无新增数据。是因为插入操作需要提交后才正式生效，而上例中并无此操作。要想实现真正的插入操作，MyBatis 提供两种提交方式，即手动提交和自动提交。

手动提交的方式为插入操作后显示调用。手动提交的代码如下。

```
session.commit();
```

自动提交的代码如下。

```
SqlSession session = ssf.openSession(true);
```

自动提交的方式：传入一个 true 值给 openSession 方法。该方式只需编写一次，相对于手动提交方式更简便，开发者较多，也是本书采用的方式。修改代码后，数据库 ccourses 表中的数据如图 2-7 所示。

id	cid	sno
1	2	3
2	2	2
3	2	3
4	2	3
5	4	5
6	3	6
7	3	2
8	1	3
9	3	3
10	2	1

图 2-6　ccourses 表中的数据

id	cid	sno
1	2	3
2	2	2
3	2	3
4	2	3
5	4	5
6	3	6
7	3	2
8	1	3
9	3	3
10	2	1
11	3	3

第 1 条记录 (共 11 条) 于第 1 页

图 2-7　ccourses 表中的数据

insert 元素还包含许多其他属性，需根据实际业务场景选择使用，通常情况下选择默认值即可。较为常用的功能是主键回填，该功能一般需要结合数据库表的自增字段使用。

例 2-20：演示使用 insert 元素实现主键回填的方式。需要改动的文件为 CcoursesMapper.java、CcoursesMapper.xml 和 TestMB.java。其余文件均沿用例 2-19 的即可。

CcoursesMapper.java 的新增代码如下。

```
int insertCcourses(Ccourses ccourses);
```

CcoursesMapper.xml 的新增代码如下。

```
<insert id="insertCcourses" parameterType="ccourses" useGeneratedKeys="true"
keyProperty="id">
    insert into ccourses(cid, sno) values(#{cid}, #{sno})
</insert>
```

useGeneratedKeys 属性设置为 true,则会获取新插入数据库 ccourses 表中自增字段的值。keyProperty 属性值为类对象的属性,用于接收 useGeneratedKeys 属性返回的值。

例 2-20 中的 CcoursesMapper. xml 里新增的语句需要传入选课表类型参数,parameterType 属性使用的是别名的方式,其中 Ccourses 类的 id 字段并未赋值。

TestMB.java 的新增代码如下,位于 testCcoursesMapper 方法中。

```
ccourses.setCid(1);
ccourses.setSno(2);
int result = mapper.insertCcourses(ccourses);
if (result==1) {
    System.out.println("选课数据插入成功");
}else
    System.out.println("选课数据插入失败");
```

执行例 2-20 前,需要注释例 2-19 的测试代码,否则会因主键重复而报错。要想观看主键回填的效果,可以使用 Debug 或直接输出 id 属性值的方式。即 insertCcourses 方法执行完毕后,查看 ccourses 对象中的 id 属性值,是否为新插入数据库表的值。

例 2-20 的运行结果如图 2-8 所示。

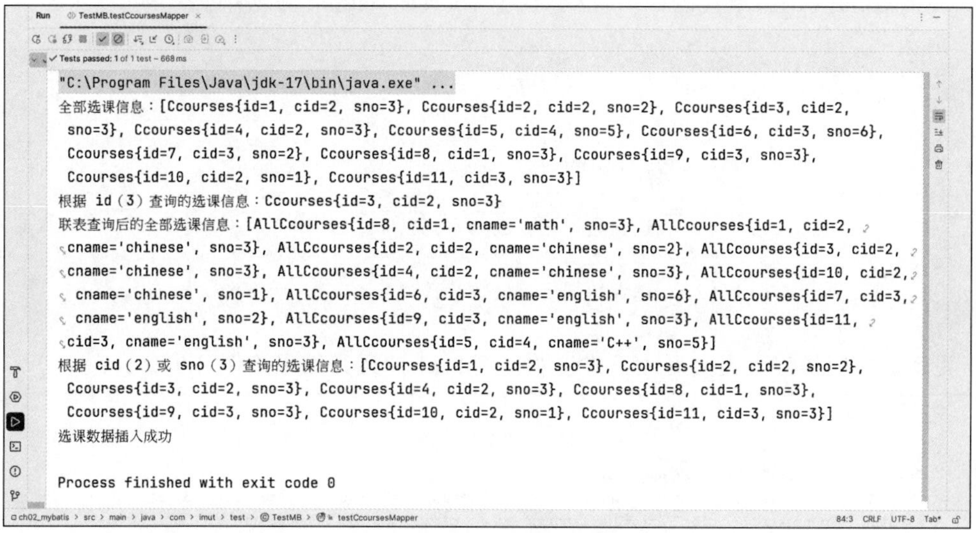

```
"C:\Program Files\Java\jdk-17\bin\java.exe" ...
全部选课信息:[Ccourses{id=1, cid=2, sno=3}, Ccourses{id=2, cid=2, sno=2}, Ccourses{id=3, cid=2,
 sno=3}, Ccourses{id=4, cid=2, sno=3}, Ccourses{id=5, cid=4, sno=5}, Ccourses{id=6, cid=3, sno=6},
 Ccourses{id=7, cid=3, sno=2}, Ccourses{id=8, cid=1, sno=3}, Ccourses{id=9, cid=3, sno=3},
 Ccourses{id=10, cid=2, sno=1}, Ccourses{id=11, cid=3, sno=3}]
根据 id (3) 查询的选课信息:Ccourses{id=3, cid=2, sno=3}
联表查询后的全部选课信息:[AllCcourses{id=8, cid=1, cname='math', sno=3}, AllCcourses{id=1, cid=2,
 cname='chinese', sno=3}, AllCcourses{id=2, cid=2, cname='chinese', sno=2}, AllCcourses{id=3, cid=2,
 cname='chinese', sno=3}, AllCcourses{id=4, cid=2, cname='chinese', sno=3}, AllCcourses{id=10, cid=2,
 cname='chinese', sno=1}, AllCcourses{id=6, cid=3, cname='english', sno=6}, AllCcourses{id=7, cid=3,
 cname='english', sno=2}, AllCcourses{id=9, cid=3, cname='english', sno=3}, AllCcourses{id=11,
 cid=3, cname='english', sno=3}, AllCcourses{id=5, cid=4, cname='C++', sno=5}]
根据 cid (2) 或 sno (3) 查询的选课信息:[Ccourses{id=1, cid=2, sno=3}, Ccourses{id=2, cid=2, sno=2},
 Ccourses{id=3, cid=2, sno=3}, Ccourses{id=4, cid=2, sno=3}, Ccourses{id=8, cid=1, sno=3},
 Ccourses{id=9, cid=3, sno=3}, Ccourses{id=10, cid=2, sno=1}, Ccourses{id=11, cid=3, sno=3}]
选课数据插入成功

Process finished with exit code 0
```

图 2-8　例 2-20 的运行结果

2.2.3 update 元素

update 元素用于更新数据,支持自动生成 SQL 语句、自定义 SQL 语句。id 属性和 parameterType 属性是最为常用的属性。id 属性用于指定语句的唯一标识,即被调用的方法名,与相对应的 Mapper 接口中的方法名一致;parameterType 属性用于指定参数类型,可以为单个参数类型,也可以为多参数封装类型,支持别名。

例 2-21:演示使用 update 元素实现根据 id 更新选课表中数据的方式。共包含 7 个文件:Ccourses. java、CcoursesMapper. java、CcoursesMapper. xml、mybatis-config. xml、db. properties、MyBatisUtil. java 和 TestMB. java。需要改动的文件为 CcoursesMapper. java、

CcoursesMapper.xml 和 TestMB.java。其余文件均沿用例 2-20 的即可。

CcoursesMapper.java 的新增代码如下。

```
int updateByPrimaryKey(Ccourses ccourses);
```

CcoursesMapper.xml 的新增代码如下。

```
<update id="updateByPrimaryKey" parameterType="com.imut.pojo.Ccourses">
    update ccourses
    set cid = #{cid,jdbcType=INTEGER}, sno = #{sno,jdbcType=INTEGER}
    where id = #{id,jdbcType=INTEGER}
</update>
```

在例子 2-21 的 CcoursesMapper.xml 文件中新增的语句里，parameterType 属性值不是 id 的类型 Integer，而是选课表的封装类。

需要注意的是，这个封装类不必与数据库表的映射类完全相同，可以只包含需要更新的字段和对应的属性，以及与查询结果一致的属性和 getter/setter 方法。这样可以实现灵活和精确地更新操作。

TestMB.java 的新增代码如下，位于 testCcoursesMapper 方法中。

```
ccourses.setId(11);
ccourses.setCid(3);
ccourses.setSno(1);
result = mapper.updateByPrimaryKey(ccourses);
if (result==1) {
    System.out.println("选课数据更新成功");
}else
    System.out.println("选课数据更新失败");
```

更新操作同样需要提交后才正式生效。例 2-21 的运行结果如图 2-9 所示。

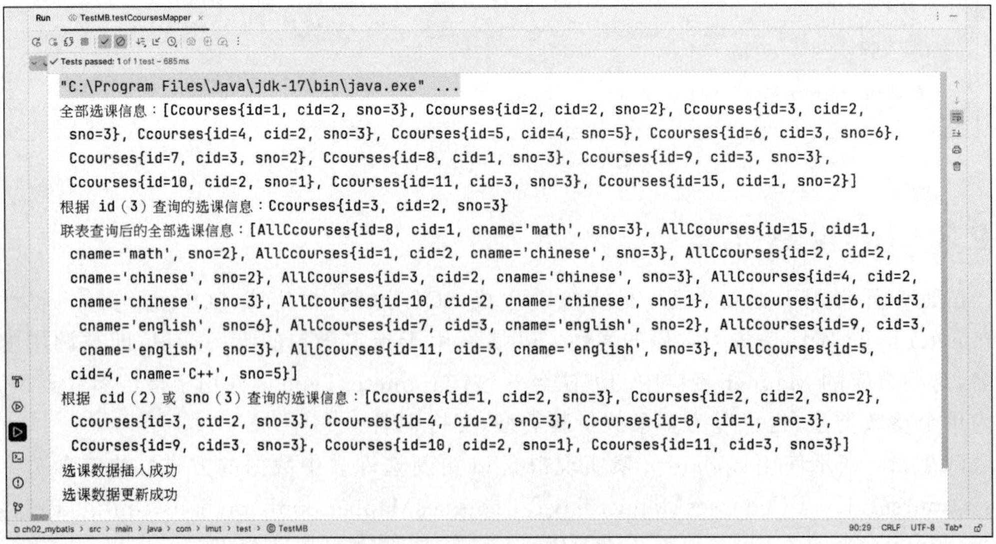

```
Run    TestMB.testCcoursesMapper ×
✓ Tests passed: 1 of 1 test – 685 ms
"C:\Program Files\Java\jdk-17\bin\java.exe" ...
全部选课信息：[Ccourses{id=1, cid=2, sno=3}, Ccourses{id=2, cid=2, sno=2}, Ccourses{id=3, cid=2,
 sno=3}, Ccourses{id=4, cid=2, sno=3}, Ccourses{id=5, cid=4, sno=5}, Ccourses{id=6, cid=3, sno=6},
 Ccourses{id=7, cid=3, sno=2}, Ccourses{id=8, cid=1, sno=3}, Ccourses{id=9, cid=3, sno=3},
 Ccourses{id=10, cid=2, sno=1}, Ccourses{id=11, cid=3, sno=3}, Ccourses{id=15, cid=1, sno=2}]
根据 id（3）查询的选课信息：Ccourses{id=3, cid=2, sno=3}
联表查询后的全部选课信息：[AllCcourses{id=8, cid=1, cname='math', sno=3}, AllCcourses{id=15, cid=1,
 cname='math', sno=2}, AllCcourses{id=1, cid=2, cname='chinese', sno=3}, AllCcourses{id=2, cid=2,
 cname='chinese', sno=2}, AllCcourses{id=3, cid=2, cname='chinese', sno=3}, AllCcourses{id=4, cid=2,
 cname='chinese', sno=3}, AllCcourses{id=10, cid=2, cname='chinese', sno=1}, AllCcourses{id=6, cid=3,
 cname='english', sno=6}, AllCcourses{id=7, cid=3, cname='english', sno=2}, AllCcourses{id=9, cid=3,
 cname='english', sno=3}, AllCcourses{id=11, cid=3, cname='english', sno=3}, AllCcourses{id=5,
 cid=4, cname='C++', sno=5}]
根据 cid（2）或 sno（3）查询的选课信息：[Ccourses{id=1, cid=2, sno=3}, Ccourses{id=2, cid=2, sno=2},
 Ccourses{id=3, cid=2, sno=3}, Ccourses{id=4, cid=2, sno=3}, Ccourses{id=8, cid=1, sno=3},
 Ccourses{id=9, cid=3, sno=3}, Ccourses{id=10, cid=2, sno=1}, Ccourses{id=11, cid=3, sno=3}]
选课数据插入成功
选课数据更新成功
ch02_mybatis › src › main › java › com › imut › test › TestMB        90:29  CRLF  UTF-8  Tab*
```

图 2-9　例 2-21 的运行结果

update 元素还包含许多其他属性，需根据实际业务场景选择使用，通常情况下选择默认值即可。

2.2.4 delete 元素

delete 元素用于删除数据，支持自动生成 SQL 语句、自定义 SQL 语句。id 属性和 parameterType 属性是最为常用的属性。id 属性用于指定语句的唯一标识，即被调用的方法名，与相对应的 Mapper 接口中的方法名一致；parameterType 属性用于指定参数类型，可以为单个参数类型，也可以为多参数封装类型，支持别名。

例 2-22：演示使用 delete 元素实现根据 id 删除选课表中数据的方式。需要改动的文件为 CcoursesMapper.java、CcoursesMapper.xml 和 TestMB.java。其余文件均沿用例 2-21 的即可。

CcoursesMapper.java 的新增代码如下。

```
int deleteByPrimaryKey(Integer cid);
```

CcoursesMapper.xml 的新增代码如下。

```
<delete id="deleteByPrimaryKey" parameterType="java.lang.Integer">
    delete from ccourses
    where id = #{id,jdbcType=INTEGER}
</delete>
```

值得注意的是，例 2-22 中的 CcoursesMapper.xml 里新增的语句需要传入 Integer 类型参数(全限定类名)id，由 Mybatis 自动生成。

TestMB.java 的新增代码如下，位于 testCcoursesMapper 方法中。

```
int id = 10;
result = mapper.deleteByPrimaryKey(id);
if (result==1) {
    System.out.println("选课数据删除成功");
}else
    System.out.println("选课数据删除失败");
```

删除操作同样需要提交后才正式生效，例 2-22 的运行结果如图 2-10 所示。

delete 元素还包含许多其他属性，需根据实际业务场景选择使用，通常情况下选择默认值即可。

2.2.5 resultMap 元素

resultMap 元素用于结果映射，包含 id 子元素和 result 子元素。id 子元素用于标识唯一列的结果映射，通常与主键相对应；result 子元素用于除主键外其他字段的结果映射，即指定某个列和与其对应的属性。两者都包含 column 属性、jdbcType 属性和 property 属性。column 属性的值为数据库表中对应的列名，不一致的列名称将导致获取值为 Null；jdbcType 属性的值为数据库中对应字段的类型，可省略；property 属性的值需要与 Java

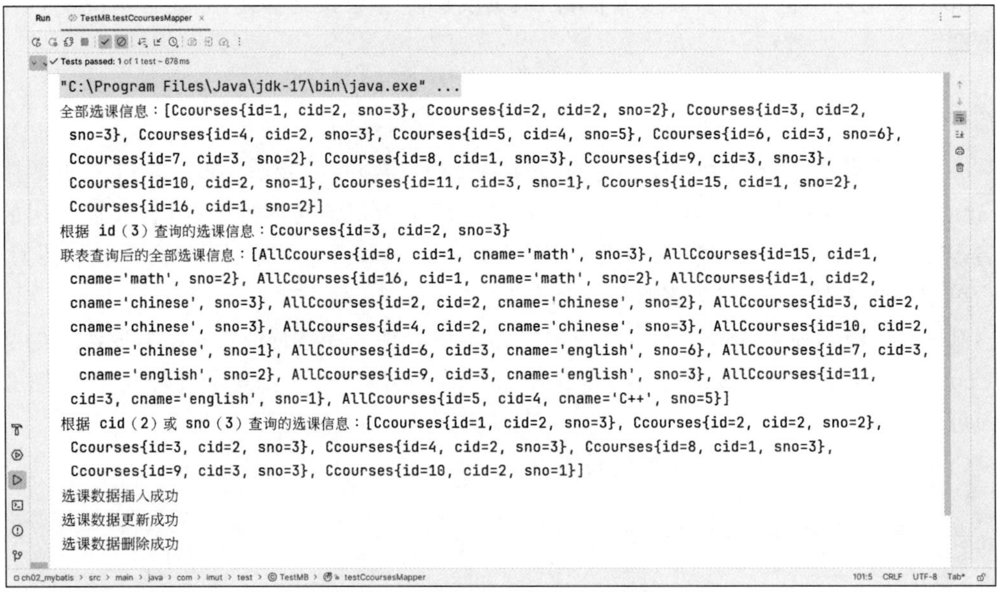
Run TestMB.testCcoursesMapper
Tests passed: 1 of 1 test – 678 ms
"C:\Program Files\Java\jdk-17\bin\java.exe" ...
全部选课信息：[Ccourses{id=1, cid=2, sno=3}, Ccourses{id=2, cid=2, sno=2}, Ccourses{id=3, cid=2, sno=3}, Ccourses{id=4, cid=2, sno=3}, Ccourses{id=5, cid=4, sno=5}, Ccourses{id=6, cid=3, sno=6}, Ccourses{id=7, cid=3, sno=2}, Ccourses{id=8, cid=1, sno=3}, Ccourses{id=9, cid=3, sno=3}, Ccourses{id=10, cid=2, sno=1}, Ccourses{id=11, cid=3, sno=1}, Ccourses{id=15, cid=1, sno=2}, Ccourses{id=16, cid=1, sno=2}]
根据 id（3）查询的选课信息：Ccourses{id=3, cid=2, sno=3}
联表查询后的全部选课信息：[AllCcourses{id=8, cid=1, cname='math', sno=3}, AllCcourses{id=15, cid=1, cname='math', sno=2}, AllCcourses{id=16, cid=1, cname='math', sno=2}, AllCcourses{id=1, cid=2, cname='chinese', sno=3}, AllCcourses{id=2, cid=2, cname='chinese', sno=2}, AllCcourses{id=3, cid=2, cname='chinese', sno=3}, AllCcourses{id=4, cid=2, cname='chinese', sno=3}, AllCcourses{id=10, cid=2, cname='chinese', sno=1}, AllCcourses{id=6, cid=3, cname='english', sno=6}, AllCcourses{id=7, cid=3, cname='english', sno=2}, AllCcourses{id=9, cid=3, cname='english', sno=3}, AllCcourses{id=11, cid=3, cname='english', sno=1}, AllCcourses{id=5, cid=4, cname='C++', sno=5}]
根据 cid（2）或 sno（3）查询的选课信息：[Ccourses{id=1, cid=2, sno=3}, Ccourses{id=2, cid=2, sno=2}, Ccourses{id=3, cid=2, sno=3}, Ccourses{id=4, cid=2, sno=3}, Ccourses{id=8, cid=1, sno=3}, Ccourses{id=9, cid=3, sno=3}, Ccourses{id=10, cid=2, sno=1}]
选课数据插入成功
选课数据更新成功
选课数据删除成功
ch02_mybatis › src › main › java › com › imut › test › TestMB › testCcoursesMapper
101:5 CRLF UTF-8 Tab*

图 2-10　例 2-22 的运行结果

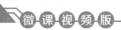

Bean(本章中可认为是 Ccourses 类，即选课实体)中定义的属性名称一致。

例 2-23：演示使用 resultMap 元素实现数据库表字段与实体类映射的方式。共包含 7 个文件：Ccourses.java、CcoursesMapper.java、CcoursesMapper.xml、mybatis-config.xml、db.properties、MyBatisUtil.java 和 TestMB.java。需要改动的文件为 CcoursesMapper.xml。其余文件均沿用例 2-22 的即可。如下代码由 Mybatis 自动生成。

CcoursesMapper.xml 的新增代码如下。

```
<resultMap id="BaseResultMap" type="com.imut.pojo.Ccourses">
    <id column="id" jdbcType="INTEGER" property="id" />
    <result column="cid" jdbcType="INTEGER" property="cid" />
    <result column="sno" jdbcType="INTEGER" property="sno" />
</resultMap>
```

例 2-23 中的查询结果包含 id、cid、sno 三列，即选课表中的所有字段。因为 id 列是主键，所以使用 id 子元素映射结果至 Ccourses 类的 id 属性。而使用 result 子元素映射其余列的值。当 SQL 语句中使用了别名，也可以使用 resultMap 元素映射结果。resultMap 元素通常不单独使用，而是与 select 元素中的 resultMap 属性共同使用，即 resultMap 元素的 id 属性值与 select 元素中的 resultMap 属性值一致。type 属性用于标识查询结果映射的实体类，值可以是全限定类名或者别名。当有多个 resultMap 元素时，通过 select 元素中的 resultMap 属性值选定使用的 resultMap。

例 2-23 的运行结果如图 2-11 所示。

例 2-24：演示多 resultMap 情形下，项目使用名称为 BaseResultMap 的 resultMap 元素的方式。共包含 7 个文件：Ccourses.java、CcoursesMapper.java、CcoursesMapper.xml、mybatis-config.xml、db.properties、MyBatisUtil.java 和 TestMB.java。需要改动的文件为 CcoursesMapper.xml。其余文件均沿用例 2-23 的即可。

```
"C:\Program Files\Java\jdk-17\bin\java.exe" ...
全部选课信息：[Ccourses{id=1, cid=2, sno=3}, Ccourses{id=2, cid=2, sno=2}, Ccourses{id=3, cid=2,
 sno=3}, Ccourses{id=4, cid=2, sno=3}, Ccourses{id=5, cid=4, sno=5}, Ccourses{id=6, cid=3, sno=6},
 Ccourses{id=7, cid=3, sno=2}, Ccourses{id=8, cid=1, sno=3}, Ccourses{id=9, cid=3, sno=3},
 Ccourses{id=11, cid=3, sno=1}, Ccourses{id=15, cid=1, sno=2}, Ccourses{id=16, cid=1, sno=2},
 Ccourses{id=17, cid=1, sno=2}]
根据 id（3）查询的选课信息：Ccourses{id=3, cid=2, sno=3}
联表查询后的全部选课信息：[AllCcourses{id=8, cid=1, cname='math', sno=3}, AllCcourses{id=15, cid=1,
 cname='math', sno=2}, AllCcourses{id=16, cid=1, cname='math', sno=2}, AllCcourses{id=17, cid=1,
 cname='math', sno=2}, AllCcourses{id=1, cid=2, cname='chinese', sno=3}, AllCcourses{id=2, cid=2,
 cname='chinese', sno=2}, AllCcourses{id=3, cid=2, cname='chinese', sno=3}, AllCcourses{id=4, cid=2,
 cname='chinese', sno=3}, AllCcourses{id=6, cid=3, cname='english', sno=6}, AllCcourses{id=7, cid=3,
 cname='english', sno=2}, AllCcourses{id=9, cid=3, cname='english', sno=3}, AllCcourses{id=11, cid=3,
 cname='english', sno=1}, AllCcourses{id=5, cid=4, cname='C++', sno=5}]
根据 cid（2）或 sno（3）查询的选课信息：[Ccourses{id=1, cid=2, sno=3}, Ccourses{id=2, cid=2, sno=2},
 Ccourses{id=3, cid=2, sno=3}, Ccourses{id=4, cid=2, sno=3}, Ccourses{id=8, cid=1, sno=3},
 Ccourses{id=9, cid=3, sno=3}]
选课数据插入成功
选课数据更新成功
选课数据删除失败
```

图 2-11　例 2-23 的运行结果

CcoursesMapper.xml 的改动代码如下。

```
<resultMap id="BaseResultMap" type="com.imut.pojo.Ccourses">
    <id column="id" jdbcType="INTEGER" property="id" />
    <result column="cid" jdbcType="INTEGER" property="cid" />
    <result column="sno" jdbcType="INTEGER" property="sno" />
</resultMap>
<resultMap id="JoinResultMap" type="allCcourses">
    <id column="id" jdbcType="INTEGER" property="id" />
    <result column="cid" jdbcType="INTEGER" property="cid" />
    <result column="cname" jdbcType="VARCHAR" property="cname" />
    <result column="sno" jdbcType="INTEGER" property="sno" />
</resultMap>
<select id="selectAll" resultMap="BaseResultMap">
    select id, cid, sno
    from ccourses
</select>
```

例 2-24 中共包含 2 个定义好的 resultMap，名字分别为 BaseResultMap 和 JoinResultMap。名为 selectAll 的 select 查询方法使用的是 id 属性值为 BaseResultMap 的 resultMap。其中只有 JoinResultMap 元素包裹的内容为新增代码，其余内容与例 2-15 和例 2-23 中的一致。例 2-24 的运行结果如图 2-12 所示。

2.2.6　sql 元素

sql 元素用于定义可被重复使用的 SQL 语句段，通常包含大段的 SQL 语句，以保证代码的可重用性和整洁性。sql 元素包含 id 属性，用于指定唯一标识。通过 include 元素使用定义好的 sql 元素，即 include 元素中的 refid 属性值为 sql 元素的 id 属性值。

```
"C:\Program Files\Java\jdk-17\bin\java.exe" ...
全部选课信息：[Ccourses{id=1, cid=2, sno=3}, Ccourses{id=2, cid=2, sno=2}, Ccourses{id=3, cid=2,
 sno=3}, Ccourses{id=4, cid=2, sno=3}, Ccourses{id=5, cid=4, sno=5}, Ccourses{id=6, cid=3, sno=6},
 Ccourses{id=7, cid=3, sno=2}, Ccourses{id=8, cid=1, sno=3}, Ccourses{id=9, cid=3, sno=3},
 Ccourses{id=11, cid=3, sno=1}, Ccourses{id=15, cid=1, sno=2}, Ccourses{id=16, cid=1, sno=2},
 Ccourses{id=17, cid=1, sno=2}, Ccourses{id=18, cid=1, sno=2}]
根据 id（3）查询的选课信息：Ccourses{id=3, cid=2, sno=3}
联表查询后的全部选课信息：[AllCcourses{id=8, cid=1, cname='math', sno=3}, AllCcourses{id=15, cid=1,
 cname='math', sno=2}, AllCcourses{id=16, cid=1, cname='math', sno=2}, AllCcourses{id=17, cid=1,
 cname='math', sno=2}, AllCcourses{id=18, cid=1, cname='math', sno=2}, AllCcourses{id=1, cid=2,
 cname='chinese', sno=3}, AllCcourses{id=2, cid=2, cname='chinese', sno=2}, AllCcourses{id=3, cid=2,
 cname='chinese', sno=3}, AllCcourses{id=4, cid=2, cname='chinese', sno=3}, AllCcourses{id=6, cid=3,
 cname='english', sno=6}, AllCcourses{id=7, cid=3, cname='english', sno=2}, AllCcourses{id=9, cid=3,
 cname='english', sno=3}, AllCcourses{id=11, cid=3, cname='english', sno=1}, AllCcourses{id=5, cid=4,
 cname='C++', sno=5}]
根据 cid（2）或 sno（3）查询的选课信息：[Ccourses{id=1, cid=2, sno=3}, Ccourses{id=2, cid=2, sno=2},
 Ccourses{id=3, cid=2, sno=3}, Ccourses{id=4, cid=2, sno=3}, Ccourses{id=8, cid=1, sno=3},
 Ccourses{id=9, cid=3, sno=3}]
选课数据插入成功
选课数据更新成功
选课数据删除失败
```

图 2-12　例 2-24 的运行结果

例 2-25：演示使用 sql 元素定义 SQL 语句段，使用 include 元素引用定义语句段查询某门课程选课情况的方式。需要改动的文件为 CcoursesMapper.java、CcoursesMapper.xml 和 TestMB.java。其余文件均沿用例 2-24 的即可。

CcoursesMapper.java 的新增代码如下。

```
List<Ccourses> selectCcourses(Integer cid);
```

CcoursesMapper.xml 的新增代码如下。

```
<sql id="query_courses">
    cid=#{cid}
</sql>
<select id="selectCcourses" parameterType="java.lang.Integer" resultMap="
BaseResultMap">
    select * from ccourses where
    <include refid="query_courses"/>
</select>
```

TestMB.java 的新增代码如下，位于 testCcoursesMapper 方法中。

```
cid = 1;
ccoursesList = mapper.selectCcourses(cid);
System.out.println("根据 cid(1)查询的选课信息:" + ccoursesList);
```

例 2-25 的运行结果如图 2-13 所示。

例 2-26：演示在非定义 XML 文件中使用 sql 元素的方式。需要改动的文件为 CcoursesMapper.java、CcoursesMapper.xml 和 TestMB.java。Courses.java 文件内容详见

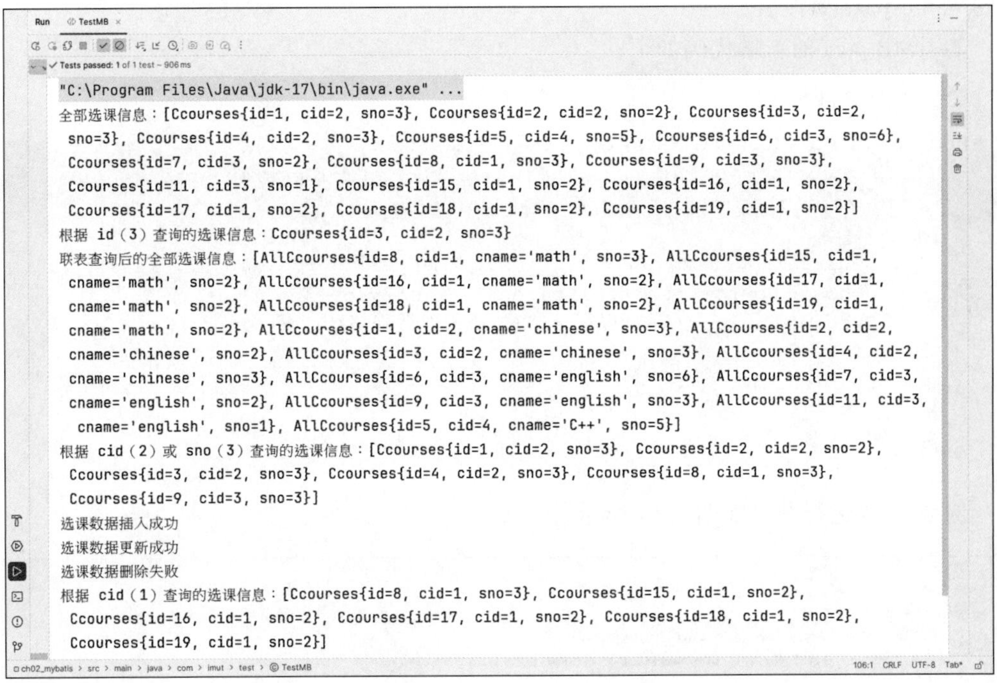

```
Run    TestMB
✓ Tests passed: 1 of 1 test – 906 ms
"C:\Program Files\Java\jdk-17\bin\java.exe" ...
全部选课信息：[Ccourses{id=1, cid=2, sno=3}, Ccourses{id=2, cid=2, sno=2}, Ccourses{id=3, cid=2,
sno=3}, Ccourses{id=4, cid=2, sno=3}, Ccourses{id=5, cid=4, sno=5}, Ccourses{id=6, cid=3, sno=6},
Ccourses{id=7, cid=3, sno=2}, Ccourses{id=8, cid=1, sno=3}, Ccourses{id=9, cid=3, sno=3},
Ccourses{id=11, cid=3, sno=1}, Ccourses{id=15, cid=1, sno=2}, Ccourses{id=16, cid=1, sno=2},
Ccourses{id=17, cid=1, sno=2}, Ccourses{id=18, cid=1, sno=2}, Ccourses{id=19, cid=1, sno=2}]
根据 id（3）查询的选课信息：Ccourses{id=3, cid=2, sno=3}
联表查询后的全部选课信息：[AllCcourses{id=8, cid=1, cname='math', sno=3}, AllCcourses{id=15, cid=1,
cname='math', sno=2}, AllCcourses{id=16, cid=1, cname='math', sno=2}, AllCcourses{id=17, cid=1,
cname='math', sno=2}, AllCcourses{id=18, cid=1, cname='math', sno=2}, AllCcourses{id=19, cid=1,
cname='math', sno=2}, AllCcourses{id=1, cid=2, cname='chinese', sno=3}, AllCcourses{id=2, cid=2,
cname='chinese', sno=2}, AllCcourses{id=3, cid=2, cname='chinese', sno=3}, AllCcourses{id=4, cid=2,
cname='chinese', sno=3}, AllCcourses{id=6, cid=3, cname='english', sno=6}, AllCcourses{id=7, cid=3,
cname='english', sno=2}, AllCcourses{id=9, cid=3, cname='english', sno=3}, AllCcourses{id=11, cid=3,
cname='english', sno=1}, AllCcourses{id=5, cid=4, cname='C++', sno=5}]
根据 cid（2）或 sno（3）查询的选课信息：[Ccourses{id=1, cid=2, sno=3}, Ccourses{id=2, cid=2, sno=2},
Ccourses{id=3, cid=2, sno=3}, Ccourses{id=4, cid=2, sno=3}, Ccourses{id=8, cid=1, sno=3},
Ccourses{id=9, cid=3, sno=3}]
选课数据插入成功
选课数据更新成功
选课数据删除失败
根据 cid（1）查询的选课信息：[Ccourses{id=8, cid=1, sno=3}, Ccourses{id=15, cid=1, sno=2},
Ccourses{id=16, cid=1, sno=2}, Ccourses{id=17, cid=1, sno=2}, Ccourses{id=18, cid=1, sno=2},
Ccourses{id=19, cid=1, sno=2}]
ch02_mybatis > src > main > java > com > imut > test > TestMB          106:1  CRLF  UTF-8  Tab*
```

图 2-13　例 2-25 的运行结果

例 1-4，其余文件均沿用例 2-25 的即可。

CoursesMapper.java 的代码如下。

```
package com.imut.mapper;
import com.imut.pojo.Courses;
public interface CoursesMapper {
    Courses selectByPrimaryKey(Integer cid);
}
```

CoursesMapper.xml 的代码如下。

```
<?xml version="1.0" encoding="UTF-8"?>
<!DOCTYPE mapper PUBLIC "-//mybatis.org//DTD Mapper 3.0//EN" "http://mybatis.
org/dtd/mybatis-3-mapper.dtd">
<mapper namespace="com.imut.mapper.CoursesMapper">
    <select id="selectByPrimaryKey" parameterType="java.lang.Integer"
resultType="courses">
        select cid, cname, tid
        from courses
        where <include refid="com.imut.mapper.CcoursesMapper.query_courses"/>
    </select>
</mapper>
```

例 2-26 是以在 CoursesMapper.xml 中使用 CcoursesMapper.xml 中定义的 sql 元素为例，引用时依旧使用 include 元素。但需注意：refid 属性值不能只是 sql 元素的 id 属性值，还需要加上 CcoursesMapper 的命名空间作为前缀，即 CcoursesMapper.xml 中 mapper 元

素的 namespace 属性值。

TestMB.java 的新增代码如下。

```
@Test
public void testCoursesMapper() throws Exception {
    SqlSessionFactory ssf = new MyBatisUtil().getSqlSessionFactory();
    SqlSession session = ssf.openSession();
    try {
        CoursesMapper mapper = session.getMapper(CoursesMapper.class);
        Courses courses = mapper.selectByPrimaryKey(3);
        System.out.println("根据 cid(3)查询的课程信息:" + courses);
    } finally {
        session.close();
    }
}
```

例 2-26 的运行结果如图 2-14 所示。

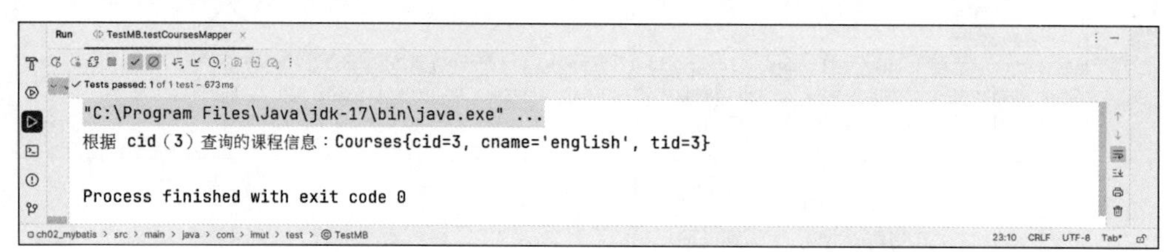

图 2-14　例 2-26 的运行结果

至此，MyBatis 的 XML 文件中的常用元素及其常用场景讲解完毕。

2.3　综合案例

本章案例主要在第 1 章的基础上增加实现“学生信息管理”模块中查询所有学生信息、增加学生信息、修改学生信息以及删除学生信息的功能，并把查询结果及日志信息输出到控制台上。

2.3.1　案例设计

本章案例划分为持久层和数据访问层。持久层由实体类 Student.java 组成。数据访问层由接口和映射文件组成，接口的名称统一以 Mapper 结尾，映射文件名称与接口文件名称相同。案例中将获取 SqlSession 的公共代码抽取出来，形成工具类。本章案例使用的主要文件如表 2-5 所示。

表 2-5　本章案例使用的主要文件

文　　件	所在包/路径	功　　能
log4j.properties	src/	日志输出配置文件
db.properties	src/	数据库连接的属性文件

续表

文　件	所在包/路径	功　能
mybatis-config.xml	src/	MyBatis 配置文件
Student.java	com.imut.pojo	封装学生信息的类
StudentMapper.java	com.imut.mapper	定义操作学生表方法的接口
StudentMapper.xml	com.imut.mapper	Student 类的映射文件
MybatisUtil.java	com.imut.utils	获取 SqlSession 对象的工具类
StudentTest.java	com.imut.test	测试操作学生信息方法的类

2.3.2 案例演示

使用 Junit4 测试执行 StudentTest.java 中的方法。执行 listStudentsTest()方法列出所有学生信息，在控制台输出的结果如图 2-15 所示。

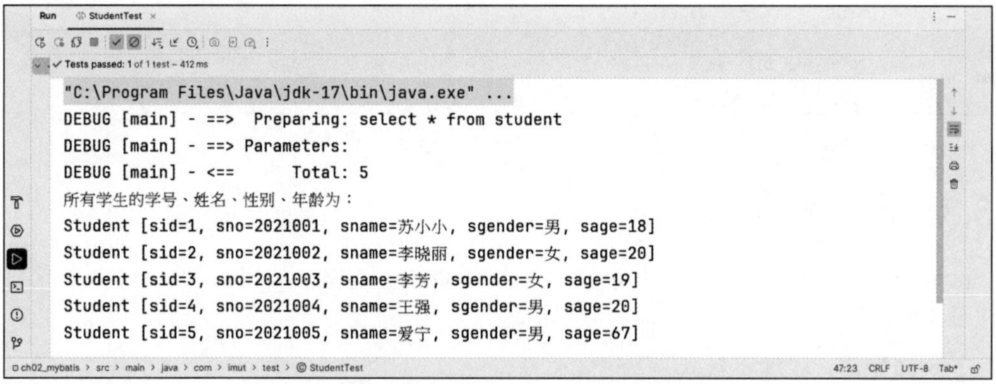
Run
StudentTest
Tests passed: 1 of 1 test – 412 ms
"C:\Program Files\Java\jdk-17\bin\java.exe" ...
DEBUG [main] - ==> Preparing: select * from student
DEBUG [main] - ==> Parameters:
DEBUG [main] - <==   Total: 5
所有学生的学号、姓名、性别、年龄为：
Student [sid=1, sno=2021001, sname=苏小小, sgender=男, sage=18]
Student [sid=2, sno=2021002, sname=李晓丽, sgender=女, sage=20]
Student [sid=3, sno=2021003, sname=李芳, sgender=女, sage=19]
Student [sid=4, sno=2021004, sname=王强, sgender=男, sage=20]
Student [sid=5, sno=2021005, sname=爱宁, sgender=男, sage=67]
ch02_mybatis > src > main > java > com > imut > test > StudentTest
47:23 CRLF UTF-8 Tab*

图 2-15　输出所有学生信息的运行结果

执行 addStudentTest()方法，增加一条姓名为“李林”的信息，在控制台输出的结果如图 2-16 所示。

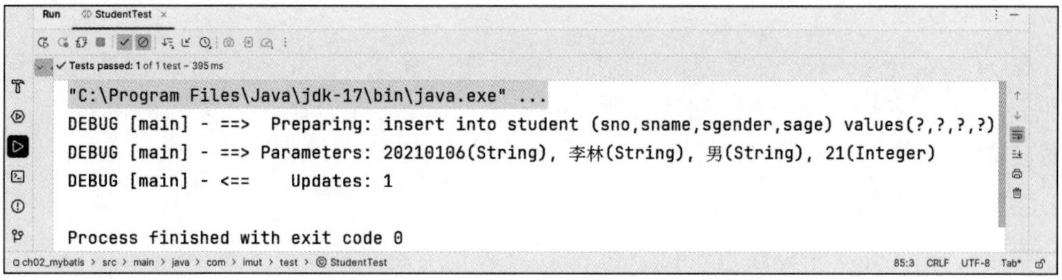
Run
StudentTest
Tests passed: 1 of 1 test – 395 ms
"C:\Program Files\Java\jdk-17\bin\java.exe" ...
DEBUG [main] - ==> Preparing: insert into student (sno,sname,sgender,sage) values(?,?,?,?)
DEBUG [main] - ==> Parameters: 20210106(String), 李林(String), 男(String), 21(Integer)
DEBUG [main] - <==   Updates: 1
Process finished with exit code 0
ch02_mybatis > src > main > java > com > imut > test > StudentTest
85:3 CRLF UTF-8 Tab*

图 2-16　执行增加学生信息的运行结果

执行 updateStudentTest()方法，将 sid 为 6 的学生姓名由“李林”修改为“吕永林”，在控制台输出的结果如图 2-17 所示。

执行 deleteStudentTest()方法，删除 sid 为 6 的学生信息，在控制台输出的结果如图 2-18 所示。

```
"C:\Program Files\Java\jdk-17\bin\java.exe" ...
DEBUG [main] - ==>  Preparing: select * from student where sid = ?
DEBUG [main] - ==> Parameters: 6(Integer)
DEBUG [main] - <==      Total: 1
DEBUG [main] - ==>  Preparing: update student set sno=?,sname=?,sgender=?,sage=? where
 sid=?
DEBUG [main] - ==> Parameters: 20210106(String), 吕永林(String), 男(String), 21(Integer),
 6(Integer)
DEBUG [main] - <==    Updates: 1
```

图 2-17　更新学生信息的运行结果

```
"C:\Program Files\Java\jdk-17\bin\java.exe" ...
DEBUG [main] - ==>  Preparing: delete from student where sid=?
DEBUG [main] - ==> Parameters: 6(Integer)
DEBUG [main] - <==    Updates: 1

Process finished with exit code 0
```

图 2-18　删除学生信息的运行结果

2.3.3　代码实现

pom.xml、log4j.properties、db.properties、mybatis-config.xml 以及 Student.java 文件内容详见 1.6.3 节。

1. 创建项目，添加依赖

创建名为 ch02_mybatis 的 Maven 工程，在 pom.xml 文件中添加 junit、log4j、mybatis、mysql-connector-java 4 个依赖，代码与 1.6.3 节中的 pom.xml 文件内容一致，这里不再赘述。

2. 创建配置文件

在/src/main/resources 下创建数据库连接属性配置文件 db.properties、日志输出配置文件 log4j.properties、mybatis 配置文件 mybatis-config.xml。这 3 个文件的内容与 1.6.3 节对应文件内容一致，这里不再赘述。

3. 创建实体类

在/src/main/java 下创建包 com.imut.pojo，在该包下新建实体类 Student.java，Student.java 文件内容详见 1.6.3 节。

4. 创建接口文件

在/src/main/java 下创建包 com.imut.mapper，在该包下创建接口文件 StudentMapper.java，代码如下。

```
package com.imut.mapper;
import java.util.List;
import com.imut.pojo.Student;
public interface StudentMapper {
    public Student getStudentById(int sid);
    public List<Student> listStudents();
    public void addStudent(Student student);
    public void updateStudent(Student student);
    public void deleteStudent(int sid);
}
```

5. 创建映射文件

在 com.imut.mapper 下创建映射器文件 StudentMapper.xml，StudentMapper.xml 的代码如下。

```
<?xml version="1.0" encoding="UTF-8" ?>
<!DOCTYPE mapper PUBLIC "-//mybatis.org//DTD Mapper 3.0//EN"
"http://mybatis.org/dtd/mybatis-3-mapper.dtd">
<mapper namespace="com.imut.mapper.StudentMapper">
    <select id="getStudentById" resultType="com.imut.pojo.Student">
        select * from student where sid = #{sid}
    </select>
    <select id="listStudents" resultType="com.imut.pojo.Student">
        select * from student
    </select>
    <insert id="addStudent">
        insert into student (sno,sname,sgender,sage)
            values(#{sno},#{sname},#{sgender},#{sage})
    </insert>
    <update id="updateStudent">
        update student set sno=#{sno},sname=#{sname},sgender=#{sgender},sage
=#{sage}
        where sid=#{sid}
    </update>
    <delete id="deleteStudent">
        delete from student where sid=#{sid}
    </delete>
</mapper>
```

6. 创建工具类

在 com.imut 目录下创建包 utils，在 com.imut.utils 包下新建工具类 MybatisUtil.java，用于创建并获取 SqlSession 对象。MybatisUtil.java 的代码如下。

```
package com.imut.utils;
import java.io.IOException;
import java.io.InputStream;
import org.apache.ibatis.io.Resources;
```

```
import org.apache.ibatis.session.SqlSession;
import org.apache.ibatis.session.SqlSessionFactory;
import org.apache.ibatis.session.SqlSessionFactoryBuilder;
public class MybatisUtil {
    private static SqlSessionFactory sqlSessionFactory;
    static {
        if(sqlSessionFactory==null) {
            try {
                String resource = "mybatis-config.xml";
                InputStream inputStream = Resources.getResourceAsStream(resource);
                sqlSessionFactory = new SqlSessionFactoryBuilder().build
(inputStream);
            } catch (IOException e) {
                e.printStackTrace();
            }
        }
    }
    public static SqlSession getSqlSession() {
        return sqlSessionFactory.openSession();
    }
}
```

7. 创建测试类

在 com.imut 目录下新建 test 目录，在 com.imut.test 目录创建测试类 StudentTest.java。测试列出所有学生信息、增加学生信息、修改学生信息以及删除学生信息的功能。StudentTest.java 的代码如下。

```
package com.imut.test;
import java.io.InputStream;
import java.util.List;
import org.apache.ibatis.io.Resources;
import org.apache.ibatis.session.SqlSession;
import org.apache.ibatis.session.SqlSessionFactory;
import org.apache.ibatis.session.SqlSessionFactoryBuilder;
import org.junit.Test;
import com.imut.mapper.StudentMapper;
import com.imut.pojo.Student;
import com.imut.utils.MybatisUtil;
public class StudentTest {
    @Test
    public void getStudentByIdTest() {
        String resource = "mybatis-config.xml";
        SqlSession sqlSession=null;
        try {
            InputStream inputStream = Resources.getResourceAsStream(resource);
            SqlSessionFactory sqlSessionFactory = new SqlSessionFactoryBuilder().
build(inputStream);
            sqlSession = sqlSessionFactory.openSession();
```

```
        StudentMapper mapper = sqlSession.getMapper(StudentMapper.class);
        Student student = mapper.getStudentById(2);
        System.out.println("查询出的编号为 2 的学生的学号、姓名、性别、年龄为:");
        System.out.println(student);
    }catch(Exception e) {
        e.printStackTrace();
    }finally {
        sqlSession.close();
    }
}
@Test
public void listStudentsTest() {
    SqlSession sqlSession = MybatisUtil.getSqlSession();
    StudentMapper mapper = sqlSession.getMapper(StudentMapper.class);
    List<Student> list = mapper.listStudents();
    System.out.println("所有学生的学号、姓名、性别、年龄为:");
    for (Student s : list) {
        System.out.println(s);
    }
    sqlSession.close();
}
@Test
public void addStudentTest() {
    SqlSession sqlSession = MybatisUtil.getSqlSession();
    StudentMapper mapper = sqlSession.getMapper(StudentMapper.class);
    Student student = new Student();
    student.setSno("20210106");
    student.setSname("李林");
    student.setSgender("男");
    student.setSage(21);
    mapper.addStudent(student);
    sqlSession.commit();
    sqlSession.close();
}
@Test
public void updateStudentTest() {
    SqlSession sqlSession = MybatisUtil.getSqlSession();
    StudentMapper mapper = sqlSession.getMapper(StudentMapper.class);
    Student student = mapper.getStudentById(6);
    student.setSname("吕永林");
    mapper.updateStudent(student);
    sqlSession.commit();
    sqlSession.close();
}
@Test
public void deleteStudentTest() {
    SqlSession sqlSession = MybatisUtil.getSqlSession();
    StudentMapper mapper = sqlSession.getMapper(StudentMapper.class);
    mapper.deleteStudent(6);
    sqlSession.commit();
```

```
        sqlSession.close();
    }
}
```

2.4 习题

1. 单选题

(1) MyBatis的两种事务管理器类型是(　　)。

A. JDBC和JNDI　　B. POOLED和UNPOOLED

C. JDBC和MANAGED　　D. MANAGED和JNDI

(2) 在MyBatis的配置文件中,不包含以下(　　)元素。

A. <sqlMapConfig>　　B. <environments>

C. <configuration>　　D. <environment>

2. 填空题

(1) select元素传入参数类型的属性是________,返回类型的属性是________。

(2) resultMap元素的子元素有________。

(3) 事务提交的方式有两种,分别是________和________。对应的代码是________。

3. 简答题

(1) 在XML映射文件中,除select、insert、update、delete元素之外,还有哪些元素?

(2) 在不同的XML映射文件中,id是否可以重复?为什么?

4. 编程题

基于XML配置方式实现新闻信息的增、删、改、查。新闻表结构如表2-6所示。

表2-6　新闻表结构

字段名	字段类型	字段含义
newsid	int	新闻id
newstitle	varchar(40)	新闻标题
newsauthor	varchar(10)	新闻作者

第3章 关联查询

本章主要讲解表与表间的关系，并详细讲解一对一、一对多、多对多关系。

3.1 一对一

一对一体现了唯一性，生活中有许多唯一标识的对象间的关系，下面以图书借阅证与学生的对应关系为例，说明其是如何完成一对一查询的。

例 3-1：演示根据学号查询图书借阅证信息的方式，显示信息需包含学生姓名。共包含 11 个文件：Students.java（学生信息实体类）、StudentsMapper.java（学生信息 Mapper 接口）、StudentsMapper.xml（学生信息 XML 映射器）、Librarycards.java（图书借阅证信息实体类）、LibrarycardsMapper.java（图书借阅证信息 Mapper 接口）、LibrarycardsMapper.xml（图书借阅证信息 XML 映射器）、SelectLibrarycardsBySno.java（学生图书借阅证信息实体类）、mybatis-config.xml、db.properties、MyBatisUtil.java 和 TestMB.java。创建实体类前，需要先创建数据库表 students，并填充表中数据。

students 表结构及数据如表 3-1 所示。

表 3-1 学生（students）表结构及数据

sno（学号，主键，自增，非空）	sname（姓名，默认为 NULL）	ssex（性别，默认为 NULL）	sage（年龄，默认为 NULL）	sclaid（班级 ID，默认为 NULL）
1	zhangsan	男	20	1
2	lisi	男	21	2
3	wangwu	女	24	3
4	zhaoliu	男	22	4

续表

sno(学号,主键,自增,非空)	sname(姓名,默认为 NULL)	ssex(性别,默认为 NULL)	sage(年龄,默认为 NULL)	sclaid(班级 ID,默认为 NULL)
5	qianqi	女	23	1
6	tiezhu	男	20	1
7	John	女	18	NULL

Students.java 的代码如下。

```
package com.imut.pojo;
import java.io.Serializable;
public class Students implements Serializable{
    private String sno;
    private String sname;
    private String ssex;
    private int sage;
    private int sclaid;
    //此处省略属性对应的 getter()、setter()和 toString()方法
}
```

librarycards 表结构及数据如表 3-2 所示。

表 3-2　图书借阅证(librarycards)表结构及数据

lid(图书借阅证 ID,主键,非空)	sno(学号,外键,默认为 NULL)
1	1
3	2
2	4

注意:建立约束前,需要确保 students 表的学号列存在索引,否则将会报错。

学号为外键,用于与学生信息关联,即与学生表中的学号关联。

根据创建的 librarycards 表建立图书借阅证实体类 Librarycards.java。与 students 表的关联需要自行编写,即添加一个 Students 类型的变量,其余内容可自动生成。Librarycards.java 的代码如下。

```
package com.imut.pojo;
public class Librarycards {
    private Integer lid;
    private Students students;
    //此处省略属性对应的 getter()、setter()和 toString()方法
}
```

实体定义完毕,接着编写相应的接口及具体实现。实现需求的方法不止一种,下面将介绍 3 种方法。

方法 1:通过两条 SQL 语句分别查询两张表的数据,将两次查询结果拼接为一条信息。

要实现上述方法，首先需要在 StudentsMapper.java 中定义一个查询学生信息的方法，该方法是根据学号查询学生信息的，通过学号即可与图书借阅证信息关联。StudentsMapper.java 的代码如下。

```
package com.imut.mapper;
import com.imut.pojo.Students;
public interface StudentsMapper {
    Students selectStudentAllBySno(int sno);
}
```

StudentsMapper.java 只是接口，具体的 SQL 语句需要在 StudentsMapper.xml 中编写。编写时需注意：①select 元素的 id 属性值需与 StudentsMapper.java 接口中定义的方法名称一致；②返回结果集类型既可以使用 select 元素中的 resultMap 属性指定，也可以使用 resultType 属性。使用 resultMap 属性，则需要额外使用 resultMap 元素定义结果映射；使用 resultType 属性，则属性值写为全限定类名或别名。以使用 resultMap 属性为例，StudentsMapper.xml 的代码如下。

```
<?xml version="1.0" encoding="UTF-8" ?>
<!DOCTYPE mapper PUBLIC "-//mybatis.org//DTD Mapper 3.0//EN"
    "http://mybatis.org/dtd/mybatis-3-mapper.dtd">
<mapper namespace="com.imut.mapper.StudentsMapper">
    <resultMap id="BaseResultMap" type="com.imut.pojo.Students">
        <id column="sno" jdbcType="INTEGER" property="sno" />
        <result column="sname" jdbcType="VARCHAR" property="sname" />
        <result column="ssex" jdbcType="VARCHAR" property="ssex" />
        <result column="sage" jdbcType="INTEGER" property="sage" />
    </resultMap>
    <select id="selectStudentAllBySno" parameterType="Integer" resultMap=
"BaseResultMap">
        select * from students where sno=#{sno}
    </select>
</mapper>
```

针对学生表的操作定义完毕，resultMap 元素中缺少一个 result 子元素，即未对 sclaid 字段映射，但是因为数据库列名与实体类属性名一致，故依旧可以获取该值。接着需要在 LibrarycardsMapper.java 中定义一个查询图书借阅证信息的方法，该方法是根据学号查询图书借阅证信息的，通过学号即可与学生信息关联。LibrarycardsMapper.java 的代码如下。

```
package com.imut.mapper;
import com.imut.pojo.Librarycards;
public interface LibrarycardsMapper {
    Librarycards selectLibrarycardBySno(int sno);
}
```

LibrarycardsMapper.java 只是接口，具体的 SQL 语句需要在 LibrarycardsMapper.xml 中编写。因为 Students 类对象定义在 Librarycards 类中，故而结果集类型应为 Librarycards 才能满

足需求。整体的思路为：第一条 SQL 语句查询 librarycards 表，第二条 SQL 语句查询 students 表，最后将需要的结果封装。

要想实现一对一关联查询，需要使用一个新的元素 association，association 元素包含多个属性，这里主要介绍 4 个属性：column 属性、javaType 属性、property 属性和 select 属性。column 属性用于指定两表关联的列名称，也就是学号，值得注意的是，此处写的必须是 librarycards 中的学号列名称，因为首先查询的是 librarycard 表，即调用的是 LibrarycardsMapper. java 中的 selectLibrarycardBySno 方法，而不是 StudentsMapper.java 中的 selectStudentAllBySno 方法。也就是说，如果使用 students 表中的学号列名称，学号这个形参的值无法传递给第二条 SQL 语句，学生信息自然无法获取；javaType 属性用于指定关联类型；property 属性用于指定 Librarycards 类中关联的 Students 类的属性名称；select 属性用于指定执行的查询语句。在本案例中，语句名称与 StudentsMapper.xml 里 select 元素的 id 属性值一致，但因不在一个 XML 文件中，故需加 namespace 前缀。resultMap 元素中定义需要显示的信息即可，除此之外仍需注意 StudentsMapper.xml 编写时提到的问题。以使用自定义 resultMap 元素为例，LibrarycardsMapper.xml 的代码如下。

```
<?xml version="1.0" encoding="UTF-8"?>
<!DOCTYPE mapper PUBLIC "-//mybatis.org//DTD Mapper 3.0//EN" "http://mybatis.
org/dtd/mybatis-3-mapper.dtd">
<mapper namespace="com.imut.mapper.LibrarycardsMapper">
    <resultMap id="lcardAndStudent" type="com.imut.pojo.Librarycards">
        <id column="lid" jdbcType="INTEGER" property="lid" />
        <association column="sno" javaType="com.imut.pojo.Students" property
="students" select="com.imut.mapper.StudentsMapper.selectStudentAllBySno" />
    </resultMap>
    <select id="selectLibrarycardBySno" parameterType="Integer" resultMap=
"lcardAndStudent">
        select * from librarycards where sno=#{sno}
    </select>
</mapper>
```

mybatis-config.xml 文件沿用例 2-14 后形成的内容，db.properties 文件内容详见例 2-2，MyBatisUtil.java 文件内容详见例 1-2。

TestMB.java 的新增代码如下。

```
@Test
public void testLibrarycardsMapper() throws Exception {
    SqlSessionFactory ssf = new MyBatisUtil().getSqlSessionFactory();
    SqlSession session = ssf.openSession();
    try {
        LibrarycardsMapper mapper = session.getMapper(LibrarycardsMapper.class);
        System.out.println("========方法 1========");
        int sno = 1;
        Librarycards librarycards = mapper.selectLibrarycardBySno(sno);
        System.out.println("根据 sno(1)查询的图书借阅证信息:" + librarycards);
    } finally {
```

```
            session.close();
        }
    }
```

通过上述方式即可实现需求。

方法 2：通过一条 SQL 语句查询两张表数据，将查询结果放在一个 resultMap 中映射。实现上述方法，依旧需要方法 1 的所有文件，其中需要改动的程序文件为 LibrarycardMapper.java、LibrarycardMapper.xml 和 TestMB.java。首先需要在 LibrarycardMapper.java 中新定义一个查询图书借阅证信息的方法，该方法还是根据学号查询图书借阅证信息的，通过学号即可与学生信息关联。LibrarycardsMapper.java 的新增代码如下。

```
Librarycards selectLibrarycardBySno1(int sno);
```

编写完毕 LibrarycardsMapper.java，接着需要在 LibrarycardsMapper.xml 中编写具体的 SQL 语句，结果集类型仍然为 Librarycards 类型。使用的依旧是 association 元素，因为只有一条 SQL 语句，故不需要使用其 select 属性，且 column 属性值可以选择两张表中的任一学号列名称。在 association 元素中需要定义返回字段映射，笔者在案例中完成 students 表中部分字段的映射，但因为 SQL 语句只查询姓名列，故返回结果其他属性值为列类型的默认值。以使用自定义 resultMap 元素为例，LibrarycardsMapper.xml 的新增代码如下。

```
<resultMap id="lcardAndStudent1" type="com.imut.pojo.Librarycards">
    <id column="lid" jdbcType="INTEGER" property="lid" />
    <association column="sno" javaType="com.imut.pojo.Students" property="students">
        <id column="sno" jdbcType="INTEGER" property="sno" />
        <result column="sname" jdbcType="VARCHAR" property="sname" />
        <result column="sage" jdbcType="INTEGER" property="sage" />
    </association>
</resultMap>
<select id="selectLibrarycardBySno1" parameterType="Integer" resultMap=
"lcardAndStudent1">
    select lc.*, s.sname
    from librarycards lc, students s
    where lc.sno=s.sno and lc.sno=#{sno}
</select>
```

TestMB.java 的新增代码如下，位于 testLibrarycardsMapper 方法中。

```
System.out.println("========方法 2========");
librarycards = mapper.selectLibrarycardBySno1(sno);
System.out.println("根据 sno(1)查询的图书借阅证信息:" + librarycards);
```

通过上述方式即可实现需求。方法 1 和方法 2 都使用与表相对应的类，方法 3 则转变了思路。

方法 3：通过一条 SQL 语句查询两张表数据，将查询结果映射为一个新的类。想要实现上述方法，首先需要手动编写一个新类 SelectLibrarycardBySno.java，作为结果的封装类，类中包含所需显示的全部属性及属性相应的 getter、setter 和 toString 方法。SelectLibrarycardBySno.java 的代码如下。

```
package com.imut.pojo;
public class SelectLibrarycardBySno {
    private Integer lid;
    private Integer sno;
    private String sname;
    //此处省略属性对应的 getter()、setter()和 toString()方法
}
```

想要实现上述方法，需要在 LibrarycardsMapper.java 中新定义一个查询图书借阅证信息的方法，该方法还是根据学号查询图书借阅证信息的，通过学号即可与学生信息关联。不同的是返回类型变为 SelectLibrarycardBySno。LibrarycardsMapper.java 的新增代码如下。

```
SelectLibrarycardBySno selectLibrarycardBySno2(int sno);
```

方法 3 的 LibrarycardsMapper.xml 编写非常简单：SQL 语句与方法 2 一致，只是 select 元素中使用的属性由 resultMap 变为 resultType，resultType 的属性值类型为 SelectLibrarycardBySno 类的全限定名。LibrarycardsMapper.xml 的新增代码如下。

```
<select id="selectLibrarycardBySno2" parameterType="Integer" resultType=
"com.imut.pojo.SelectLibrarycardBySno">
    select lc.* , s.sname
    from librarycards lc, students s
    where lc.sno=s.sno and lc.sno=#{sid}
</select>
```

TestMB.java 的新增代码如下，位于 testLibrarycardsMapper 方法中。

```
System.out.println("========方法 3========");
SelectLibrarycardBySno selectLibrarycardBySno = mapper. selectLibrarycardBySno2
(sno);
System.out.println("根据 sno(1)查询的图书借阅证信息:" + selectLibrarycardBySno);
```

通过上述方式即可实现需求，至此 3 种方法都讲解完毕。其运行结果如图 3-1 所示。

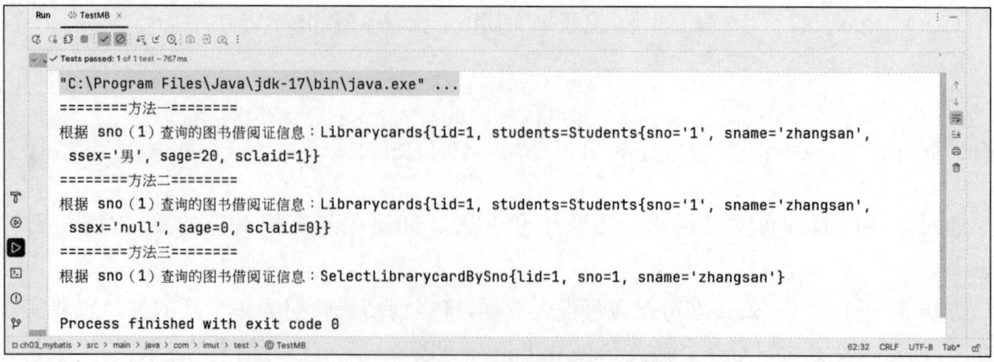

```
"C:\Program Files\Java\jdk-17\bin\java.exe" ...
========方法一========
根据 sno (1) 查询的图书借阅证信息：Librarycards{lid=1, students=Students{sno='1', sname='zhangsan',
 ssex='男', sage=20, sclaid=1}}
========方法二========
根据 sno (1) 查询的图书借阅证信息：Librarycards{lid=1, students=Students{sno='1', sname='zhangsan',
 ssex='null', sage=0, sclaid=0}}
========方法三========
根据 sno (1) 查询的图书借阅证信息：SelectLibrarycardBySno{lid=1, sno=1, sname='zhangsan'}

Process finished with exit code 0
```

图 3-1　一对一 3 种方法运行结果

3.2 一对多

生活中包含许多一对多的关系，如客户与订单、部门与员工、班级与学生等。下面以班级与学生的对应关系为例，说明其是如何完成一对多查询的。

例 3-2：演示根据班级 ID 查询学生信息的方式。共包含 11 个文件：Students.java、StudentsMapper. java、StudentsMapper. xml、Allclasses. java（班级信息实体类）、AllclassesMapper.java（班级信息 Mapper 接口）、AllclassesMapper.xml（班级信息 XML 映射器）、SelectClassStudentsByClaid.java（班级学生信息实体类）、mybatis-config.xml、db.properties、MyBatisUtil.java 和 TestMB.java。创建实体类前，需要先创建数据库表 allclasses，并填充表中数据。

allclasses 表结构及数据如表 3-3 所示。

表 3-3　班级(allclasses)表结构及数据

claid(班级 ID，主键，非空)	claname(班级名称，默认为 NULL)
1	一班
2	二班
3	三班
4	四班

根据创建的 allclasses 表建立班级实体类 Allclasses.java。与 students 表的关联需要自行编写，即添加一个 Students 类型的变量，因为一个班级通常有多名学生，故使用集合存储。Allclasses.java 代码如下。

```
package com.imut.pojo;
import java.util.List;
public class Allclasses {
    private String claname;
    private List<Students> studentsList;
    ...
        //此处省略属性对应的 getter()、setter()和 toString()方法
    ...
}
```

实体类定义完毕，接着编写相应的接口及具体实现。实现需求的方法不止一种，下面同样介绍 3 种方式。

方法 1：通过两条 SQL 语句分别查询两张表数据，将两次查询结果拼接为一条信息。实现上述方法，首先需要在 StudentsMapper.java 中定义一个查询学生信息的方法，该方法是根据班级 ID 查询学生信息的，通过班级 ID 即可与班级信息关联。StudentsMapper.java 的新增代码如下。

```
List<Students> selectStudentByClaid(int claid);
```

StudentsMapper.java 只是接口文件，具体的 SQL 语句需要在 StudentsMapper.xml 中编写。编写时需注意的问题在 3.1 节已讲过，这里不再赘述。以使用 resultType 属性为例，StudentsMapper.xml 的新增代码如下。

```
<select id="selectStudentByClaid" resultType="com.imut.pojo.Students"
parameterType="Integer">
    select * from students where sclaid=#{claid}
</select>
```

针对学生表的操作编写完毕，接着需要在 AllclassesMapper.java 中定义一个查询班级信息的方法，该方法是根据班级 ID 查询班级信息的，通过班级 ID 即可与学生信息关联。AllclassesMapper.java 的代码如下。

```
package com.imut.mapper;
import com.imut.pojo.Allclasses;
public interface AllclassesMapper {
    Allclasses selectClassStudentsByClaid(int claid);
}
```

AllclassesMapper.java 只是接口，具体的 SQL 语句需要在 AllclassesMapper.xml 中编写。因为 Students 类对象定义在 Allclasses 类中，故而结果集类型应为 Allclasses 才能够满足需求。整体的思路为：第 1 条 SQL 语句查询 allclasses 表，第 2 条 SQL 语句查询 student 表，最后将需要的结果封装。

要想实现一对多关联查询，需要使用一个新的元素 collection。collection 元素包含多个属性，这里主要介绍 4 个属性：column 属性、ofType 属性、property 属性和 select 属性。column 属性用于指定两表关联的列名称，也就是班级 ID，值得注意的是，这里写的必须是 allclasses 中的班级 ID 列名称，因为首先查询的是 allclasses 表，即调用的是 AllclassesMapper.java 中的 selectClassStudentsByClaid 方法，而不是 StudentsMapper.java 中的 selectStudentByClaid 方法，即如果使用 students 表中的班级 ID 列名称，班级 ID 这个形参的值无法传递给第 2 条 SQL 语句，学生信息自然无法获取；ofType 属性用于指定集合中的元素类型；property 属性用于指定 Allclasses 类中关联的 Students 类的属性名称；select 属性用于指定执行的查询语句。在本案例中，语句名称与 StudentsMapper.xml 里 select 元素的 id 属性值一致，但因不在一个 XML 文件中，故需加 namespace 前缀。resultMap 元素中定义需要显示的信息即可，除此之外仍需注意 StudentsMapper.xml 编写时提到的问题。以使用自定义 resultMap 元素为例，AllclassesMapper.xml 的代码如下。

```
<?xml version="1.0" encoding="UTF-8" ?>
<!DOCTYPE mapper PUBLIC "-//mybatis.org//DTD Mapper 3.0//EN" "http://mybatis.
org/dtd/mybatis-3-mapper.dtd" >
<mapper namespace="com.imut.mapper.AllclassesMapper" >
    <resultMap id="classAndStudents" type="com.imut.pojo.Allclasses" >
```

```
        <id column="claid" property="claid" jdbcType="INTEGER" />
        <result column="claname" property="claname" jdbcType="VARCHAR" />
        <collection property="studentsList" ofType="com.imut.pojo.Students"
column="claid" select="com.imut.mapper.StudentsMapper.selectStudentByClaid" />
    </resultMap>
    <select id="selectClassStudentsByClaid" parameterType="Integer"
            resultMap="classAndStudents">
        select * from allclasses where claid = #{claid}
    </select>
</mapper>
```

TestMB.java 的新增代码如下。

```
@Test
public void testAllclassesMapper() throws Exception {
    SqlSessionFactory ssf = new MyBatisUtil().getSqlSessionFactory();
    SqlSession session = ssf.openSession();
    try {
        AllclassesMapper mapper = session.getMapper(AllclassesMapper.class);
        System.out.println("========方法 1========");
        int claid = 1;
        Allclasses allclasses = mapper.selectClassStudentsByClaid(claid);
        System.out.println("根据 claid(1)查询的班级信息:" + allclasses);
    } finally {
        session.close();
    }
}
```

通过上述方式即可实现需求。

方法 2：通过一条 SQL 语句查询两张表数据，将查询结果放在一个 resultMap 中进行映射。实现该方法，依旧需要方法 1 的所有文件，其中需要改动的程序为 AllclassesMapper.java、AllclassesMapper.xml 和 TestMB.java。首先需要在 AllclassesMapper.java 中新定义一个查询班级及学生信息的方法，该方法还是根据班级 ID 查询班级信息的，通过班级 ID 即可与学生信息关联。AllclassesMapper.java 的新增代码如下。

```
Allclasses selectClassStudentsByClaid1(int claid);
```

编写完毕 AllclassesMapper.java，接着需要在 AllclassesMapper.xml 中编写具体的 SQL 语句，结果集类型仍然为 Allclasses。使用的依旧是 collection 元素，因为只有一条 SQL 语句，故不需要使用其 select 属性，且 column 属性值可以选择两张表中的任一班级 ID 列名称。collection 元素中需要定义返回字段映射，案例中配置了 students 表中学号、姓名、年龄字段的映射。以使用自定义 resultMap 元素为例，AllclassesMapper.xml 的新增代码如下。

```
<resultMap id="classAndStudents1" type="com.imut.pojo.Allclasses" >
    <id column="claid" property="claid" jdbcType="INTEGER" />
```

```
        <result column="claname" property="claname" jdbcType="VARCHAR" />
        <collection property="studentsList" ofType="com.imut.pojo.Students">
            <id column="sno" jdbcType="INTEGER" property="sno" />
            <result column="sname" jdbcType="VARCHAR" property="sname" />
            <result column="sage" jdbcType="INTEGER" property="sage" />
        </collection>
    </resultMap>
    <select id="selectClassStudentsByClaid1" parameterType="Integer" resultMap=
    "classAndStudents1">
        select a.*, s.sno, s.sname, s.sage
        from allclasses a, students s
        where a.claid = s.sclaid and a.claid=#{claid}
    </select>
```

TestMB.java 的新增代码如下，位于 testAllclassesMapper 方法中。

```
System.out.println("========方法 2========");
allclasses = mapper.selectClassStudentsByClaid1(claid);
System.out.println("根据 claid(1)查询的班级信息:" + allclasses);
```

通过上述方式即可实现需求。方法 1 和方法 2 都是使用与表相对应的类，方法 3 则转变了思路。

方法 3：通过一条 SQL 语句查询两张表数据，将查询结果映射为一个新类。实现该方法，首先需要手动编写一个新类 SelectClassStudentsByClaid.java，作为结果的封装类，类中包含所需显示的全部属性及属性相应的 getter、setter 和 toString 方法。根据需求，SelectClassStudentsByClaid.java 的代码如下。

```
package com.imut.pojo;
public class SelectClassStudentsByClaid {
    private Integer claid;
    private String claname;
    private Integer sno;
    private String sname;
    private Integer sage;
        //此处省略属性对应的 getter()、setter()和 toString()方法
}
```

想要实现上述方法，需要在 AllclassesMapper.java 中新定义一个查询班级及学生信息的方法，该方法还是根据班级 ID 查询班级信息的，通过班级 ID 即可与学生信息关联。不同的是返回类型变为 SelectClassStudentsByClaid。AllclassesMapper.java 的新增代码如下。

```
List<SelectClassStudentsByClaid> selectClassStudentsByClaid2(int claid);
```

方法 3 中的 AllclassesMapper.xml 编写非常简单：SQL 语句与方法 2 一致，只是 select 元素中使用的属性由 resultMap 变为 resultType，resultType 的属性值为 SelectClassStudentsByClaid 类的全限定名。AllclassesMapper.xml 的新增代码如下。

```
<select id="selectClassStudentsByClaid2" parameterType="Integer" resultType
="com.imut.pojo.SelectClassStudentsByClaid">
    select a.*, s.sno, s.sname, s.sage
    from allclasses a, students s
    where a.claid = s.sclaid and a.claid=#{claid}
</select>
```

TestMB.java 的新增代码如下,位于 testAllclassesMapper 方法中。

```
System.out.println("========方法 3========");
List < SelectClassStudentsByClaid > selectClassStudentsByClaidList = mapper.
selectClassStudentsByClaid2(claid);
System.out.println("根据 claid(1)查询的班级信息:" + selectClassStudentsByClaidList);
```

通过上述方式即可实现需求,至此 3 种方法笔者都讲解完毕。其运行结果如图 3-2 所示。

```
Run   TestMB ×
✓ Tests passed: 1 of 1 test – 840 ms
"C:\Program Files\Java\jdk-17\bin\java.exe" ...
========方法一========
根据 claid (1) 查询的班级信息：Allclasses{claid=1, claname='一班', studentsList=[Students{sno='1',
 sname='zhangsan', ssex='男', sage=20, sclaid=1}, Students{sno='5', sname='qianqi', ssex='女',
 sage=23, sclaid=1}, Students{sno='6', sname='tiezhu', ssex='男', sage=20, sclaid=1}]}
========方法二========
根据 claid (1) 查询的班级信息：Allclasses{claid=1, claname='一班', studentsList=[Students{sno='1',
 sname='zhangsan', ssex='null', sage=20, sclaid=0}, Students{sno='5', sname='qianqi', ssex='null',
 sage=23, sclaid=0}, Students{sno='6', sname='tiezhu', ssex='null', sage=20, sclaid=0}]}
========方法三========
根据 claid (1) 查询的班级信息：[SelectClassStudentsByClaid{claid=1, claname='一班', sno=1,
 sname='zhangsan', sage=20}, SelectClassStudentsByClaid{claid=1, claname='一班', sno=5,
 sname='qianqi', sage=23}, SelectClassStudentsByClaid{claid=1, claname='一班', sno=6, sname='tiezhu',
 sage=20}]
ch03_mybatis > src > main > java > com > imut > test > TestMB        78:19  CRLF  UTF-8  Tab*
```

图 3-2 一对多 3 种方法运行结果

3.3 多对多

生活中包含大量多对多的关系,如教师与学生、订单与商品、电影与顾客、课程与学生等。下面以课程与学生的对应关系为例,说明其是如何完成多对多查询的。多对多其实可以看作两个一对多,即一门课程可以被多个学生选,一个学生可以选择多门课程。

例 3-3:演示查询所有课程及相应选课学生的信息的方式。共包含 8 个文件:Students. java、Courses. java、CoursesMapper. java、CoursesMapper. xml、mybatis-config. xml、db.properties、MyBatisUtil.java 和 TestMB.java。

先前已经创建过学生实体类 Students,需要在先前的基础上添加与之关联的 Courses 类属性及相应的 getter()和 setter()方法。为了使输出结果合理、简洁,故 toString()方法中不需要添加 Courses 类属性。Students.java 的新增代码如下。

```
import java.util.List;
public class Students implements Serializable{
    private List<Courses> coursesList;
        //此处省略属性对应的 getter()、setter()和 toString()方法
}
```

courses 表的 cid 用于与学生选课信息关联，即与选课表中的课程 ID 关联。

继续沿用例 1-4 所创建的 Courses 类，并添加与之关联的 Students 类属性及该属性的 getter()、setter()方法和 toString()方法。Courses.java 的新增代码如下。

```
import java.util.List;
public class Courses {
    private List<Students> studentsList;
        //此处省略属性对应的 getter()、setter()和 toString()方法
}
```

要想实现多对多，还需要一张中间表，通过课程、学生和中间表(选课表 ccourses)的关系实现多对多拆分成一对多的情况。

注意：建立约束前，需要确保 courses 表的课程 ID 列、students 表的学号列已经添加索引，否则将会报出错误。

不需要为选课记录创建实体。其余的实体定义完毕，接着编写相应的接口及具体实现。思路如下：通过一条 SQL 语句查询 3 张表数据，将查询结果封装成为需要的样式，即课程信息与选择该课学生的信息。实现上述方法，需要在 CoursesMapper.java 中定义一个查询所有课程及选课学生信息的方法，1.5.1 节已创建过 CoursesMapper.java，故只列出新增方法。CoursesMapper.java 的新增代码如下。

```
List<Courses> selectAllCoursesAndStudent();
```

编写完毕 CoursesMapper.java，需要在 CoursesMapper.xml 中编写具体的 SQL 语句，结果集类型为 Courses。使用的依旧是 collection 元素，因为只有一条 SQL 语句，故不需要使用其 select 属性。因为课程表和学生表并无直接关联，故不需要 column 属性。collection 元素中需要定义返回字段映射，笔者在案例中完成 students 表中全部字段的映射。以使用自定义 resultMap 元素为例，CoursesMapper.xml 的新增代码如下。

```
<resultMap id="CourseWithStudentResult" type="com.imut.pojo.Courses">
    <id column="cid" jdbcType="INTEGER" property="cid" />
    <result column="cname" jdbcType="VARCHAR" property="cname" />
    <result column="tid" jdbcType="INTEGER" property="tid" />
    <collection ofType="com.imut.pojo.Students"  property="studentsList">
        <id column="sno" jdbcType="INTEGER" property="sno" />
        <result column="sname" jdbcType="VARCHAR" property="sname" />
        <result column="ssex" jdbcType="VARCHAR" property="ssex" />
        <result column="sage" jdbcType="INTEGER" property="sage" />
        <result column="sclaid" jdbcType="INTEGER" property="sclaid" />
    </collection>
```

```
</resultMap>
<select id="selectAllCoursesAndStudent" parameterType="Integer" resultMap="
CourseWithStudentResult">
    select * from courses c, students s, ccourses cc
    where cc.cid=c.cid
    and cc.sno=s.sno
</select>
```

TestMB.java 的新增代码如下，位于 testCoursesMapper()方法中。

```
List<Courses> coursesList = mapper.selectAllCoursesAndStudent();
System.out.println("所有课程及相应选课学生的信息" + coursesList);
```

通过上述方式即可实现需求，运行结果如图 3-3 所示。

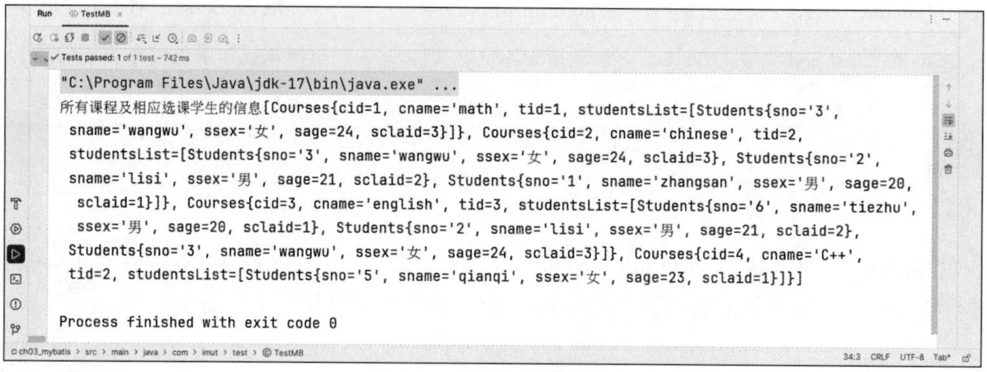

```
Run   TestMB ×
✓ Tests passed: 1 of 1 test – 742 ms
"C:\Program Files\Java\jdk-17\bin\java.exe" ...
所有课程及相应选课学生的信息[Courses{cid=1, cname='math', tid=1, studentsList=[Students{sno='3',
 sname='wangwu', ssex='女', sage=24, sclaid=3}]}, Courses{cid=2, cname='chinese', tid=2,
 studentsList=[Students{sno='3', sname='wangwu', ssex='女', sage=24, sclaid=3}, Students{sno='2',
 sname='lisi', ssex='男', sage=21, sclaid=2}, Students{sno='1', sname='zhangsan', ssex='男', sage=20,
  sclaid=1}]}, Courses{cid=3, cname='english', tid=3, studentsList=[Students{sno='6', sname='tiezhu',
  ssex='男', sage=20, sclaid=1}, Students{sno='2', sname='lisi', ssex='男', sage=21, sclaid=2},
 Students{sno='3', sname='wangwu', ssex='女', sage=24, sclaid=3}]}, Courses{cid=4, cname='C++',
 tid=2, studentsList=[Students{sno='5', sname='qianqi', ssex='女', sage=23, sclaid=1}]}]

Process finished with exit code 0
ch03_mybatis > src > main > java > com > imut > test > TestMB        34:3  CRLF  UTF-8  Tab*
```

图 3-3 多对多运行结果

3.4 综合案例

本章案例主要实现"学生信息管理"模块中根据学生 SID 查询该学生的选课信息的功能，实现"课程信息管理"模块中根据课程 CID 查询该课程的选课信息、根据学生 SID 查询该学生没有选修的课程等功能。

3.4.1 案例设计

本章案例划分为持久层和数据访问层。持久层由实体类 Student.java、Course.java、Ccourse.java 组成。数据访问层由接口和映射文件组成，接口的名称统一以 Mapper 结尾，映射文件名称与接口文件名称相同。本章案例使用的主要文件如表 3-4 所示。

表 3-4 本章案例使用的主要文件

文　件	所在包/路径	功　能
log4j.properties	src/	日志输出配置文件
db.properties	src/	数据库连接的属性文件

续表

文　　件	所在包/路径	功　　能
mybatis-config.xml	src/	MyBatis 配置文件
Student.java	com.imut.pojo	封装学生信息的类
Course.java	com.imut.pojo	封装课程信息的类
Ccourse.java	com.imut.pojo	封装选课信息的类
StudentMapper.java	com.imut.mapper	定义操作学生表方法的接口
CourseMapper.java	com.imut.mapper	定义操作课程表方法的接口
StudentMapper.xml	com.imut.mapper	Student 类的映射文件
CourseMapper.xml	com.imut.mapper	Course 类的映射文件
MybatisUtil.java	com.imut.utils	获取 SqlSession 对象的工具类
StudentTest.java	com.imut.test	测试操作学生信息方法的类
CourseTest.java	com.imut.test	测试操作课程信息方法的类

3.4.2 案例演示

使用 Junit4 测试执行 StudentTest.java 中的方法。执行 getCcoursesByIdTest()方法根据学生 SID 查询该学生的选课信息，在控制台输出的结果如图 3-4 所示。

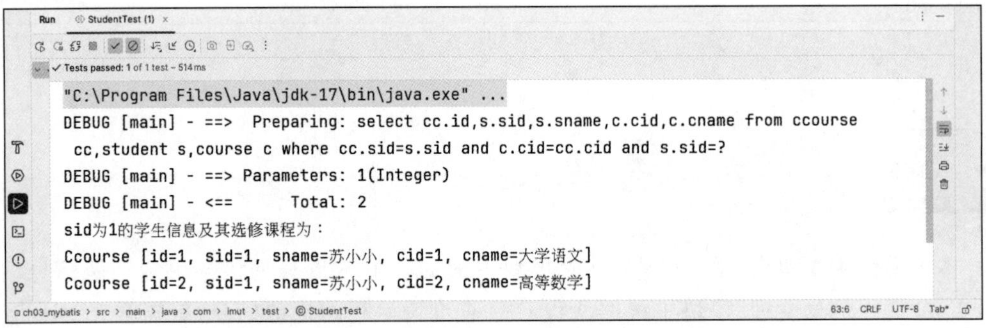

```
"C:\Program Files\Java\jdk-17\bin\java.exe" ...
DEBUG [main] - ==>  Preparing: select cc.id,s.sid,s.sname,c.cid,c.cname from ccourse
 cc,student s,course c where cc.sid=s.sid and c.cid=cc.cid and s.sid=?
DEBUG [main] - ==> Parameters: 1(Integer)
DEBUG [main] - <==      Total: 2
sid为1的学生信息及其选修课程为：
Ccourse [id=1, sid=1, sname=苏小小, cid=1, cname=大学语文]
Ccourse [id=2, sid=1, sname=苏小小, cid=2, cname=高等数学]
```

图 3-4　SID 为 1 的学生的选课信息

使用 Junit4 测试执行 CourseTest.java 中的方法。执行 getCcourseByIdTest()方法根据课程 CID 查询该课程的选课信息，在控制台输出的结果如图 3-5 所示。

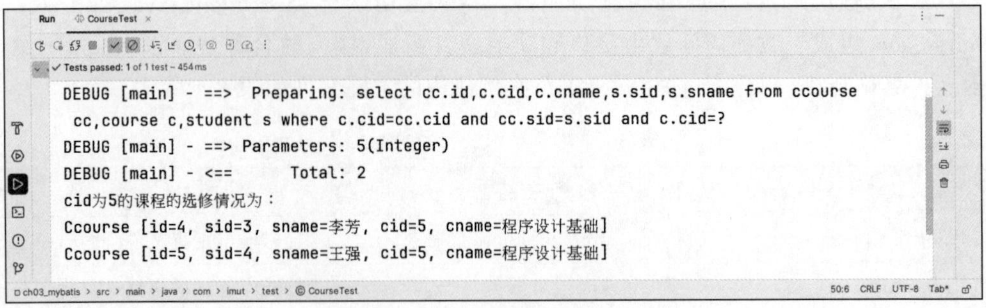

```
DEBUG [main] - ==>  Preparing: select cc.id,c.cid,c.cname,s.sid,s.sname from ccourse
 cc,course c,student s where c.cid=cc.cid and cc.sid=s.sid and c.cid=?
DEBUG [main] - ==> Parameters: 5(Integer)
DEBUG [main] - <==      Total: 2
cid为5的课程的选修情况为：
Ccourse [id=4, sid=3, sname=李芳, cid=5, cname=程序设计基础]
Ccourse [id=5, sid=4, sname=王强, cid=5, cname=程序设计基础]
```

图 3-5　CID 为 5 的课程的选修信息

执行 listSelectTest()方法根据学生 SID 查询该学生没有选修的课程，在控制台输出的结果如图 3-6 所示。

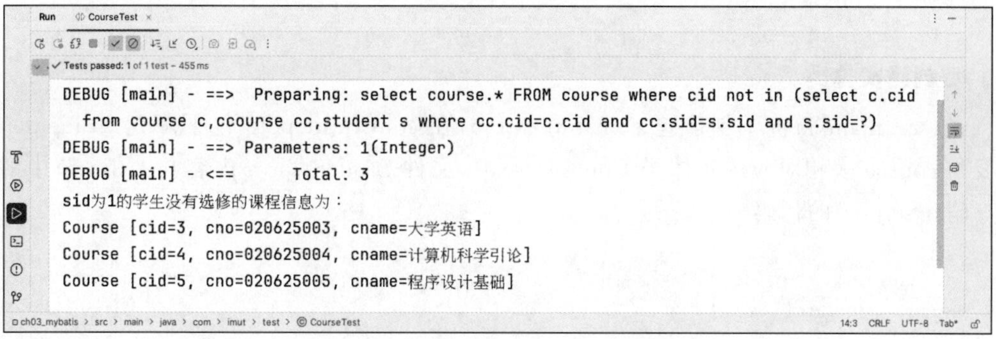
Run
CourseTest
Tests passed: 1 of 1 test – 455 ms
DEBUG [main] - ==> Preparing: select course.* FROM course where cid not in (select c.cid from course c,ccourse cc,student s where cc.cid=c.cid and cc.sid=s.sid and s.sid=?)
DEBUG [main] - ==> Parameters: 1(Integer)
DEBUG [main] - <==   Total: 3
sid为1的学生没有选修的课程信息为：
Course [cid=3, cno=020625003, cname=大学英语]
Course [cid=4, cno=020625004, cname=计算机科学引论]
Course [cid=5, cno=020625005, cname=程序设计基础]
ch03_mybatis > src > main > java > com > imut > test > CourseTest
14:3 CRLF UTF-8 Tab*

图 3-6　SID 为 1 的学生没有选修的课程信息

3.4.3　代码实现

1. 创建数据库表

在名为 db_student 的数据库下创建 course 表和 ccourse 表，并插入相应数据。course 表和 ccourse 表的具体信息分别如表 1-4 和表 1-5 所示。

2. 创建项目，添加依赖

创建名为 ch03_mybatis 的 Maven 工程，文件 pom.xml 内容与 2.3.3 节中的 pom.xml 文件内容一致。

3. 创建配置文件

在/src/main/resources 下创建配置文件 log4j.properties、db.properties 和 mybatis-config.xml。log4j.properties、db.properties 与 2.3.3 节中的对应文件内容一致。文件 mybatis-config.xml 的代码如下。

```
<?xml version="1.0" encoding="UTF-8" ?>
<!DOCTYPE configuration PUBLIC "-//mybatis.org//DTD Config 3.0//EN"
"http://mybatis.org/dtd/mybatis-3-config.dtd">
<configuration>
    <properties resource="db.properties"/>
    <environments default="development">
        <environment id="development">
            <transactionManager type="JDBC" />
            <dataSource type="POOLED">
                <property name="driver" value="${jdbc.driver}" />
                <property name="url" value="${jdbc.url}" />
                <property name="username" value="${jdbc.username}" />
                <property name="password" value="${jdbc.password}" />
            </dataSource>
        </environment>
    </environments>
    <mappers>
```

```
        <package name="com.imut.mapper"/>
    </mappers>
</configuration>
```

4. 创建实体类

在/src/main/java 下创建包 com.imut.pojo，在 com.imut.pojo 包下新建 Student.java 以及与 course 表相对应的实体类 Course .java。文件 Student.java 内容与 2.3.3 节中的一致。Course.java 的代码如下。

```
package com.imut.pojo;
import java.util.List;
public class Course {
    private int cid;
    private String cno;
    private String cname;
    private List<Student> students;
    //此处省略 getter()、setter()t 和 toString()方法
}
```

5. 创建接口文件

在 com.imut 下新建 mapper 包，在该包下创建接口文件 StudentMapper.java 和 CourseMapper.java。StudentMapper.java 的代码如下。

```
package com.imut.mapper;
import java.util.List;
import com.imut.pojo.Ccourse;
import com.imut.pojo.Student;
public interface StudentMapper {
    public Student getStudentById(int sid);
    public List<Student> listStudents();
    public List<Ccourse> getCcoursesById(int sid);
    public void addStudent(Student student);
    public void updateStudent(Student student);
    public void deleteStudent(int sid);
}
```

CourseMapper.java 的代码如下。

```
package com.imut.mapper;
import java.util.List;
import com.imut.pojo.Ccourse;
import com.imut.pojo.Course;
public interface CourseMapper {
    public List<Ccourse> getCcourseById(int cid);
    public List<Course> listSelect(int sid);
}
```

6. 创建映射文件

在 com.imut.mapper 目录下创建映射器文件 StudentMapper.xml，StudentMapper.xml 的代码如下。

```
<?xml version="1.0" encoding="UTF-8" ?>
<!DOCTYPE mapper
PUBLIC "-//mybatis.org//DTD Mapper 3.0//EN"
"http://mybatis.org/dtd/mybatis-3-mapper.dtd">
<mapper namespace="com.imut.mapper.StudentMapper">
    <select id="getStudentById" resultType="com.imut.pojo.Student">
        select * from student where sid = #{sid}
    </select>
    <select id="listStudents" resultType="com.imut.pojo.Student">
        select * from student
    </select>
    <select id="getCcoursesById" resultType="com.imut.pojo.Ccourse" >
         select cc.id, s.sid, s.sname, c.cid, c.cname from ccourse cc, student s,
course c
            where cc.sid=s.sid and c.cid=cc.cid and s.sid=#{sid}
    </select>
    <insert id="addStudent">
        insert into student (sno,sname,sgender,sage)
            values(#{sno},#{sname},#{sgender},#{sage})
    </insert>
    <update id="updateStudent">
        update student set sno=#{sno},sname=#{sname},sgender=#{sgender},sage
=#{sage}
        where sid=#{sid}
    </update>
    <delete id="deleteStudent">
        delete from student where sid=#{sid}
    </delete>
</mapper>
```

在 com.imut.mapper 目录下创建映射器文件 CourseMapper.xml，CourseMapper.xml 的代码如下。

```
<?xml version="1.0" encoding="UTF-8" ?>
<!DOCTYPE mapper
PUBLIC "-//mybatis.org//DTD Mapper 3.0//EN"
"http://mybatis.org/dtd/mybatis-3-mapper.dtd">
<mapper namespace="com.imut.mapper.CourseMapper">
    <select id="getCcourseById" resultType="com.imut.pojo.Ccourse">
         select cc.id, c.cid, c.cname, s.sid, s.sname from ccourse cc, course c,
student s
            where c.cid=cc.cid and cc.sid=s.sid and c.cid=#{cid}
    </select>
    <select id="listSelect" resultType="com.imut.pojo.Course">
```

```
        select course.* FROM course where cid not in
            (select c.cid from course c,ccourse cc,student s
                where cc.cid=c.cid and cc.sid=s.sid and s.sid=#{sid})
    </select>
</mapper>
```

7. 创建测试类

在 com.imut 目录下新建 test 目录，在 com.imut.test 目录创建测试类 StudentTest.java。StudentTest.java 的代码如下。

```
package com.imut.test;
import java.io.InputStream;
import java.util.List;
import org.apache.ibatis.io.Resources;
import org.apache.ibatis.session.SqlSession;
import org.apache.ibatis.session.SqlSessionFactory;
import org.apache.ibatis.session.SqlSessionFactoryBuilder;
import org.junit.Test;
import com.imut.mapper.StudentMapper;
import com.imut.pojo.Ccourse;
import com.imut.pojo.Course;
import com.imut.pojo.Student;
import com.imut.utils.MybatisUtil;
public class StudentTest {
    @Test
    public void getStudentByIdTest() {
        String resource = "mybatis-config.xml";
        SqlSession sqlSession = null;
        try {
            InputStream inputStream = Resources.getResourceAsStream(resource);
            SqlSessionFactory sqlSessionFactory = new SqlSessionFactoryBuilder().
build(inputStream);
            sqlSession = sqlSessionFactory.openSession();
            StudentMapper mapper = sqlSession.getMapper(StudentMapper.class);
            Student student = mapper.getStudentById(2);
            System.out.println("查询出的编号为 2 的学生的学号、姓名、性别、年龄为:");
            System.out.println(student);
        } catch (Exception e) {
            e.printStackTrace();
        } finally {
            sqlSession.close();
        }
    }
    @Test
    public void getCcoursesByIdTest() {
        SqlSession sqlSession = MybatisUtil.getSqlSession();
        StudentMapper mapper = sqlSession.getMapper(StudentMapper.class);
        List<Ccourse> list= mapper.getCcoursesById(1);
```

```
        System.out.println("SID 为 1 的学生信息及其选修课程为:");
        for (Ccourse cc : list) {
            System.out.println(cc);
        }
        sqlSession.close();
    }
    @Test
    public void listStudentsTest() {
        SqlSession sqlSession = MybatisUtil.getSqlSession();
        StudentMapper mapper = sqlSession.getMapper(StudentMapper.class);
        List<Student> list = mapper.listStudents();
        System.out.println("所有学生的学号、姓名、性别、年龄为:");
        for (Student s : list) {
            System.out.println(s);
        }
        sqlSession.close();
    }
    @Test
    public void addStudentTest() {
        SqlSession sqlSession = MybatisUtil.getSqlSession();
        StudentMapper mapper = sqlSession.getMapper(StudentMapper.class);
        Student student = new Student();
        student.setSno("20210106");
        student.setSname("李林");
        student.setSgender("男");
        student.setSage(21);
        mapper.addStudent(student);
        sqlSession.commit();
        sqlSession.close();
    }
    @Test
    public void updateStudentTest() {
        SqlSession sqlSession = MybatisUtil.getSqlSession();
        StudentMapper mapper = sqlSession.getMapper(StudentMapper.class);
        Student student = mapper.getStudentById(2);
        student.setSname("李晓丽");
        mapper.updateStudent(student);
        sqlSession.commit();
        sqlSession.close();
    }
    @Test
    public void deleteStudentTest() {
        SqlSession sqlSession = MybatisUtil.getSqlSession();
        StudentMapper mapper = sqlSession.getMapper(StudentMapper.class);
        mapper.deleteStudent(6);
        sqlSession.commit();
        sqlSession.close();
    }
}
```

在 com.imut.test 目录创建测试类 CourseTest.java。CourseTest.java 的代码如下。

```
package com.imut.test;
import java.util.List;
import org.apache.ibatis.session.SqlSession;
import org.junit.Test;
import com.imut.mapper.CourseMapper;
import com.imut.pojo.Ccourse;
import com.imut.pojo.Course;
import com.imut.utils.MybatisUtil;
public class CourseTest {
    @Test
    public void getCcourseByIdTest() {
        SqlSession sqlSession = MybatisUtil.getSqlSession();
        CourseMapper mapper = sqlSession.getMapper(CourseMapper.class);
        List<Ccourse> list=mapper.getCcourseById(5);
        System.out.println("CID 为 5 的课程的选修情况为:");
        for(Ccourse cc:list) {
            System.out.println(cc);
        }
        sqlSession.close();
    }
    @Test
    public void listSelectTest() {
        SqlSession sqlSession = MybatisUtil.getSqlSession();
        CourseMapper mapper = sqlSession.getMapper(CourseMapper.class);
        List<Course> list=mapper.listSelect(1);
        System.out.println("SID 为 1 的学生没有选修的课程信息为:");
        for(Course c:list) {
            System.out.println(c);
        }
        sqlSession.close();
    }
}
```

3.5 习题

1. 单选题

(1) 一对一、多对多使用的元素是(　　)。

A. association 和 association　　B. association 和 collection

C. collection 和 collection　　D. collection 和 association

(2) 下面关于数据库中多表之间的关联关系,说法错误的是(　　)。

A. 一对一关联关系可以在任意一方引入对方主键作为外键

B. 一对多关联关系在"一"的一方,添加"多"的一方的主键作为外键

C. 多对多关联关系会产生中间关系表,引入两张表的主键作为外键

D. 多对多关联关系的两张表的主键可以成为联合主键或使用新的字段作为主键

2. 填空题

(1) association 元素的属性有________个,分别是________。

(2) collection 元素的属性有________个,分别是________。

3. 简答题

(1) 使用 MyBatis 的 mapper 接口调用时,需要注意什么?

(2) MyBatis 是如何将 SQL 执行结果封装为目标对象并返回的?

(3) 列举生活中其他的一对一、一对多、多对多的例子。

第4章 动态SQL

许多读者都遇到过 SQL 拼接的场景，过程非常痛苦。缺失空格、未判断变量是否为空、未添加引号等问题不时发生，使得编写人员不得不返工。为了解决上述问题，动态 SQL 诞生。它是 MyBatis 重要的特性之一，包含 if、choose、when、otherwise 等多种元素。本章将一一讲解。

4.1 if 元素

if 元素的命名比较容易地让人联想到 Java 中的 if 条件判断语句。事实上，if 元素确实是用于判断，在多数场景中用于判断参数是否为空。if 元素只包含 1 个属性：test。test 属性值即需要判断的条件，支持多个条件组合。

例 4-1：演示根据课程 ID 查询选课记录的方式，以使用 1 个单条件查询为例。共包含 7 个文件：Ccourses.java、CcoursesMapper.java、CcoursesMapper.xml、mybatis-config.xml、db.properties、MyBatisUtil.java 和 TestMB.java。需要改动的文件为 CcoursesMapper.java、CcoursesMapper.xml 和 TestMB.java。其余文件均沿用例 2-25 的即可。

CcoursesMapper.java 的代码如下。

```
package com.imut.mapper;
import com.imut.pojo.Ccourses;
import java.util.List;
public interface CcoursesMapper {
    List<Ccourses> selectCcoursesByCid(Integer cid);
}
```

CcoursesMapper.xml 的代码如下。

```
<?xml version="1.0" encoding="UTF-8"?>
<!DOCTYPE mapper PUBLIC "-//mybatis.org//DTD Mapper 3.0//EN" "http://mybatis.
org/dtd/mybatis-3-mapper.dtd">
<mapper namespace="com.imut.mapper.CcoursesMapper">
    <select id="selectCcoursesByCid" resultType="com.imut.pojo.Ccourses">
        select id, cid, sno
        from ccourses where
        <if test="cid!=null">
            cid = #{cid,jdbcType=INTEGER}
        </if>
    </select>
</mapper>
```

if 元素的 test 属性判断课程 ID 是否为 null，如果为 null，则不会执行赋值课程 ID 的语句，且因未拼接赋值语句，导致 SQL 语句语法格式不对，进而报错；如果不为 null，则语句成功执行。

if 元素除了可以在 select 元素中使用，还可以在 insert、update、delete 等元素中使用，并且使用次数无限制。

TestMB.java 的代码如下。

```
package com.imut.test;
import com.imut.mapper.CcoursesMapper;
import com.imut.pojo.*;
import com.imut.utils.MyBatisUtil;
import org.apache.ibatis.session.SqlSession;
import org.apache.ibatis.session.SqlSessionFactory;
import org.junit.Test;
import java.util.List;
public class TestMB {
    @Test
    public void testCcoursesMapper() throws Exception {
        SqlSessionFactory ssf = new MyBatisUtil().getSqlSessionFactory();
        SqlSession session = ssf.openSession();
        try {
            CcoursesMapper mapper = session.getMapper(CcoursesMapper.class);
            int cid = 1;
            List<Ccourses> ccoursesList = mapper.selectCcoursesByCid(cid);
            System.out.println("根据 cid(1)查询的选课信息:" + ccoursesList);
        } finally {
            session.close();
        }
    }
}
```

例 4-1 的运行结果如图 4-1 所示。

例 4-2：演示根据课程 ID、学号查询选课记录的方式，以使用多个单条件查询为例。需要改动的文件为 CcoursesMapper.java、CcoursesMapper.xml 和 TestMB.java。其余文件均

```
"C:\Program Files\Java\jdk-17\bin\java.exe" ...
根据 cid (1) 查询的选课信息: [Ccourses{id=8, cid=1, sno=3}, Ccourses{id=12, cid=1, sno=2}, Ccourses{id=13, cid=1, sno=2}, Ccourses{id=14, cid=1, sno=2}]

Process finished with exit code 0
```

图 4-1　例 4-1 的运行结果

沿用例 2-25 的即可。

CcoursesMapper.java 的新增代码如下。

```
List<Ccourses> selectCcoursesByID(int cid, Integer sid);
```

注意：传入参数的个数会影响语句是否成功执行。如果传入多个参数，可能会抛出 org. apache. ibatis. binding. BindingException: Parameter 'cid' not found. Available parameters are [arg1, arg0, param1, param2]异常。2.2.1 节讲过，是因为 MyBatis 无法与正确的参数对应。想要实现该需求，共有 2 种解决方案。除例 2-18 提到的添加@Param 注解标明属性外，还可以通过封装多个参数为一个对象的方式传递。即传入参数类型变为包含课程 ID 属性和学号的实体类。MyBatis 便能够自动从实体类对象中解析所需属性，代码如下。

```
List<Ccourses> selectCcoursesByID(Ccourses ccourses);
```

读者可自行选择两种方式中的一种。此处选择添加@Param 注解的方式。

CcoursesMapper.xml 的新增代码如下。

```
<select id="selectCcoursesByID" resultType="com.imut.pojo.Ccourses">
    select id, cid, sno
    from ccourses where
    <if test="cid!=null">
        cid = #{cid,jdbcType=INTEGER}
    </if>
    <if test="sno!=null">
        and sno = #{sno,jdbcType=INTEGER}
    </if>
</select>
```

温馨提示：为避免课程 ID 或学号为 null 的情况发生，应注意 if 元素包裹 SQL 语句的位置或补充适当的 SQL 语句。

TestMB.java 的新增代码如下，位于 testCcoursesMapper 方法中。

```
int sno = 2;
ccoursesList = mapper.selectCcoursesByID(cid, sno);
System.out.println("根据 cid(1)和 sno(1)查询的选课信息:" + ccoursesList);
```

例 4-2 的运行结果如图 4-2 所示。

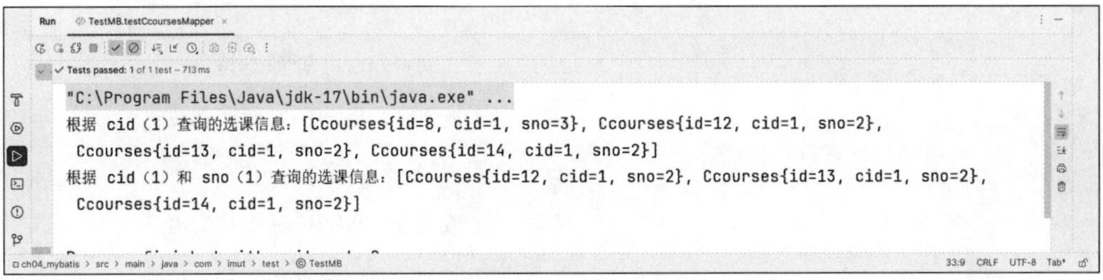

```
Run   TestMB.testCcoursesMapper
Tests passed: 1 of 1 test – 713 ms
"C:\Program Files\Java\jdk-17\bin\java.exe" ...
根据 cid (1) 查询的选课信息: [Ccourses{id=8, cid=1, sno=3}, Ccourses{id=12, cid=1, sno=2},
 Ccourses{id=13, cid=1, sno=2}, Ccourses{id=14, cid=1, sno=2}]
根据 cid (1) 和 sno (1) 查询的选课信息: [Ccourses{id=12, cid=1, sno=2}, Ccourses{id=13, cid=1, sno=2},
 Ccourses{id=14, cid=1, sno=2}]
ch04_mybatis > src > main > java > com > imut > test > TestMB          33:9  CRLF  UTF-8  Tab*
```

图 4-2　例 4-2 的运行结果

上述为单条件判断的场景，实际场景亦包含多条件判断。多条件组合即 test 中的语句不止一条，语句间通过 and、or 或||逻辑符连接。

例 4-3：演示根据课程 ID、学号且课程 ID 大于 0 查询选课记录的方式，以使用多条件组合查询为例。需要改动的文件为 CcoursesMapper.xml。CcoursesMapper.java 和 TestMB.java 文件内容详见例 4-2，其余文件均沿用例 2-25 的即可。

CcoursesMapper.xml 的新增代码如下。

```
<select id="selectCcoursesByID" resultType="com.imut.pojo.Ccourses">
    select id, cid, sno
    from ccourses where
    <if test="cid!=null and cid>0">
        cid = #{cid,jdbcType=INTEGER}
    </if>
    <if test="sno!=null">
        and sno = #{sno,jdbcType=INTEGER}
    </if>
</select>
```

注意：运行本例前，需要将 CcoursesMapper.xml 中例 4-2 的内容进行注释，否则会报错。另外，if 元素的 test 属性值不能包含<字符(关系运算符)，故建议使用相应缩写。如果是在一般 SQL 中使用缩写，需要在前面加 & 符号，后面加；符号，即将 gt 变为 >。

例 4-3 的运行结果如图 4-3 所示。

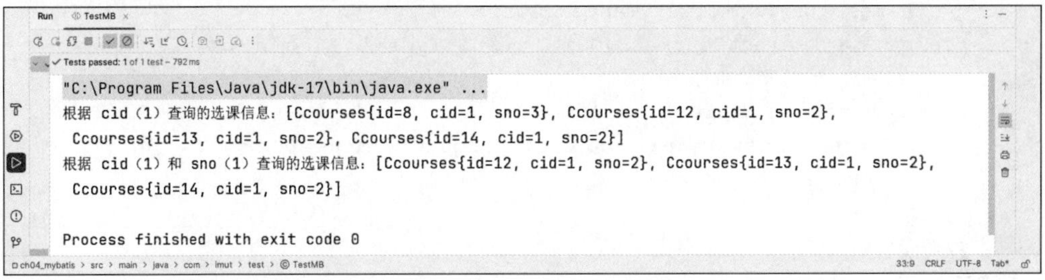

```
Run   TestMB
Tests passed: 1 of 1 test – 792 ms
"C:\Program Files\Java\jdk-17\bin\java.exe" ...
根据 cid (1) 查询的选课信息: [Ccourses{id=8, cid=1, sno=3}, Ccourses{id=12, cid=1, sno=2},
 Ccourses{id=13, cid=1, sno=2}, Ccourses{id=14, cid=1, sno=2}]
根据 cid (1) 和 sno (1) 查询的选课信息: [Ccourses{id=12, cid=1, sno=2}, Ccourses{id=13, cid=1, sno=2},
 Ccourses{id=14, cid=1, sno=2}]

Process finished with exit code 0
ch04_mybatis > src > main > java > com > imut > test > TestMB          33:9  CRLF  UTF-8  Tab*
```

图 4-3　例 4-3 的运行结果

关系运算符有其他的表达方式，对应关系如表 4-1 所示。

表 4-1　关系运算符的对应关系

符　号	缩　写	备　注
>	gt	大于。MyBatis 两者都支持
>=	gte	大于或等于。MyBatis 两者都支持
<	lt	小于。MyBatis 支持后者
<=	lte	小于或等于。MyBatis 支持后者
==	eq	等于。MyBatis 两者都支持
!=	neq	不等于。MyBatis 两者都支持

参数包含数字类型、字符串类型。针对字符串类型参数，本书介绍以下 4 种情形。前 3 种情形使用的都是 indexOf 方法。

情形 1：判断参数是否以某字符开头。鉴于选课记录中没有 String 类型属性，故使用课程实体的课程名称。

例 4-4：演示根据课程名称查询课程信息的方式，以使用 if 元素判断课程名称是否不为 null，且以 m 开头为例。需要改动的文件为 CoursesMapper.java、CoursesMapper.xml 和 TestMB.java。Students.java 和 Courses.java 文件内容详见例 3-3，其余文件均沿用例 2-25 的即可。

CoursesMapper.java 的代码如下。

```
package com.imut.mapper;
import com.imut.pojo.Courses;
import java.util.List;
public interface CoursesMapper {
    List<Courses> selectCoursesByName(String cname);
}
```

CoursesMapper.xml 的代码如下。

```
<?xml version="1.0" encoding="UTF-8"?>
<!DOCTYPE mapper PUBLIC "-//mybatis.org//DTD Mapper 3.0//EN" "http://mybatis.
org/dtd/mybatis-3-mapper.dtd">
<mapper namespace="com.imut.mapper.CoursesMapper">
    <select id="selectCoursesByName" resultType="com.imut.pojo.Courses">
        select cid, cname, tid
        from courses where
        <if test="cname!=null and cname.indexOf('m')==0">
            cname = #{cname,jdbcType=VARCHAR}
        </if>
    </select>
</mapper>
```

TestMB.java 的新增代码如下。

```
@Test
public void testCoursesMapper() throws Exception {
    SqlSessionFactory ssf = new MyBatisUtil().getSqlSessionFactory();
    SqlSession session = ssf.openSession();
    try {
        CoursesMapper mapper = session.getMapper(CoursesMapper.class);
        String cname = "math";
        List<Courses> coursesList = mapper.selectCoursesByName(cname);
        System.out.println("根据 cname(math)查询的课程信息:" + coursesList);
    } finally {
        session.close();
    }
}
```

例 4-4 的运行结果如图 4-4 所示。

```
"C:\Program Files\Java\jdk-17\bin\java.exe" ...
根据 cname (math) 查询的课程信息：[Courses{cid=1, cname='math', tid=1, studentsList=null}]

Process finished with exit code 0
```

图 4-4　例 4-4 的运行结果

情形 2：判断参数是否包含某字符。

例 4-5：演示根据课程名称查询课程信息的方式，以使用 if 元素判断课程名称是否不为 null，且包含 ma 为例。需要改动的文件为 CoursesMapper.java、CoursesMapper.xml 和 TestMB.java。Students.java 和 Courses.java 文件内容详见例 3-3，其余文件均沿用例 2-25 的即可。

CoursesMapper.java 的新增代码如下。

```
List<Courses> selectCoursesByName1(String cname);
```

CoursesMapper.xml 的新增代码如下。

```
<select id="selectCoursesByName1" resultType="com.imut.pojo.Courses">
    select cid, cname, tid
    from courses where
    <if test="cname!=null and cname.indexOf('ma')>=0">
        cname = #{cname,jdbcType=VARCHAR}
    </if>
</select>
```

TestMB.java 的新增代码如下，位于 testCoursesMapper 方法中。

```
cname = "mtah";
coursesList = mapper.selectCoursesByName1(cname);
System.out.println("根据 cname(mtah)查询的课程信息:" + coursesList);
```

例 4-5 的运行结果如图 4-5 所示。

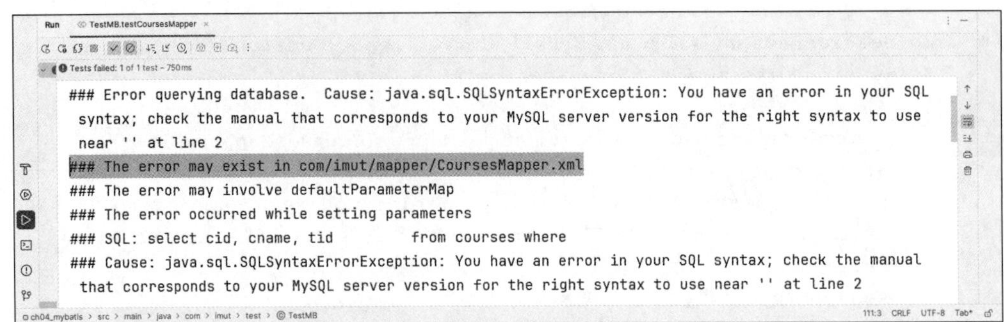

图 4-5　例 4-5 的运行结果

测试方法使用了反面测试的方式，即不满足 if 元素的判断条件，故程序因 SQL 语法不正确而报错。

情形 3：判断参数是否以某字符结尾。

例 4-6：演示根据课程名称查询课程信息的方式，以使用 if 元素判断课程名称是否不为 null，且以 h 结尾为例。需要改动的文件为 CoursesMapper.java、CoursesMapper.xml 和 TestMB.java。Students.java 和 Courses.java 文件内容详见例 3-3，其余文件均沿用例 2-25 的即可。

CoursesMapper.java 的新增代码如下。

```
List<Courses> selectCoursesByName2(String cname);
```

CoursesMapper.xml 的新增代码如下。

```
<select id="selectCoursesByName2" resultType="com.imut.pojo.Courses">
    select cid, cname, tid
    from courses where
    <if test="cname!=null and cname.indexOf('h')==(cname.length-1)">
        cname = #{cname,jdbcType=VARCHAR}
    </if>
</select>
```

TestMB.java 的新增代码如下，位于 testCoursesMapper 方法中。

```
cname = "chinese";
coursesList = mapper.selectCoursesByName2(cname);
System.out.println("根据 cname(chinese)查询的课程信息:" + coursesList);
```

例 4-6 的运行结果如图 4-6 所示。

测试方法使用了反面测试的方式，即不满足 if 元素的判断条件，故程序因 SQL 语法不正确而报错。

情形 4：判断参数是否等于某字符。

例 4-7：演示根据课程名称查询课程信息的方式，以使用 if 元素判断课程名称是否不为 null，且是否等于 math 为例。需要改动的文件为 CoursesMapper.java、CoursesMapper.xml

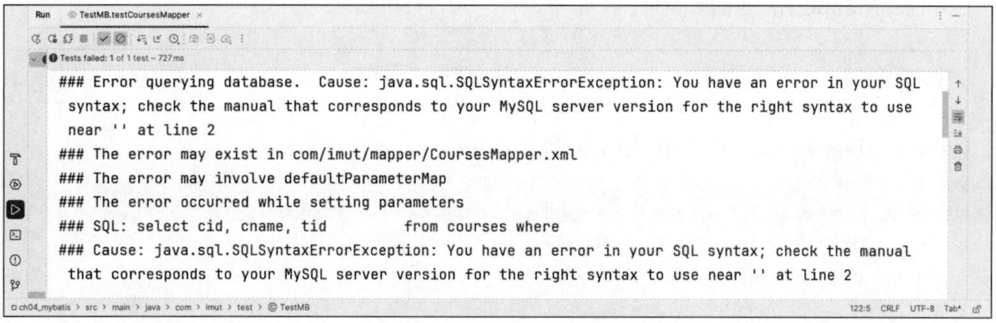

```
### Error querying database.  Cause: java.sql.SQLSyntaxErrorException: You have an error in your SQL
 syntax; check the manual that corresponds to your MySQL server version for the right syntax to use
 near '' at line 2
### The error may exist in com/imut/mapper/CoursesMapper.xml
### The error may involve defaultParameterMap
### The error occurred while setting parameters
### SQL: select cid, cname, tid          from courses where
### Cause: java.sql.SQLSyntaxErrorException: You have an error in your SQL syntax; check the manual
 that corresponds to your MySQL server version for the right syntax to use near '' at line 2
```

图 4-6　例 4-6 的运行结果

和 TestMB.java。Students.java 和 Courses.java 文件内容详见例 3-3，其余文件均沿用例 2-25 的即可。

CoursesMapper.java 的新增代码如下。

```
List<Courses> selectCoursesByName3(String cname);
```

CoursesMapper.xml 的新增代码如下。

```
<select id="selectCoursesByName3" resultType="com.imut.pojo.Courses">
    select cid, cname, tid
    from courses where
    <if test="cname!=null and cname=='math'">
        cname = #{cname,jdbcType=VARCHAR}
    </if>
</select>
```

TestMB.java 的新增代码如下，位于 testCoursesMapper 方法中。

```
cname = "math";
coursesList = mapper.selectCoursesByName3(cname);
System.out.println("根据 cname(math)查询的课程信息:" + coursesList);
```

例 4-7 的运行结果如图 4-7 所示。

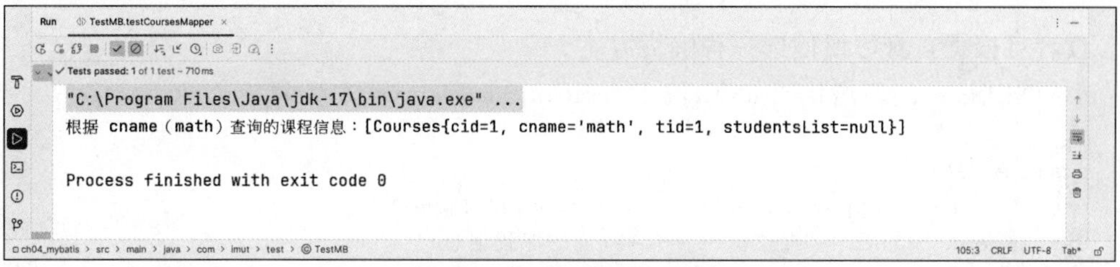

```
"C:\Program Files\Java\jdk-17\bin\java.exe" ...
根据 cname（math）查询的课程信息：[Courses{cid=1, cname='math', tid=1, studentsList=null}]

Process finished with exit code 0
```

图 4-7　例 4-7 的运行结果

例 4-8：演示根据课程名称（单个字符）查询课程信息的方式。需要改动的文件为 CoursesMapper.java、CoursesMapper.xml 和 TestMB.java。Students.java 和 Courses.java 文件内容详见例 3-3，其余文件均沿用例 2-25 的即可。

CoursesMapper.java 的新增代码如下。

```
List<Courses> selectCoursesByName4(String cname);
```

CoursesMapper.xml 的新增代码如下。

```
<select id="selectCoursesByName4" resultType="com.imut.pojo.Courses">
    select cid, cname, tid
    from courses where
    <if test="cname!=null and cname=='a'">
        cname = #{cname,jdbcType=VARCHAR}
    </if>
</select>
```

值得注意的是,if 元素中的表达式是由 OGNL 处理的。如果单引号内为单个字符,会将其识别为 Java 中的 char 类型,而参数是 String 类型。两者做==运算,返回 False,表达式自然不成立。如果要实现该需求,共有 3 个解决方案。

解决方案 1:将原本的单引号变为双引号,双引号变为单引号。

```
<if test='cname!=null and cname=="a"'>
    cname = #{cname,jdbcType=VARCHAR}
</if>
```

解决方案 2:将其转换为 String 类型,即调用 toString 方法。

```
<if test="cname!=null and cname=='a'.toString()">
    cname = #{cname,jdbcType=VARCHAR}
</if>
```

解决方案 3:使用转义字符。

```
<if test="cname!=null and cname==&quot;a&quot;">
    cname = #{cname,jdbcType=VARCHAR}
</if>
```

读者可根据自身习惯任选一种解决方案。

TestMB.java 的新增代码如下,位于 testCoursesMapper 方法中。

```
cname = "a";
coursesList = mapper.selectCoursesByName4(cname);
System.out.println("根据 cname(a)查询的课程信息:" + coursesList);
```

如果未选用上述 3 种解决方案,则例 4-8 的运行结果如图 4-8 所示。

如果选用上述 3 种解决方案中的一种,则例 4-8 的运行结果如图 4-9 所示。

支持的字符串参数可调用的方法还有许多,这里不再一一介绍,感兴趣的读者可以自行尝试。另外,if 元素支持通过自定义方法进行判断。

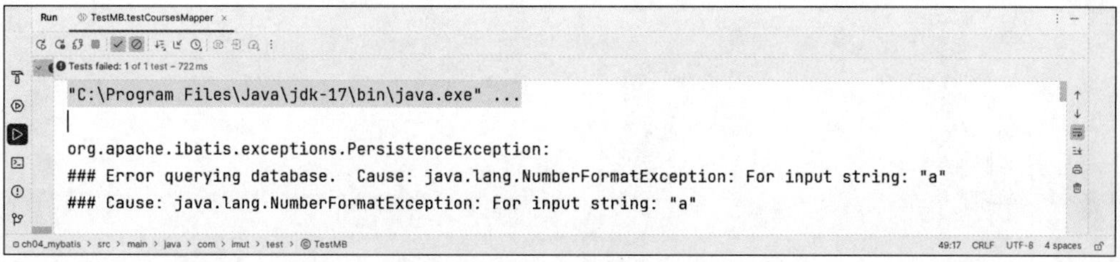

```
Run    TestMB.testCoursesMapper
Tests failed: 1 of 1 test – 722 ms
"C:\Program Files\Java\jdk-17\bin\java.exe" ...

org.apache.ibatis.exceptions.PersistenceException:
### Error querying database.  Cause: java.lang.NumberFormatException: For input string: "a"
### Cause: java.lang.NumberFormatException: For input string: "a"
ch04_mybatis > src > main > java > com > imut > test > TestMB        49:17  CRLF  UTF-8  4 spaces
```

图 4-8　例 4-8 的错误运行结果

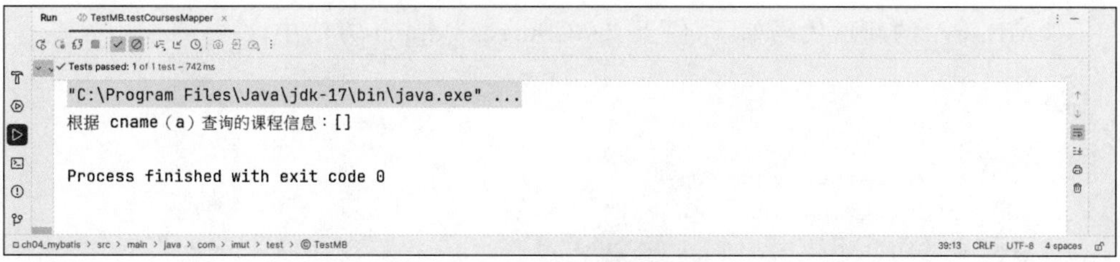

```
Run    TestMB.testCoursesMapper
Tests passed: 1 of 1 test – 742 ms
"C:\Program Files\Java\jdk-17\bin\java.exe" ...
根据 cname（a）查询的课程信息：[]

Process finished with exit code 0
ch04_mybatis > src > main > java > com > imut > test > TestMB        39:13  CRLF  UTF-8  4 spaces
```

图 4-9　例 4-8 的正确运行结果

4.2　choose、when、otherwise 元素

在 Java 中，当有多个条件选择时，可以选用 switch 语句。MyBatis 中也提供类似的功能，即 choose 元素。when 元素相当于 switch 语句中的 case 关键字，otherwise 元素相当于 switch 语句中的 default 关键字，三者一起使用。choose 元素和 otherwise 元素不包含任何属性；when 元素包含 1 个属性：test。test 属性值即需要判断的条件，支持多个条件组合。

例 4-9：演示根据课程名称(不为 null 且长度为 7)或授课教师 ID(不为 null)或无限制条件(1＝1 永远为真)查询课程信息的方式。需要改动的文件为 CoursesMapper.java、CoursesMapper.xml 和 TestMB.java。Students.java 和 Courses.java 文件内容详见例 3-3，其余文件均沿用例 2-25 的即可。

CoursesMapper.java 的新增代码如下。

```
List<Courses> selectCoursesByCnameOrTid(Courses courses);
```

CoursesMapper.xml 的新增代码如下。

```
<select id="selectCoursesByCnameOrTid" resultType="com.imut.pojo.Courses">
    select cid, cname, tid
    from courses where
    <choose>
        <when test="cname!=null and cname.length()==7">
            cname = #{cname,jdbcType=VARCHAR}
        </when>
        <when test="tid!=null">
            tid = #{tid,jdbcType=INTEGER}
```

```
        </when>
        <otherwise>
            1 = 1
        </otherwise>
    </choose>
</select>
```

传入参数按照从上到下的顺序依次判断，当满足其中一个 when 元素的判断条件时，后续 when 元素失效。如果所有 when 元素中的判断条件都不满足，会拼接 otherwise 元素中的语句。

TestMB.java 的新增代码如下，位于 testCoursesMapper 方法中。

```
cname = "math";
int tid = 1;
Courses courses = new Courses();
courses.setCname(cname);
courses.setTid(tid);
coursesList = mapper.selectCoursesByCnameOrTid(courses);
System.out.println("根据 tid(1)查询的课程信息:" + coursesList);
```

例 4-9 的运行结果如图 4-10 所示。

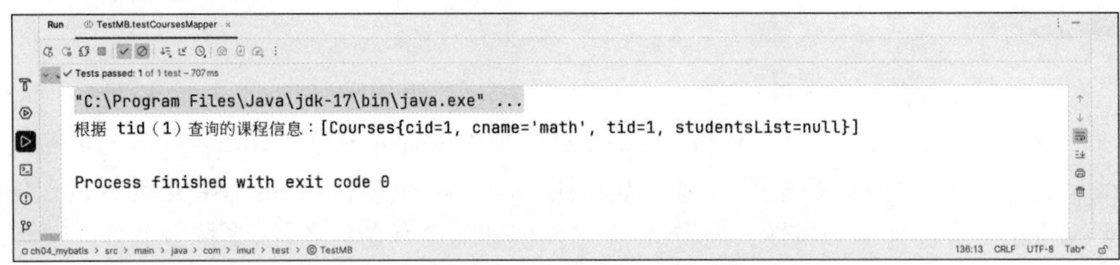

图 4-10　例 4-9 的运行结果

测试方法传入的 cname 值不满足第一个 when 元素的判断条件，而 tid 满足第二个 when 元素的判断条件，故执行的 SQL 语句为 select cid, cname, tid from courses where tid = ?。

4.3　where、set、trim 元素

where、set 和 trim 元素多与其他元素搭配使用，且三者中 trim 元素的功能最为强大。三个元素的使用能够有效地解决多字段传值、值为空的问题，避免因值为空造成 SQL 语句语法格式错误的情况。

4.3.1　where 元素

where 元素相当于 SQL 语句中的 where 关键字，用于指定选择条件。既然 SQL 语句有 where 关键字，为什么还需要 where 元素呢？4.1 节讲解多个单条件查询代码时曾讲过，当 if 元素不满足，会出现 SQL 语句结构不完整，从而导致查询失败的情况。虽然能够通过移动 if 元素位置或补充 SQL 语句的方式解决，但这种方式不能百分之百地解决问题，以

例 4-10 为例说明。

例 4-10：演示根据课程 ID(为 null)、学号(不为 null)查询选课信息的方式。需要改动的文件为 CcoursesMapper.java、CcoursesMapper.xml 和 TestMB.java。其余文件均沿用例 2-25 的即可。

CcoursesMapper.java 的新增代码如下。

```
List<Ccourses> selectCcoursesByID1(@Param("cid") Integer cid, @Param("sno")
Integer sno);
```

CcoursesMapper.xml 的新增代码如下。

```
<select id="selectCcoursesByID1" resultType="com.imut.pojo.Ccourses">
    select id, cid, sno
    from ccourses
    <if test="cid!=null">
        where cid = #{cid,jdbcType=INTEGER}
    </if>
    <if test="sno!=null">
        and sno = #{sno,jdbcType=INTEGER}
    </if>
</select>
```

TestMB.java 的新增代码如下，位于 testCcoursesMapper 方法中。

```
sno = 3;
ccoursesList = mapper.selectCcoursesByID1(null, sno);
System.out.println("根据 cid(null)和 sno(3)查询的选课信息:" + ccoursesList);
```

例 4-10 的运行结果如图 4-11 所示。

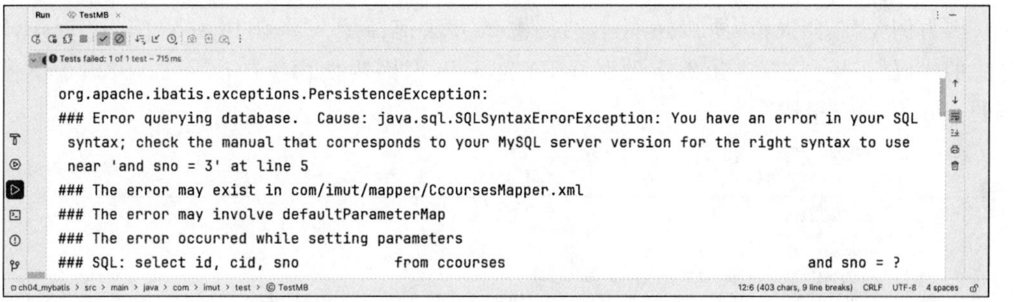

```
Run   TestMB
Tests failed: 1 of 1 test – 715 ms
org.apache.ibatis.exceptions.PersistenceException:
### Error querying database.  Cause: java.sql.SQLSyntaxErrorException: You have an error in your SQL
 syntax; check the manual that corresponds to your MySQL server version for the right syntax to use
 near 'and sno = 3' at line 5
### The error may exist in com/imut/mapper/CcoursesMapper.xml
### The error may involve defaultParameterMap
### The error occurred while setting parameters
### SQL: select id, cid, sno          from ccourses                              and sno = ?
ch04_mybatis > src > main > java > com > imut > test > TestMB      12:6 (403 chars, 9 line breaks)  CRLF  UTF-8  4 spaces
```

图 4-11　例 4-10 的运行结果

当课程 ID 为 null，学号不为 null，where 关键字包含于第一个 if 元素中，故不会出现。但包含于第二个 if 元素中的 and 关键字会出现，SQL 语句语法结构依然不对。因此 where 元素的诞生正是为了解决这类问题。where 元素没有属性，写入 SQL 的位置即 where 关键字允许出现的位置，如果由 where 元素包裹的 SQL 部分无内容输出，则 SQL 语句中不会出现 where 关键字；如果有执行结果返回且开头为 and、or 关键字，MyBatis 将会删除第一个 and、or 关键字，除此之外，MyBatis 还会补充应有的空格，如 where 关键字前缺失的空格。

而这一切对用户来说是透明的，使用便捷。

例 4-11：演示使用 where 元素后根据课程 ID（为 null）、学号（不为 null）查询选课信息的方式。需要改动的文件为 CcoursesMapper.java、CcoursesMapper.xml 和 TestMB.java。其余文件均沿用例 2-25 的即可。

CcoursesMapper.java 的新增代码如下。

```
List<Ccourses> selectCcoursesByID2(@Param("cid") Integer cid, @Param("sno")
Integer sno);
```

CcoursesMapper.xml 的新增代码如下。

```
<select id="selectCcoursesByID2" resultType="com.imut.pojo.Ccourses">
select id, cid, sno
    from ccourses
    <where>
        <if test="cid!=null and cid > 0">
            cid = #{cid,jdbcType=INTEGER}
        </if>
        <if test="sno!=null">
            and sno = #{sno,jdbcType=INTEGER}
        </if>
    </where>
</select>
```

TestMB.java 的新增代码如下，位于 testCcoursesMapper 方法中。

```
sno = 2;
ccoursesList = mapper.selectCcoursesByID2(null, sno);
System.out.println("根据 cid(null)和 sno(2)查询的选课信息:" + ccoursesList);
```

因为例 4-10 会报错，所以运行本例的测试方法前需要先将调用 selectCcoursesByID1 方法的代码行进行注释。虽然本例课程 ID 为 null，但因为使用 where 元素，去掉了学号列前的 and，故并未报错。

例 4-11 的运行结果如图 4-12 所示。

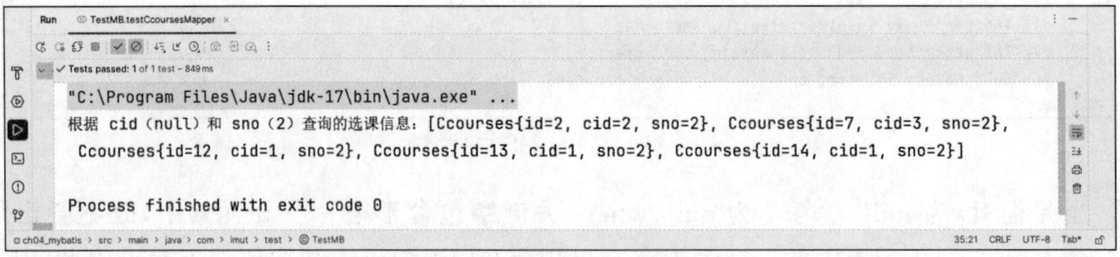

```
Run   TestMB.testCcoursesMapper
✓ Tests passed: 1 of 1 test – 849 ms
"C:\Program Files\Java\jdk-17\bin\java.exe" ...
根据 cid (null) 和 sno (2) 查询的选课信息: [Ccourses{id=2, cid=2, sno=2}, Ccourses{id=7, cid=3, sno=2},
 Ccourses{id=12, cid=1, sno=2}, Ccourses{id=13, cid=1, sno=2}, Ccourses{id=14, cid=1, sno=2}]

Process finished with exit code 0
ch04_mybatis > src > main > java > com > imut > test > TestMB      35:21  CRLF  UTF-8  Tab*
```

图 4-12　例 4-11 的运行结果

4.3.2　set 元素

set 元素相当于 SQL 语句中的 set 关键字，主要用于更新语句。既然 SQL 语句有 set

关键字,为什么还需要 set 元素呢?在只包含一个判断条件的 SQL 语句中,可以选择使用 set 关键字或 set 元素,但当包含多个 if 元素时,则可能会出现 SQL 语句结构不完整而导致执行失败的情况。虽然能够通过移动 if 元素位置或补充 SQL 语句的方式解决,但这并不能百分之百地解决问题。以例 4-12 为例说明。

例 4-12:演示根据课程 ID(不为 null)更新课程信息的方式,其中课程名称不为 null,教师 ID 为 null。需要改动的文件为 CoursesMapper.java、CoursesMapper.xml 和 TestMB.java。Students.java 和 Courses.java 文件内容详见例 3-3,其余文件均沿用例 2-25 的即可。

CoursesMapper.java 的新增代码如下。

```
int updateByPrimaryKey(Courses courses);
```

CoursesMapper.xml 的新增代码如下。

```
<update id="updateByPrimaryKey" parameterType="com.imut.pojo.Courses">
    update courses
    set
    <if test="cname!=null">
        cname = #{cname,jdbcType=VARCHAR},
    </if>
    <if test="tid!=null">
        tid = #{tid,jdbcType=INTEGER}
    </if>
    where cid = #{cid,jdbcType=INTEGER}
</update>
```

TestMB.java 的新增代码如下,位于 testCoursesMapper 方法中。

```
courses = new Courses();
courses.setCid(4);
courses.setCname("Java");
int result = mapper.updateByPrimaryKey(courses);
session.commit();
if (result==1){
    System.out.println("数据更新成功");
}else
    System.out.println("数据更新失败");
```

当授课教师 ID 为 null 时,SQL 语句就会变为"update courses set cname = ?, where cid = ?",可以看到 where 关键字前方有一个逗号,不符合语法规则,进而导致查询失败。因此 set 元素的诞生正是为了解决此类问题。set 元素没有属性,写入 SQL 的位置即 set 关键字允许出现的位置,如果由 set 元素包裹的 SQL 部分无内容输出,则 SQL 语句中不会出现 set 关键字,会导致查询出错;如果有 SQL 内容返回且结尾为逗号,MyBatis 将会删除最后一个逗号,除此之外,MyBatis 还会补充应有的空格,如 set 关键字前缺失的空格。set 元素保证只更新值变化的字段。而这一切对用户来说也是透明的,使用便捷。

例 4-12 的运行结果如图 4-13 所示。

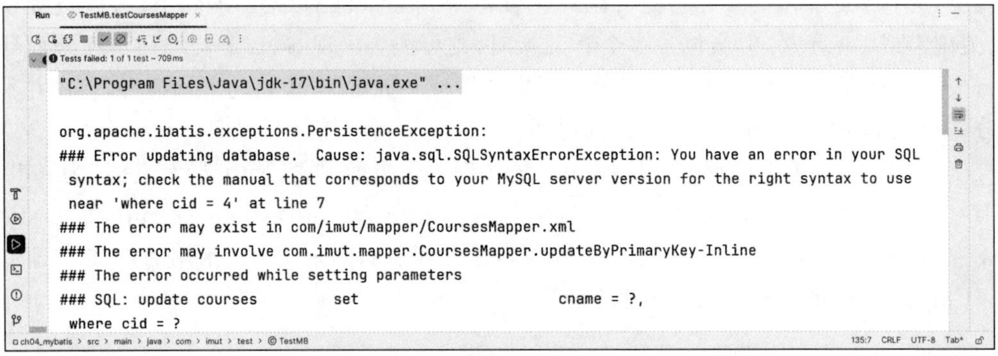

```
"C:\Program Files\Java\jdk-17\bin\java.exe" ...

org.apache.ibatis.exceptions.PersistenceException:
### Error updating database.  Cause: java.sql.SQLSyntaxErrorException: You have an error in your SQL
 syntax; check the manual that corresponds to your MySQL server version for the right syntax to use
 near 'where cid = 4' at line 7
### The error may exist in com/imut/mapper/CoursesMapper.xml
### The error may involve com.imut.mapper.CoursesMapper.updateByPrimaryKey-Inline
### The error occurred while setting parameters
### SQL: update courses         set                   cname = ?,
 where cid = ?
```

图 4-13　例 4-12 的运行结果

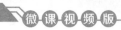

例 4-13：演示使用 set 元素后根据课程 ID(不为 null)更新课程信息的方式，其中课程名称不为 null，教师 ID 为 null。需要改动的文件为 CoursesMapper.java、CoursesMapper.xml 和 TestMB.java。Students.java 和 Courses.java 文件内容详见例 3-3，其余文件均沿用例 2-25 的即可。

CoursesMapper.java 的新增代码如下。

```
int updateByPrimaryKey1(Courses courses);
```

CoursesMapper.xml 的新增代码如下。

```
<update id="updateByPrimaryKey1" parameterType="com.imut.pojo.Courses">
    update courses
    <set>
        <if test="cname!=null">
            cname = #{cname,jdbcType=VARCHAR},
        </if>
        <if test="tid!=null">
            tid = #{tid,jdbcType=INTEGER}
        </if>
    </set>
    where cid = #{cid,jdbcType=INTEGER}
</update>
```

TestMB.java 的新增代码如下，位于 testCoursesMapper 方法中。

```
courses = new Courses();
courses.setCid(4);
courses.setCname("Java");
int result = mapper.updateByPrimaryKey1(courses);
session.commit();
if (result==1){
    System.out.println("数据更新成功");
}else
    System.out.println("数据更新失败");
```

注意：因为执行的是更新操作，而 MyBatis 的 SqlSession 默认的行为是不自动提交事务。要想更新操作执行成功，一是手动提交，即调用 commit 方法；二是开启自动提交，即将 openSession 方法传入的参数设为 true。本例使用的是手动提交的方式。

例 4-13 的运行结果如图 4-14 所示。

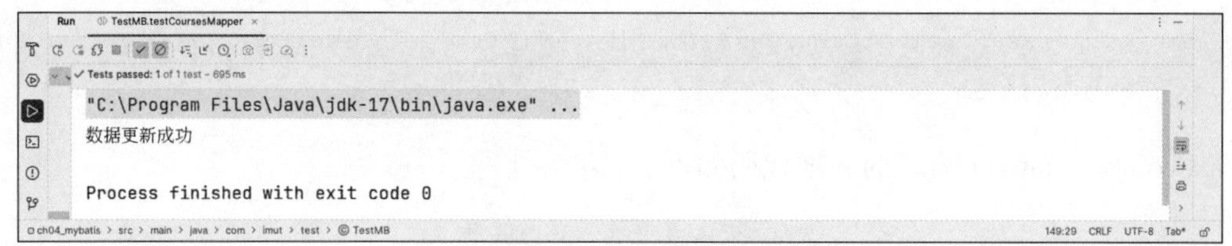

图 4-14　例 4-13 的运行结果

数据库中 courses 表更新后的数据如图 4-15 所示。

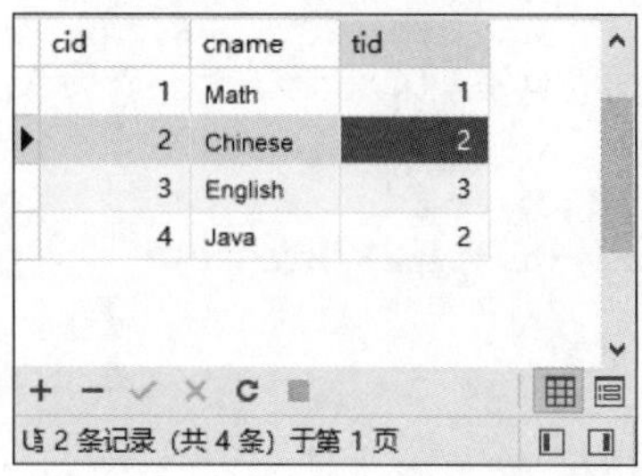

cid	cname	tid
1	Math	1
2	Chinese	2
3	English	3
4	Java	2

第 2 条记录 (共 4 条) 于第 1 页

图 4-15　courses 表更新后的数据

4.3.3　trim 元素

trim 元素类似于 Java 中的 trim 方法，甚至更加灵活。trim 元素包含 4 个属性：prefix 属性、prefixOverrides 属性、suffix 属性和 suffixOverrides 属性。prefix 属性用于指定插入的前缀；prefixOverrides 属性用于指定被剔除的前缀内容，多个字符串以 | 作为分隔符；suffix 属性用于指定插入的后缀；suffixOverrides 属性用于指定被剔除的后缀内容，多个字符串以 | 作为分隔符。写入 SQL 的位置通常为 where/set 关键字允许出现的位置，如果由 trim 元素包裹的 SQL 部分无内容输出，则视情况而定 SQL 语句是否报错；如果有 SQL 内容返回且开头为 prefixOverrides 属性指定内容，或结尾为 suffixOverrides 属性指定内容，则会被删除，接着加入 prefix 属性和 suffix 属性的值。除此之外，MyBatis 还会补充应有的空格，如 prefix 属性值和 suffix 属性值前缺失的空格。

trim 元素格式化 SQL 语句非常便捷，适用于多种场景，这里主要介绍 3 种。场景 1：实现 where 元素功能。即 prefix 属性值设为 where，prefixOverrides 属性值设为 AND | OR（不够严谨，还应考虑 AND\n、OR\n、AND\r、OR\r、AND\t、OR\t 的情况）。场景 2：实现 set 元素功能。即 prefix 属性值设为 set，suffixOverrides 属性值设为逗号。场景 3：实现更灵活的查询功能。即 prefix 属性值设为 where，prefixOverrides 属性值设为 AND | OR，suffix 属性值设为 Limit 3。分别对应于例 4-14、例 4-15、例 4-16。笔者建议：测试应该多设计不同的情况，包含正例和反例，才能更加严谨地验证结果。

例 4-14：演示使用 trim 元素实现 where 元素的方式，其中课程 ID 为 null，学号不为 null。共包含 7 个文件：Ccourses. java、CcoursesMapper. java、CcoursesMapper. xml、 mybatis-config.xml、db.properties、MyBatisUtil.java 和 TestMB.java。需要改动的文件为 CcoursesMapper.java、CcoursesMapper.xml 和 TestMB.java。其余文件均沿用例 2-25 的即可。

CcoursesMapper.java 的新增代码如下。

```
List<Ccourses> selectCcoursesByID3(@Param("cid") Integer cid, @Param("sno")
Integer sno);
```

CcoursesMapper.xml 的新增代码如下。

```
<select id="selectCcoursesByID3" resultType="com.imut.pojo.Ccourses">
    select id, cid, sno
    from ccourses
    <trim prefix="where" prefixOverrides="AND|OR" suffix="" suffixOverrides="">
        <if test="cid!=null and cid > 0">
            cid = #{cid,jdbcType=INTEGER}
        </if>
        <if test="sno!=null">
            and sno = #{sno,jdbcType=INTEGER}
        </if>
    </trim>
</select>
```

TestMB.java 的新增代码如下，位于 testCcoursesMapper 方法中。

```
sno = 4;
ccoursesList = mapper.selectCcoursesByID3(null, sno);
System.out.println("根据 cid(null)和 sno(4)查询的选课信息:" + ccoursesList);
```

例 4-14 的运行结果如图 4-16 所示。

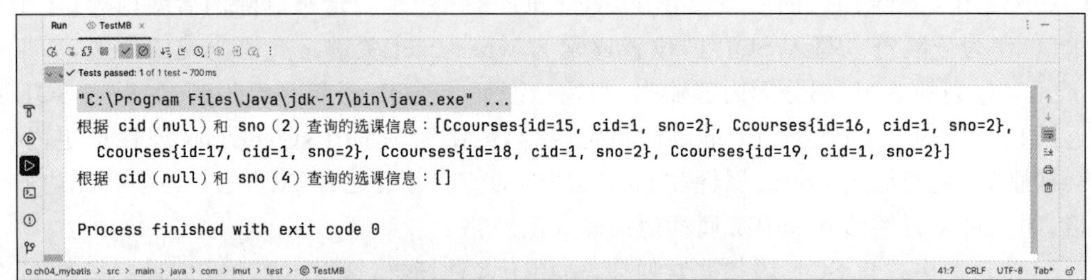

图 4-16　例 4-14 的运行结果

执行的 SQL 语句为“select id, cid, sno from ccourses where sno = ?”，与 where 元素效果一致。

例 4-15：演示使用 trim 元素实现 set 元素的方式，其中课程名称不为 null，教师 ID 为 null。需要改动的文件为 CoursesMapper.java、CoursesMapper.xml 和 TestMB.java。Students.java 和 Courses.java 文件内容详见例 3-3，其余文件均沿用例 2-25 的即可。

CoursesMapper.java 的新增代码如下。

```
int updateByPrimaryKey2(Courses courses);
```

CoursesMapper.xml 的新增代码如下。

```
<update id="updateByPrimaryKey2" parameterType="com.imut.pojo.Courses">
    update courses
    <trim prefix="set" suffixOverrides=",">
        <if test="cname!=null">
            cname = #{cname,jdbcType=VARCHAR},
        </if>
        <if test="tid!=null">
            tid = #{tid,jdbcType=INTEGER}
        </if>
    </trim>
    where cid = #{cid,jdbcType=INTEGER}
</update>
```

TestMB.java 的新增代码如下，位于 testCoursesMapper 方法中。

```
courses = new Courses();
courses.setCid(4);
courses.setCname("C++");
result = mapper.updateByPrimaryKey2(courses);
session.commit();
if (result==1){
    System.out.println("数据更新成功");
}else
    System.out.println("数据更新失败");
```

例 4-15 的运行结果如图 4-17 所示。

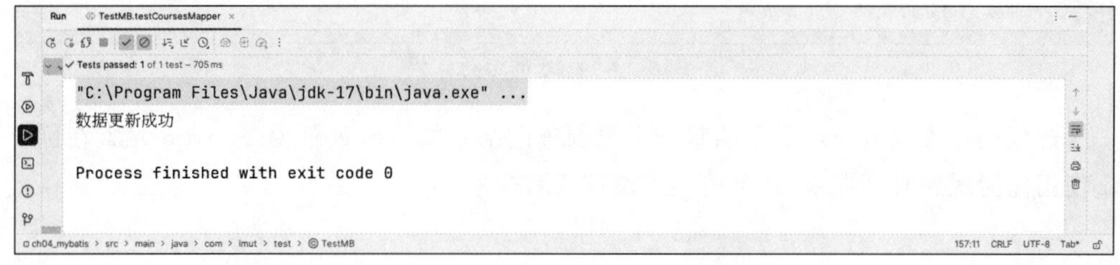

图 4-17 例 4-15 的运行结果

执行的 SQL 语句为“update courses set cname = ? where cid = ?”，与 set 元素效果一致。

例 4-16：演示使用 trim 元素实现更灵活的查询方式，其中课程 ID 为 null，学号不为 null。需要改动的文件为 CcoursesMapper.java、CcoursesMapper.xml 和 TestMB.java。其余文件均沿用例 2-25 的即可。

CcoursesMapper.java 的新增代码如下。

```
List<Ccourses> selectCcoursesByID4(@Param("cid") Integer cid, @Param("sno")
Integer sno);
```

CcoursesMapper.xml 的新增代码如下。

```
<select id="selectCcoursesByID4" resultType="com.imut.pojo.Ccourses">
    select id, cid, sno
    from ccourses
    <trim prefix="where" prefixOverrides="AND|OR" suffix="Limit 3" suffixOverrides
="">
        <if test="cid!=null and cid > 0">
            cid = #{cid,jdbcType=INTEGER}
        </if>
        <if test="sno!=null">
            or sno = #{sno,jdbcType=INTEGER}
        </if>
    </trim>
</select>
```

TestMB.java 的新增代码如下，位于 testCcoursesMapper 方法中。

```
sno = 3;
ccoursesList = mapper.selectCcoursesByID4(null, sno);
System.out.println("根据 cid(null)和 sno(3)查询的选课信息:" + ccoursesList);
```

例 4-16 的运行结果如图 4-18 所示。

```
Run    TestMB.testCcoursesMapper
Tests passed: 1 of 1 test – 905 ms
"C:\Program Files\Java\jdk-17\bin\java.exe" ...
根据 cid (null) 和 sno (2) 查询的选课信息：[Ccourses{id=15, cid=1, sno=2}, Ccourses{id=16, cid=1, sno=2},
 Ccourses{id=17, cid=1, sno=2}, Ccourses{id=18, cid=1, sno=2}, Ccourses{id=19, cid=1, sno=2}]
根据 cid (null) 和 sno (4) 查询的选课信息：[]
根据 cid (null) 和 sno (3) 查询的选课信息：[Ccourses{id=1, cid=2, sno=3}, Ccourses{id=3, cid=2, sno=3},
 Ccourses{id=4, cid=2, sno=3}]

Process finished with exit code 0
ch04_mybatis > src > main > java > com > imut > test > TestMB        48:32  CRLF  UTF-8  Tab*
```

图 4-18　例 4-16 的运行结果

上述代码在实现 where 元素的基础上限制查询结果的输出数量为 3。trim 元素在例 4-16 中不使用任何属性时会报错，其余的场景留待读者探索。

4.4 foreach 元素

foreach 元素类似于 Java 中的 foreach 循环，甚至更加灵活，通常情况下用于包含 IN 关键字的 SQL 语句中。foreach 元素包含 6 个属性：collection 属性、index 属性、item 属性、open 属性、separator 属性、close 属性。collection 属性用于指定可迭代对象，包括 Array、List、Set、Map 等，属性值为 collection、list、array 中的一种，是必填属性；index 属性用于指定遍历对象的索引；item 属性用于指定当前遍历对象的名称；open 属性用于指定拼接的前缀内容，且只在迭代开始前拼接一次；separator 属性用于指定拼接的分隔符，每次迭代都会拼接；close 属性用于指定拼接的后缀内容，且只在迭代结束后拼接一次。写入 SQL 的位置不固定，如果由 foreach 元素包裹的 SQL 部分无内容输出，则视情况而定 SQL 语句是否报错。

foreach 元素功能强大，适用多种场景，这里主要介绍 4 种。场景 1：使用 Set 作为参数，根据 Set 中包含的课程 ID 查询选课记录。场景 2：使用 Array(数组)作为参数，根据数组中包含的课程 ID 查询选课记录。场景 3(未使用 in 关键字)：使用 Map 作为参数，根据 Map 中包含的课程 ID 查询选课记录。场景 4：使用 List 作为参数，根据 List 中包含的信息批量插入选课记录。

例 4-17：演示根据多个课程 ID 查询选课记录的方式，其中多个课程 ID 存在 Set 中。需要改动的文件为 CcoursesMapper.java、CcoursesMapper.xml 和 TestMB.java。其余文件均沿用例 2-25 的即可。

CcoursesMapper.java 的新增代码如下。

```
List<Ccourses> selectCcoursesByCids(Set<Integer> cids);
```

CcoursesMapper.xml 的新增代码如下。

```
<select id="selectCcoursesByCids" resultType="com.imut.pojo.Ccourses">
    select id, cid, sno
    from ccourses
    <where>
        <foreach collection="collection" item="cid" open="cid in (" separator
=","close=")">
            #{cid}
        </foreach>
    </where>
</select>
```

Set 类型为封装数据类型 INTEGER，元素为多个课程 ID。collection 属性的值一定是 collection，如果错写为 set，会抛出异常。item 属性的值无指定规则，但需与 foreach 元素包裹的#{}中的名称一致。open 属性包含了部分 SQL 语句，避免造成查询失败并报错的情况。即当其写在 foreach 元素外部，而又未使用 where 元素，并且传入内容为 null 时报错。上述代码使用两种元素组合的方式，有效降低了查询出错的概率。

TestMB.java 的新增代码如下，位于 testCcoursesMapper 方法中。

```
Set<Integer> set = new HashSet<>();
set.add(1);
set.add(2);
set.add(3);
ccoursesList = mapper.selectCcoursesByCids(set);
System.out.println("根据 cid(1,2,3)查询的选课信息:" + ccoursesList);
```

例 4-17 的运行结果如图 4-19 所示。

执行测试方法，如果已进行日志类相关的配置，可以由控制台知晓执行的 SQL 语句为

```
"C:\Program Files\Java\jdk-17\bin\java.exe" ...
根据 cid（null）和 sno（2）查询的选课信息：[Ccourses{id=15, cid=1, sno=2}, Ccourses{id=16, cid=1, sno=2},
 Ccourses{id=17, cid=1, sno=2}, Ccourses{id=18, cid=1, sno=2}, Ccourses{id=19, cid=1, sno=2}]
根据 cid（null）和 sno（4）查询的选课信息：[]
根据 cid（null）和 sno（3）查询的选课信息：[Ccourses{id=1, cid=2, sno=3}, Ccourses{id=3, cid=2, sno=3},
 Ccourses{id=4, cid=2, sno=3}]
根据 cid（1,2,3）查询的选课信息：[Ccourses{id=1, cid=2, sno=3}, Ccourses{id=2, cid=2, sno=2},
 Ccourses{id=3, cid=2, sno=3}, Ccourses{id=4, cid=2, sno=3}, Ccourses{id=6, cid=3, sno=6},
 Ccourses{id=7, cid=3, sno=2}, Ccourses{id=8, cid=1, sno=3}, Ccourses{id=9, cid=3, sno=3},
 Ccourses{id=11, cid=3, sno=1}, Ccourses{id=15, cid=1, sno=2}, Ccourses{id=16, cid=1, sno=2},
 Ccourses{id=17, cid=1, sno=2}, Ccourses{id=18, cid=1, sno=2}, Ccourses{id=19, cid=1, sno=2}]

Process finished with exit code 0
```

图 4-19　例 4-17 的运行结果

“select id, cid, sno from ccourses WHERE cid in (? , ? , ?)。?”由传入的变量值 1、2、3 代替。

例 4-18：演示根据多个课程 ID 查询选课记录的方式，其中多个课程 ID 存在 Array 中。需要改动的文件为 CcoursesMapper.java、CcoursesMapper.xml 和 TestMB.java。其余文件均沿用例 2-25 的即可。

CcoursesMapper.java 的新增代码如下。

```
List<Ccourses> selectCcoursesByCids1(Integer[] cids);
```

CcoursesMapper.xml 的新增代码如下。

```
<select id="selectCcoursesByCids1" resultType="com.imut.pojo.Ccourses">
    select id, cid, sno
    from ccourses
    <where>
        <foreach collection="array" item="cid" open="cid in (" separator=","
close=")">
            <if test="cid!=null">
                #{cid}
            </if>
        </foreach>
    </where>
</select>
```

Array 类型为封装数据类型 Integer，Array 包含元素为多个课程 ID。上述代码使用了 where 元素结合 foreach 元素加 if 元素组合的方式，进一步降低查询出错的概率。

TestMB.java 的新增代码如下，位于 testCcoursesMapper 方法中。

```
Integer[] arr = {1,2,3};
ccoursesList = mapper.selectCcoursesByCids1(arr);
System.out.println("根据 cid(1,2,3)查询的选课信息:" + ccoursesList);
```

例 4-18 的运行结果如图 4-20 所示。

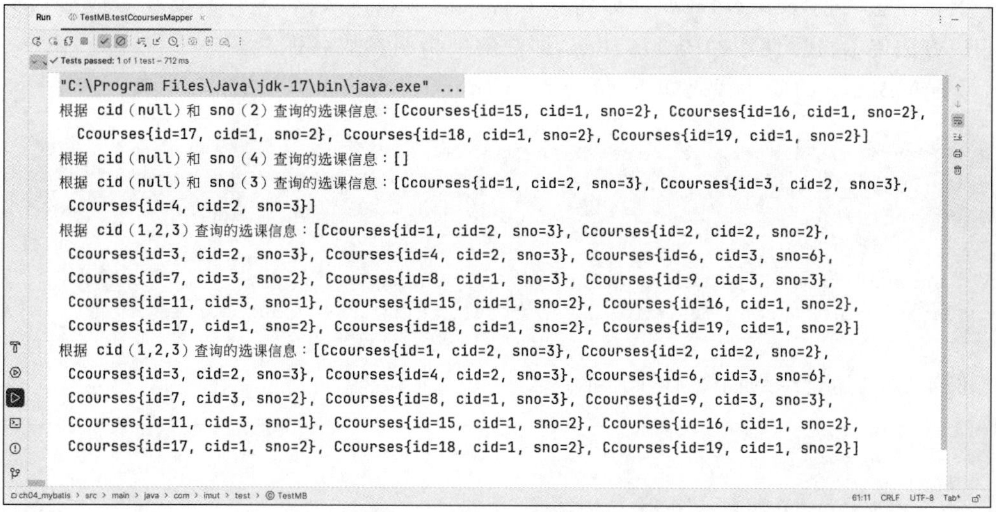

```
"C:\Program Files\Java\jdk-17\bin\java.exe" ...
根据 cid（null）和 sno（2）查询的选课信息：[Ccourses{id=15, cid=1, sno=2}, Ccourses{id=16, cid=1, sno=2},
 Ccourses{id=17, cid=1, sno=2}, Ccourses{id=18, cid=1, sno=2}, Ccourses{id=19, cid=1, sno=2}]
根据 cid（null）和 sno（4）查询的选课信息：[]
根据 cid（null）和 sno（3）查询的选课信息：[Ccourses{id=1, cid=2, sno=3}, Ccourses{id=3, cid=2, sno=3},
 Ccourses{id=4, cid=2, sno=3}]
根据 cid（1,2,3）查询的选课信息：[Ccourses{id=1, cid=2, sno=3}, Ccourses{id=2, cid=2, sno=2},
 Ccourses{id=3, cid=2, sno=3}, Ccourses{id=4, cid=2, sno=3}, Ccourses{id=6, cid=3, sno=6},
 Ccourses{id=7, cid=3, sno=2}, Ccourses{id=8, cid=1, sno=3}, Ccourses{id=9, cid=3, sno=3},
 Ccourses{id=11, cid=3, sno=1}, Ccourses{id=15, cid=1, sno=2}, Ccourses{id=16, cid=1, sno=2},
 Ccourses{id=17, cid=1, sno=2}, Ccourses{id=18, cid=1, sno=2}, Ccourses{id=19, cid=1, sno=2}]
根据 cid（1,2,3）查询的选课信息：[Ccourses{id=1, cid=2, sno=3}, Ccourses{id=2, cid=2, sno=2},
 Ccourses{id=3, cid=2, sno=3}, Ccourses{id=4, cid=2, sno=3}, Ccourses{id=6, cid=3, sno=6},
 Ccourses{id=7, cid=3, sno=2}, Ccourses{id=8, cid=1, sno=3}, Ccourses{id=9, cid=3, sno=3},
 Ccourses{id=11, cid=3, sno=1}, Ccourses{id=15, cid=1, sno=2}, Ccourses{id=16, cid=1, sno=2},
 Ccourses{id=17, cid=1, sno=2}, Ccourses{id=18, cid=1, sno=2}, Ccourses{id=19, cid=1, sno=2}]
```

图 4-20　例 4-18 的运行结果

例 4-19：演示根据多个课程 ID 查询选课记录的方式，其中多个课程 ID 存在 Map 中，未使用 IN 关键字。需要改动的文件为 CcoursesMapper.java、CcoursesMapper.xml 和 TestMB.java。其余文件均沿用例 2-25 的即可。

CcoursesMapper.java 的新增代码如下。

```
List<Ccourses> selectCcoursesByCids2(@Param("cidmap") Map<String, Integer>
cidmap);
```

CcoursesMapper.xml 的新增代码如下。

```
<select id="selectCcoursesByCids2" resultType="com.imut.pojo.Ccourses">
    select id, cid, sno
    from ccourses
    <where>
        <foreach collection="cidmap.values" item="cid" open="(" separator="or"
close=")">
            <if test="cid!=null">
                cid=#{cid}
            </if>
        </foreach>
    </where>
</select>
```

Map 键的类型为 String，值的类型为封装数据类型 Integer，Map 元素为多个课程 ID。注意：使用 Map 时，与其他方式不相同。index 属性为 Map 中的键，collection 属性为 Map 中的值，故 collection 属性值应为传入参数名.values，即类似于 Java 中获取 Map 中所有值时调用的方法 values。本例 open 属性值与上例不同，变为“(”，separator 属性变为 or，即改变了拼接内容，形成同样功能的不同 SQL 语句，足见其功能之强大。并且 foreach 元素包裹了 if 元素，if 元素中的判断条件是 cid 不为空。

值得注意的是,collection 属性如何获取传入的 Map 参数名称?需要在对应接口文件的相应方法传参位置使用@Param 注解,即注解中指明参数名称。

TestMB.java 的新增代码如下,位于 testCcoursesMapper 方法中。

```
Map<String, Integer> map = new HashMap<>();
map.put("1", 1);
map.put("2", 2);
map.put("3", 3);
ccoursesList = mapper.selectCcoursesByCids2(map);
System.out.println("根据 cid(1,2,3)查询的选课信息:" + ccoursesList);
```

例 4-19 的运行结果如图 4-21 所示。

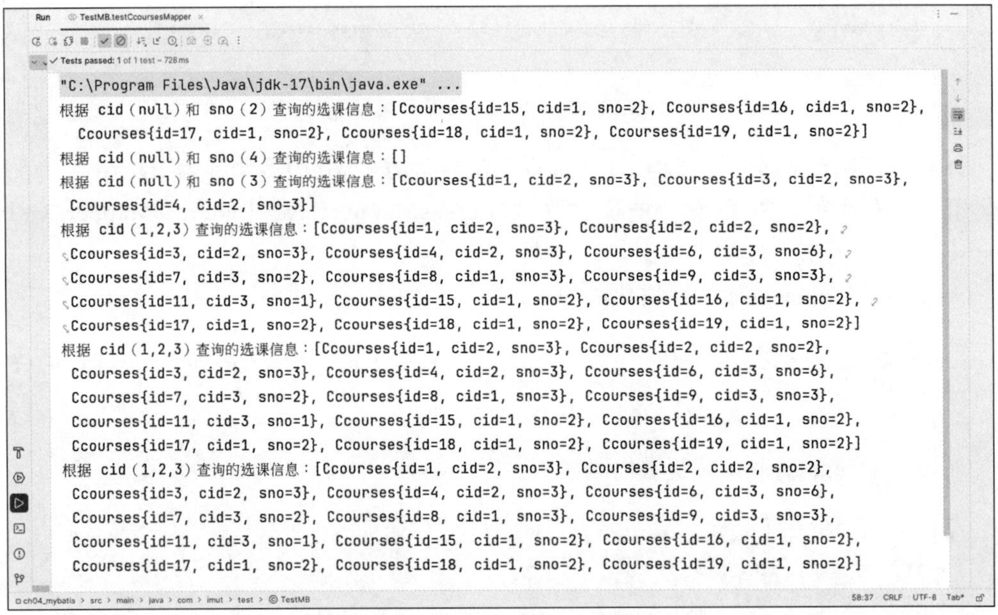

```
Run   TestMB.testCcoursesMapper
✓ Tests passed: 1 of 1 test – 728 ms
"C:\Program Files\Java\jdk-17\bin\java.exe" ...
根据 cid(null)和 sno(2)查询的选课信息:[Ccourses{id=15, cid=1, sno=2}, Ccourses{id=16, cid=1, sno=2},
 Ccourses{id=17, cid=1, sno=2}, Ccourses{id=18, cid=1, sno=2}, Ccourses{id=19, cid=1, sno=2}]
根据 cid(null)和 sno(4)查询的选课信息:[]
根据 cid(null)和 sno(3)查询的选课信息:[Ccourses{id=1, cid=2, sno=3}, Ccourses{id=3, cid=2, sno=3},
 Ccourses{id=4, cid=2, sno=3}]
根据 cid(1,2,3)查询的选课信息:[Ccourses{id=1, cid=2, sno=3}, Ccourses{id=2, cid=2, sno=2},
 Ccourses{id=3, cid=2, sno=3}, Ccourses{id=4, cid=2, sno=3}, Ccourses{id=6, cid=3, sno=6},
 Ccourses{id=7, cid=3, sno=2}, Ccourses{id=8, cid=1, sno=3}, Ccourses{id=9, cid=3, sno=3},
 Ccourses{id=11, cid=3, sno=1}, Ccourses{id=15, cid=1, sno=2}, Ccourses{id=16, cid=1, sno=2},
 Ccourses{id=17, cid=1, sno=2}, Ccourses{id=18, cid=1, sno=2}, Ccourses{id=19, cid=1, sno=2}]
根据 cid(1,2,3)查询的选课信息:[Ccourses{id=1, cid=2, sno=3}, Ccourses{id=2, cid=2, sno=2},
 Ccourses{id=3, cid=2, sno=3}, Ccourses{id=4, cid=2, sno=3}, Ccourses{id=6, cid=3, sno=6},
 Ccourses{id=7, cid=3, sno=2}, Ccourses{id=8, cid=1, sno=3}, Ccourses{id=9, cid=3, sno=3},
 Ccourses{id=11, cid=3, sno=1}, Ccourses{id=15, cid=1, sno=2}, Ccourses{id=16, cid=1, sno=2},
 Ccourses{id=17, cid=1, sno=2}, Ccourses{id=18, cid=1, sno=2}, Ccourses{id=19, cid=1, sno=2}]
根据 cid(1,2,3)查询的选课信息:[Ccourses{id=1, cid=2, sno=3}, Ccourses{id=2, cid=2, sno=2},
 Ccourses{id=3, cid=2, sno=3}, Ccourses{id=4, cid=2, sno=3}, Ccourses{id=6, cid=3, sno=6},
 Ccourses{id=7, cid=3, sno=2}, Ccourses{id=8, cid=1, sno=3}, Ccourses{id=9, cid=3, sno=3},
 Ccourses{id=11, cid=3, sno=1}, Ccourses{id=15, cid=1, sno=2}, Ccourses{id=16, cid=1, sno=2},
 Ccourses{id=17, cid=1, sno=2}, Ccourses{id=18, cid=1, sno=2}, Ccourses{id=19, cid=1, sno=2}]
ch04_mybatis > src > main > java > com > imut > test > TestMB          58:37  CRLF  UTF-8  Tab*
```

图 4-21　例 4-19 的运行结果

执行测试方法,如果已进行日志类相关的配置,可以由控制台知晓执行的 SQL 语句为"select id, cid, sno from ccourses WHERE (cid=? or cid=? or cid=?)。?"由传入的变量值 1、2、3 代替。

例 4-20:演示批量插入选课记录的方式,其中选课信息存在 List 中,未使用 IN 关键字。需要改动的文件为 CcoursesMapper.java、CcoursesMapper.xml 和 TestMB.java。其余文件均沿用例 2-25 的即可。

CcoursesMapper.java 的新增代码如下。

```
int insert(List<Ccourses> ccoursesList);
```

CcoursesMapper.xml 的新增代码如下。

```
<insert id="insert" parameterType="com.imut.pojo.Ccourses">
    insert into ccourses (id, cid, sno) values
```

```
    <foreach collection="list" item="ccourses" separator=",">
        (#{ccourses.id}, #{ccourses.cid}, #{ccourses.sno})
    </foreach>
</insert>
```

List 类型为选课记录实体类型 Ccourses,元素为多条选课信息。collection 属性的值一定是 list,如果错写为其他,会抛出异常。item 属性的值无指定规则,但需与 foreach 元素包裹的#{}中的.前面名称一致。因笔者使用的批量插入语句无须在循环开始和结束时拼接内容,故未使用 open 属性和 close 属性。foreach 元素包裹内容获取的值是 item 属性传入的值,为 Ccourses 类型属性名称。上述代码仅使用了 foreach 元素,故批量插入出错的概率较高。

TestMB.java 的新增代码如下,位于 testCcoursesMapper 方法中。

```
ccoursesList = new ArrayList<>();
Ccourses ccourses = new Ccourses();
ccourses.setId(15);
ccourses.setCid(2);
ccourses.setSno(3);
ccoursesList.add(ccourses);
ccourses = new Ccourses();
ccourses.setId(16);
ccourses.setCid(3);
ccourses.setSno(1);
ccoursesList.add(ccourses);
int result = mapper.insert(ccoursesList);
session.commit();
if (result==2)
    System.out.println("数据插入成功");
else
    System.out.println("数据插入失败");
```

例 4-20 的运行结果如图 4-22 所示。

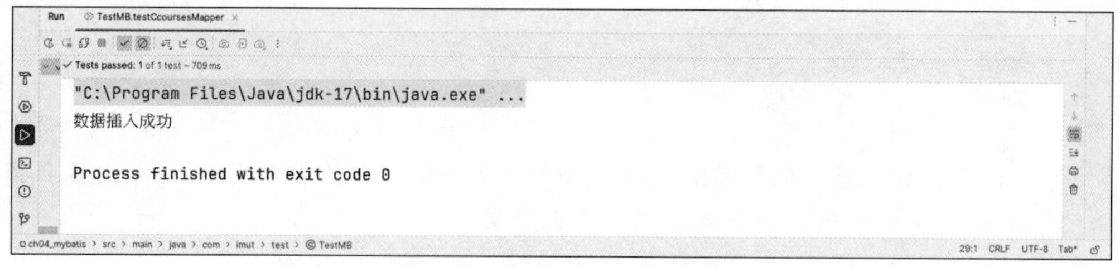

图 4-22　例 4-20 的运行结果

建议 Ccourses 实体类最好添加构造方法,以满足不同的赋值方式。执行测试方法,如果已进行日志类相关的配置,可以由控制台知晓执行的 SQL 为 insert into ccourses (id, cid, sno) values (?, ?, ?) , (?, ?, ?)。

4.5 bind 元素

bind 元素创建一个供别处使用的变量，变量的值为 OGNL 表达式，通常情况下用于模糊查询。bind 元素包含 2 个属性：name 属性、value 属性。name 属性用于指定 bind 元素的名称，作为调用的唯一标识，是必填属性；value 属性用于指定自定义的 OGNL 表达式，也是必填属性。写入 SQL 的位置不固定，bind 元素使用方式便捷，定义完毕后，通过 #{name 属性值}的方式调用。

这里主要介绍模糊查询的 4 种 bind 定义方式。方式 1：使用 MyBatis 内置参数方式。方式 2：使用 MyBatis 内置参数.参数名方式。方式 3：直接使用属性名的方式。方式 4：使用 MyBatis 内置参数.方法名方式。示例场景均为：根据课程名称查询课程信息。

例 4-21：演示根据课程名称查询课程信息记录的方式，其中参数名为使用 MyBatis 内置参数名称。共包含 8 个文件：Students.java、Courses.java、CoursesMapper.java、CoursesMapper.xml、mybatis-config.xml、db.properties、MyBatisUtil.java 和 TestMB.java。需要改动的文件为 CoursesMapper.java、CoursesMapper.xml 和 TestMB.java。Students.java 和 Courses.java 文件内容详见例 3-3，其余文件均沿用例 2-25 的即可。

CoursesMapper.java 的新增代码如下。

```
List<Courses> selectCoursesByCname(String cname);
```

CoursesMapper.xml 的新增代码如下。

```
<select id="selectCoursesByCname" resultType="com.imut.pojo.Courses">
select cid, cname, tid
    from courses
    <where>
        <if test="_parameter!=null">
            <bind name="forcname" value="'%'+_parameter+'%'"/>
            name like #{forcname}
        </if>
    </where>
</select>
```

注意：if 元素 test 属性值的参数位置也必须为 MyBatis 内置参数_parameter，否则将会报错。传入参数个数须为 1，否则无法得到查询结果。

TestMB.java 的新增代码如下，位于 testCoursesMapper 方法中。

```
cname = "e";
coursesList = mapper.selectCoursesByCname(cname);
System.out.println("根据 cname(e)模糊查询的课程信息:" + coursesList);
```

例 4-21 的运行结果如图 4-23 所示。

执行测试方法，如果已进行日志类相关的配置，可以由控制台知晓执行的 SQL 为

```
"C:\Program Files\Java\jdk-17\bin\java.exe" ...
根据 cname（e）模糊查询的课程信息：[Courses{cid=2, cname='chinese', tid=2, studentsList=null},
 Courses{cid=3, cname='english', tid=3, studentsList=null}]

Process finished with exit code 0
```

图 4-23　例 4-21 的运行结果

“select cid, cname, tid from courses WHERE cname like ?,?”由%e%替代。“%”由 CoursesMapper.xml 中的 bind 元素拼接，是通配符，可以匹配任意字符。

例 4-22：演示根据课程名称查询课程信息记录的方式，其中参数形式为 MyBatis 内置参数.参数名称。需要改动的文件为 CoursesMapper.java、CoursesMapper.xml 和 TestMB.java。Students.java 和 Courses.java 文件内容详见例 3-3，其余文件均沿用例 2-25 的即可。

CoursesMapper.java 的新增代码如下。

```
List<Courses> selectCoursesByCname1(@Param("cname") String cname);
```

注意：CoursesMapper.java 接口中的对应方法传入的参数必须使用@Param 注解标识，否则会抛出异常。

CoursesMapper.xml 的新增代码如下。

```
<select id="selectCoursesByCname1" resultType="com.imut.pojo.Courses">
    select cid, cname, tid
    from courses
    <where>
      <if test="cname!=null">
          <bind name="forcname" value="'%'+_parameter.cname+'%'"/>
          cname like #{forcname}
      </if>
    </where>
</select>
```

TestMB.java 的新增代码如下，位于 testCoursesMapper 方法中。

```
cname = "h";
coursesList = mapper.selectCoursesByCname1(cname);
System.out.println("根据 cname(h)模糊查询的课程信息:" + coursesList);
```

例 4-22 的运行结果如图 4-24 所示。

例 4-23：演示根据课程名称查询课程信息记录的方式，其中参数名为属性名称。需要改动的文件为 CoursesMapper.java、CoursesMapper.xml 和 TestMB.java。Students.java 和 Courses.java 文件内容详见例 3-3，其余文件均沿用例 2-25 的即可。

CoursesMapper.java 的新增代码如下。

```
List<Courses> selectCoursesByCname2(String cname);
```

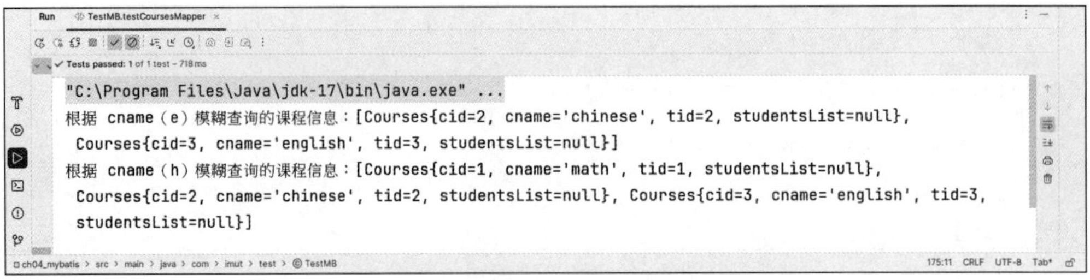

```
Run   TestMB.testCoursesMapper
Tests passed: 1 of 1 test – 718 ms
"C:\Program Files\Java\jdk-17\bin\java.exe" ...
根据 cname（e）模糊查询的课程信息：[Courses{cid=2, cname='chinese', tid=2, studentsList=null},
 Courses{cid=3, cname='english', tid=3, studentsList=null}]
根据 cname（h）模糊查询的课程信息：[Courses{cid=1, cname='math', tid=1, studentsList=null},
 Courses{cid=2, cname='chinese', tid=2, studentsList=null}, Courses{cid=3, cname='english', tid=3,
 studentsList=null}]
ch04_mybatis > src > main > java > com > imut > test > TestMB          175:11  CRLF  UTF-8  Tab*
```

图 4-24　例 4-22 的运行结果

CoursesMapper.xml 的新增代码如下。

```
<select id="selectCoursesByCname2" resultType="com.imut.pojo.Courses">
    select cid, cname, tid
    from courses
    <where>
        <if test="cname!=null">
            <bind name="forcname" value="'%'+cname+'%'"/>
            cname like #{forcname}
        </if>
    </where>
</select>
```

TestMB.java 的新增代码如下,位于 testCoursesMapper 方法中。

```
cname = "c";
coursesList = mapper.selectCoursesByCname2(cname);
System.out.println("根据 cname(c)模糊查询的课程信息:" + coursesList);
```

例 4-23 的运行结果如图 4-25 所示。

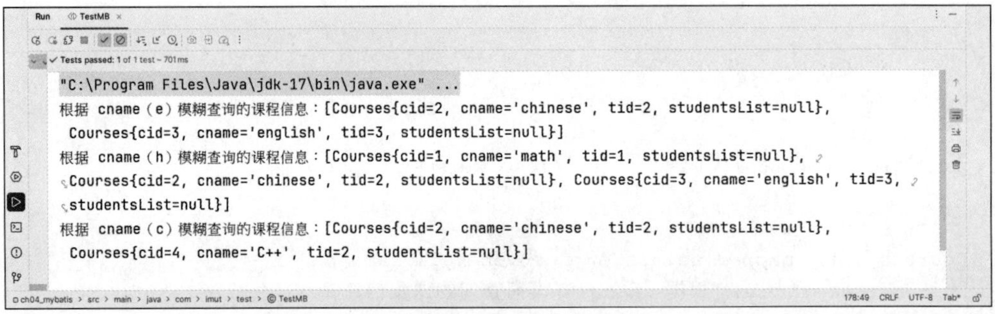

```
Run   TestMB
Tests passed: 1 of 1 test – 701 ms
"C:\Program Files\Java\jdk-17\bin\java.exe" ...
根据 cname（e）模糊查询的课程信息：[Courses{cid=2, cname='chinese', tid=2, studentsList=null},
 Courses{cid=3, cname='english', tid=3, studentsList=null}]
根据 cname（h）模糊查询的课程信息：[Courses{cid=1, cname='math', tid=1, studentsList=null},
 Courses{cid=2, cname='chinese', tid=2, studentsList=null}, Courses{cid=3, cname='english', tid=3,
 studentsList=null}]
根据 cname（c）模糊查询的课程信息：[Courses{cid=2, cname='chinese', tid=2, studentsList=null},
 Courses{cid=4, cname='C++', tid=2, studentsList=null}]
ch04_mybatis > src > main > java > com > imut > test > TestMB          178:49  CRLF  UTF-8  Tab*
```

图 4-25　例 4-23 的运行结果

例 4-24：演示根据课程名称查询课程信息记录的方式,其中参数形式为 MyBatis 内置参数.方法名称。需要改动的文件为 CoursesMapper.java、CoursesMapper.xml 和 TestMB.java。Students.java 和 Courses.java 文件内容详见例 3-3,其余文件均沿用例 2-25 的即可。

CoursesMapper.java 的新增代码如下。

```
List<Courses> selectCoursesByCname3(Courses courses);
```

CoursesMapper.xml 的新增代码如下。

```
<select id="selectCoursesByCname3" resultType="com.imut.pojo.Courses">
    select cid, cname, tid
    from courses
    <where>
        <if test="cname!=null">
            <bind name="forcname" value="'%'+_parameter.getCname()+'%'"/>
            cname like #{forcname}
        </if>
    </where>
</select>
```

注意：CoursesMapper.java 接口中的对应方法传入的参数是 Courses 类型的变量。方法不一定必须为 getter 方法，其他方法也可以。

TestMB.java 的新增代码如下，位于 testCoursesMapper 方法中。

```
courses.setCname("i");
coursesList = mapper.selectCoursesByCname3(courses);
System.out.println("根据 cname(i)模糊查询的课程信息:" + coursesList);
```

例 4-24 的运行结果如图 4-26 所示。

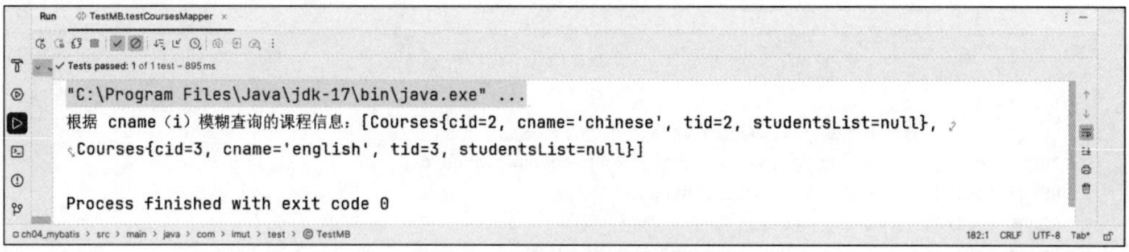

```
"C:\Program Files\Java\jdk-17\bin\java.exe" ...
根据 cname（i）模糊查询的课程信息：[Courses{cid=2, cname='chinese', tid=2, studentsList=null}, Courses{cid=3, cname='english', tid=3, studentsList=null}]

Process finished with exit code 0
```

图 4-26　例 4-24 的运行结果

bind 元素的使用可在一定程度上避免不同数据库的适配问题，还能够有效防止 SQL 注入。

至此本章讲解完毕，希望读者能够动手练习，多组合不同元素，碰撞出更多的火花。

4.6 综合案例

本章案例主要实现“课程信息管理”模块中根据课程 cid 查询课程信息、查询所有课程信息、修改课程信息以及删除课程信息的功能。

4.6.1 案例设计

本章案例划分为持久层和数据访问层。持久层由实体类组成。数据访问层由接口和映射文件组成，接口的名称统一以 Mapper 结尾，映射文件名称与接口文件名称相同。本章案例使用的主要文件如表 4-2 所示。

表 4-2　本章案例使用的主要文件

文　　件	所在包/路径	功　　能
log4j.properties	src/	日志输出配置文件
db.properties	src/	数据库连接的属性文件
mybatis-config.xml	src/	MyBatis 配置文件
Course.java	com.imut.pojo	封装课程信息的类
Ccourse.java	com.imut.pojo	封装选课信息的类
Student.java	com.imut.pojo	封装学生信息的类
CourseMapper.java	com.imut.mapper	定义操作课程表方法的接口
CourseMapper.xml	com.imut.mapper	Course 类的映射文件
MybatisUtil.java	com.imut.utils	获取 SqlSession 对象的工具类
CourseTest.java	com.imut.test	测试操作课程信息方法的类

4.6.2　案例演示

使用 Junit4 测试执行 CourseTest.java 中的方法。执行 getCourseByIdTest()方法根据课程 cid 查询课程信息，在控制台输出的结果如图 4-27 所示。

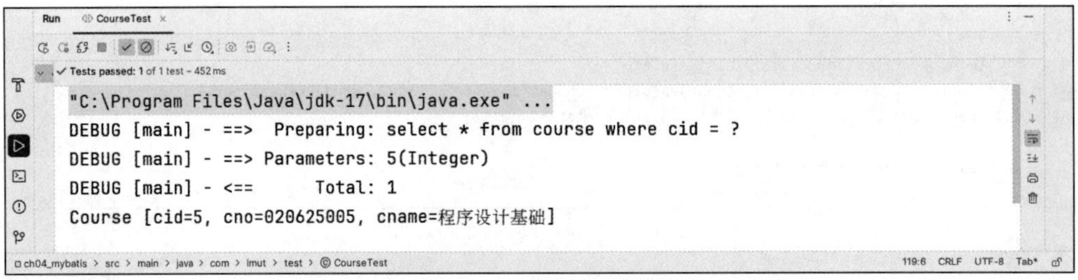

```
"C:\Program Files\Java\jdk-17\bin\java.exe" ...
DEBUG [main] - ==>  Preparing: select * from course where cid = ?
DEBUG [main] - ==> Parameters: 5(Integer)
DEBUG [main] - <==      Total: 1
Course [cid=5, cno=020625005, cname=程序设计基础]
```

图 4-27　查询 cid 为 5 的课程信息的执行结果

执行 listCoursesTest()方法输出所有课程信息，在控制台输出的结果如图 4-28 所示。

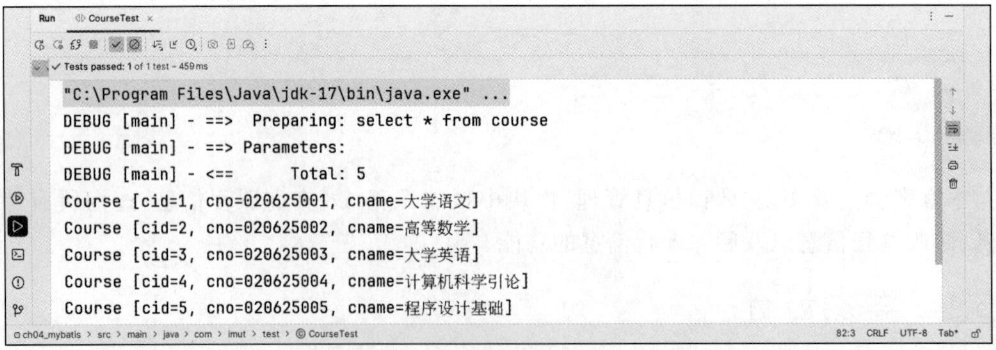

```
"C:\Program Files\Java\jdk-17\bin\java.exe" ...
DEBUG [main] - ==>  Preparing: select * from course
DEBUG [main] - ==> Parameters: 
DEBUG [main] - <==      Total: 5
Course [cid=1, cno=020625001, cname=大学语文]
Course [cid=2, cno=020625002, cname=高等数学]
Course [cid=3, cno=020625003, cname=大学英语]
Course [cid=4, cno=020625004, cname=计算机科学引论]
Course [cid=5, cno=020625005, cname=程序设计基础]
```

图 4-28　输出所有课程信息的运行结果

执行 addCourseTest()方法，增加一条名为“数据结构”的课程信息，在控制台输出的结果如图 4-29 所示。

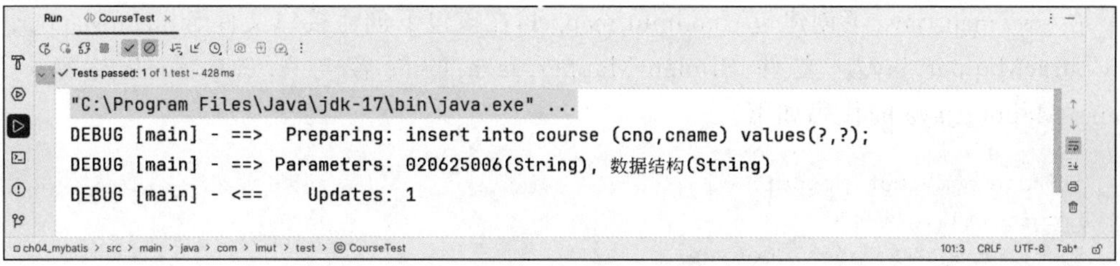
Run CourseTest
Tests passed: 1 of 1 test – 428 ms
"C:\Program Files\Java\jdk-17\bin\java.exe" ...
DEBUG [main] - ==> Preparing: insert into course (cno,cname) values(?,?);
DEBUG [main] - ==> Parameters: 020625006(String), 数据结构(String)
DEBUG [main] - <==   Updates: 1
ch04_mybatis > src > main > java > com > imut > test > CourseTest
101:3 CRLF UTF-8 Tab*

图 4-29　执行增加课程信息的运行结果

执行 updateCourseTest()方法，将 cid 为 6 的课程名称由“数据结构”修改为“数据结构与算法设计”，在控制台输出的结果如图 4-30 所示。

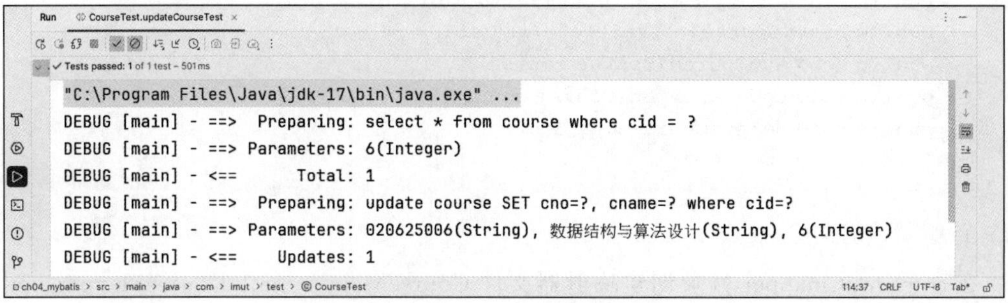
Run CourseTest.updateCourseTest
Tests passed: 1 of 1 test – 501 ms
"C:\Program Files\Java\jdk-17\bin\java.exe" ...
DEBUG [main] - ==> Preparing: select * from course where cid = ?
DEBUG [main] - ==> Parameters: 6(Integer)
DEBUG [main] - <==   Total: 1
DEBUG [main] - ==> Preparing: update course SET cno=?, cname=? where cid=?
DEBUG [main] - ==> Parameters: 020625006(String), 数据结构与算法设计(String), 6(Integer)
DEBUG [main] - <==   Updates: 1
ch04_mybatis > src > main > java > com > imut > test > CourseTest
114:37 CRLF UTF-8 Tab*

图 4-30　修改课程信息的运行结果

执行 deleteCourseTest()方法删除 cid 为 6 的课程信息，在控制台输出的结果如图 4-31 所示。

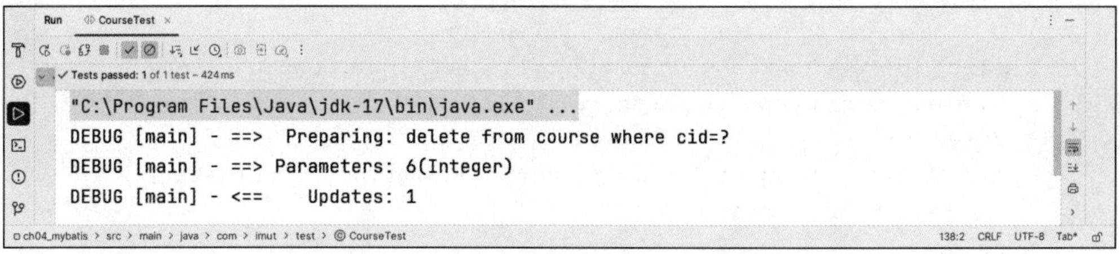
Run CourseTest
Tests passed: 1 of 1 test – 424 ms
"C:\Program Files\Java\jdk-17\bin\java.exe" ...
DEBUG [main] - ==> Preparing: delete from course where cid=?
DEBUG [main] - ==> Parameters: 6(Integer)
DEBUG [main] - <==   Updates: 1
ch04_mybatis > src > main > java > com > imut > test > CourseTest
138:2 CRLF UTF-8 Tab*

图 4-31　删除课程信息的运行结果

4.6.3　代码实现

1. 创建项目、添加依赖

创建名为 ch04_mybatis 的 Maven 工程，编写 pom.xml，内容与 3.4.3 节中的一致。

2. 创建配置文件

创建配置文件 log4j.properties、db.properties 与 mybatis-config.xml，内容与 3.4.3 节中的一致。

3. 创建实体类

创建实体类文件 Course.java、Ccourse.java、Student.java 以及 MybatisUtil.java，内容与 3.4.3 节中对应的文件内容一致。

4. 创建接口文件

在/src/main/java 下创建包 com.imut.mapper，在该包下创建接口文件 StudentMapper.java 和 CourseMapper.java。文件 StudentMapper.java 的内容与 4.6.3 节的内容一致。CourseMapper.java 的代码如下。

```
package com.imut.mapper;
import java.util.List;
import com.imut.pojo.Ccourse;
import com.imut.pojo.Course;
public interface CourseMapper {
    public Course getCourseById(int cid);
    public List<Course> listCourses();
    public List<Ccourse> getCcourseById(int cid);
    public List<Course> listSelect(int sid);
    public void addCourse(Course course);
    public void updateCourse(Course course);
    public void deleteCourse(int cid);
}
```

5. 创建映射文件

在 com.imut.mapper 目录创建映射器文件 CourseMapper.xml，CourseMapper.xml 的代码如下。

```
<?xml version="1.0" encoding="UTF-8" ?>
<!DOCTYPE mapper PUBLIC "-//mybatis.org//DTD Mapper 3.0//EN"
"http://mybatis.org/dtd/mybatis-3-mapper.dtd">
<mapper namespace="com.imut.mapper.CourseMapper">
    <select id="getCourseById" resultType="com.imut.pojo.Course">
        select * from course where cid = #{cid}
    </select>
    <select id="listCourses" resultType="com.imut.pojo.Course">
        select * from course
    </select>
    <select id="getCcourseById" resultType="com.imut.pojo.Ccourse">
        select cc.* from course c, ccourse cc
            where c.cid=cc.cid and c.cid=#{cid}
    </select>
    <select id="listSelect" resultType="com.imut.pojo.Course">
        select course.* FROM course where cid not in
            (select c.cid from course c,ccourse cc,student s
                where cc.cid=c.cid and cc.sid=s.sid and s.sid=#{sid})
    </select>
    <insert id="addCourse">
        insert into course (cno,cname)
            values(#{cno},#{cname});
    </insert>
    <update id="updateCourse">
        update course
```

```
        <set>
            <if test="cno!=null and cno!=''">cno=#{cno},</if>
            <if test="cname!=null and cname!=''">cname=#{cname}</if>
        </set>
        where cid=#{cid}
    </update>
    <delete id="deleteCourse">
        delete from course where cid=#{cid}
    </delete>
</mapper>
```

6. 创建测试类

在 com.imut 目录下新建 test 目录，在 com.imut.test 目录创建测试类 CourseTest.java。新增测试根据课程 cid 查询课程信息、查询所有课程信息、添加课程信息、更新课程信息以及删除课程信息的方法。CourseTest.java 的代码如下。

```
package com.imut.test;
import java.util.List;
import org.apache.ibatis.session.SqlSession;
import org.junit.Test;
import com.imut.mapper.CourseMapper;
import com.imut.pojo.Ccourse;
import com.imut.pojo.Course;
import com.imut.utils.MybatisUtil;
public class CourseTest {
    @Test
    public void getCcourseByIdTest() {
        SqlSession sqlSession = MybatisUtil.getSqlSession();
        CourseMapper mapper = sqlSession.getMapper(CourseMapper.class);
        List<Ccourse> list = mapper.getCcourseById(5);
        for (Ccourse cc : list) {
            System.out.println(cc);
        }
        sqlSession.close();
    }
    @Test
    public void listSelectTest() {
        SqlSession sqlSession = MybatisUtil.getSqlSession();
        CourseMapper mapper = sqlSession.getMapper(CourseMapper.class);
        List<Course> list = mapper.listSelect(1);
        for (Course c : list) {
            System.out.println(c);
        }
        sqlSession.close();
    }
    @Test
    public void getCourseByIdTest() {
        SqlSession sqlSession = MybatisUtil.getSqlSession();
```

```
        CourseMapper mapper = sqlSession.getMapper(CourseMapper.class);
        Course course = mapper.getCourseById(5);
        System.out.println(course);
        sqlSession.close();
    }
    @Test
    public void listCoursesTest() {
        SqlSession sqlSession = MybatisUtil.getSqlSession();
        CourseMapper mapper = sqlSession.getMapper(CourseMapper.class);
        List<Course> list = mapper.listCourses();
        for (Course c : list) {
            System.out.println(c);
        }
        sqlSession.close();
    }
    @Test
    public void addCourseTest() {
        SqlSession sqlSession = MybatisUtil.getSqlSession();
        CourseMapper mapper = sqlSession.getMapper(CourseMapper.class);
        Course course = new Course();
        course.setCno("020625006");
        course.setCname("数据结构");
        mapper.addCourse(course);
        sqlSession.commit();
        sqlSession.close();
    }
    @Test
    public void updateCourseTest() {
        SqlSession sqlSession = MybatisUtil.getSqlSession();
        CourseMapper mapper = sqlSession.getMapper(CourseMapper.class);
        Course course=mapper.getCourseById(6);
        course.setCname("数据结构与算法设计");
        mapper.updateCourse(course);
        sqlSession.commit();
        sqlSession.close();
    }
    @Test
    public void deleteCourseTest() {
        SqlSession sqlSession = MybatisUtil.getSqlSession();
        CourseMapper mapper = sqlSession.getMapper(CourseMapper.class);
        mapper.deleteCourse(6);
        sqlSession.commit();
        sqlSession.close();
    }
}
```

4.7 习题

1. 单选题

(1) 当包含多个选项,需要从多个选项中选择一个执行时,使用的动态 SQL 元素是(　　)。

A. IF　　B. CHOOSE

C. SET　　D. CHOOSE WHEN OTHERWISE

(2) foreach 元素包含的属性有(　　)。

A. id 和 name　　B. name 和 value

C. collection 和 index　　D. index 和 value

(3) 动态 SQL 中的主要元素不包含(　　)。

A. if　　B. trim　　C. update　　D. bind

2. 填空题

(1) test 属性的作用是________。

(2) trim 元素的属性有________个,分别是________。

3. 简答题

(1) set、where 和 trim 的异同点是什么?

(2) foreach 元素的 collection 属性的值有哪些? 每种代表什么含义?

第5章 缓存和存储过程

在日常生活中,需要经常查询信息。如果每次查询都要做一些无用的重复工作,非常浪费时间。故产生如下需求:同样的查询,如果不用每次都从数据库获取结果该有多好!缓存由此出现,既可以提升数据访问效率,又可以降低数据库压力。正如其他 ORM 框架,MyBatis 亦提供缓存机制,分别是一级缓存和二级缓存。

存储过程是为了完成特定需求的大段 SQL 语句,通常会在大型的数据库系统中使用。存储过程具有运行速度快、稳定的特点,能够在一定程度上减轻客户端压力。

5.1 一级缓存

一级缓存又名本地缓存(local cache),默认是开启状态,即用户无须额外配置。根据其作用域,细分为基于 Session 的缓存和基于 Statement 的缓存。这里将以基于 Session 的缓存为例讲解。

1.1.3 节曾讲过,SqlSession 会通过 Configuration 创建 Executor,并将具体的任务交给 Executor 执行。既然是本地缓存,肯定不适合用全局的属性存储,结合 1.4 节的图 1-3 可知,剩下的选择是将缓存放入 Executor 维护。接着详细看看 Executor 接口(与先前提及的宽泛的 Executor 不同)与其实现类之间的关系图,如图 5-1 所示。注:图中仅列出部分重点属性、方法和参数。

Executor 是一个接口,共有两个直接实现类:CachingExecutor 和 BaseExecutor(抽象类)。CachingExecutor 主要用于二级缓存,故重点关注 BaseExecutor。BaseExecutor 有一个名为 localCache 的属性,由名字可知,是用于本地缓存的,localCache 类型为 PerpetualCache(cache 接口的实现类)。PerpetualCache 类包含一个 Map<Object, Object>类型的 cache 属性,由此可知,一级缓存使用 Map 结构存储。BaseExecutor 中定义了与数据库交互的方法,包括查询、更新、提交数据至数据库。其具有 4 个子类:SimpleExecutor、BatchExecutor、ReuseExecutor 和

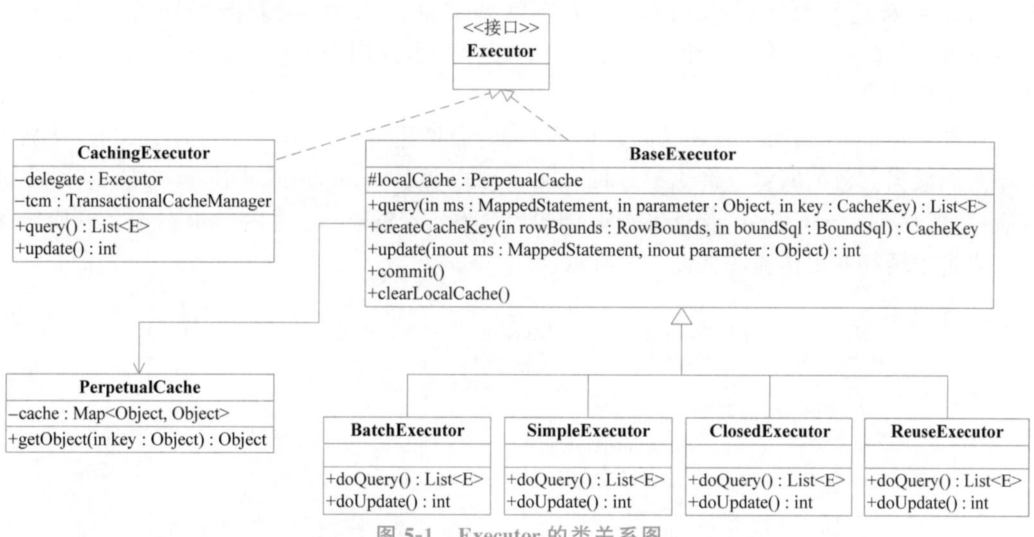

图 5-1　Executor 的类关系图

ClosedExecutor，分别实现与数据库不同需求的交互。

了解完整体结构，接着学习一级缓存的工作流程。查询有两种结果：①击中缓存；②未击中缓存。击中缓存的工作流程如图 5-2 所示。

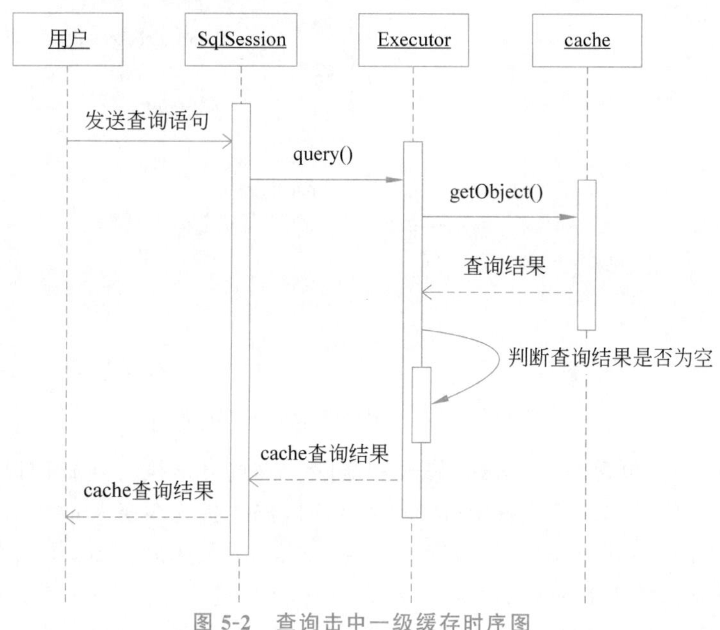

图 5-2　查询击中一级缓存时序图

用户根据自身需求发送查询请求，程序接收到用户的查询请求，判断需要执行的方法，SqlSession 职责下放至 Executor，即通过调用 Executor 的 query 方法令 Executor 执行具体任务。Executor 则通过调用 cache 的 getObject 方法获取缓存中的数据，该方法需要一个 CacheKey 类型的参数 key。key 由 MappedStatement 的 ID、SQL 的 offset、SQL 的 limit、SQL 语句以及相关参数组成。cache 具体指的是其中一个实现类 PerpetualCache，其中的 getObject 方法根据键获取 HashMap 中存放的缓存数据，并返回结果至 Executor。

Executor判断返回数据是否为空，不为空则处理输出参数，接着判断缓存级别，若是statement级别，则清空缓存，否则将获取到的缓存数据返回至SqlSession，再由SqlSession返回给用户。

如果Executor判断cache返回数据为空，则会调用queryFromDatabase方法，从数据库中查询数据。数据库将查询结果返回，Executor调用cache的putObject方法，将数据放入缓存，即存入localCache，接着将查询结果反馈给SqlSession，再由SqlSession返回给用户。未击中缓存的工作流程如图5-3所示。

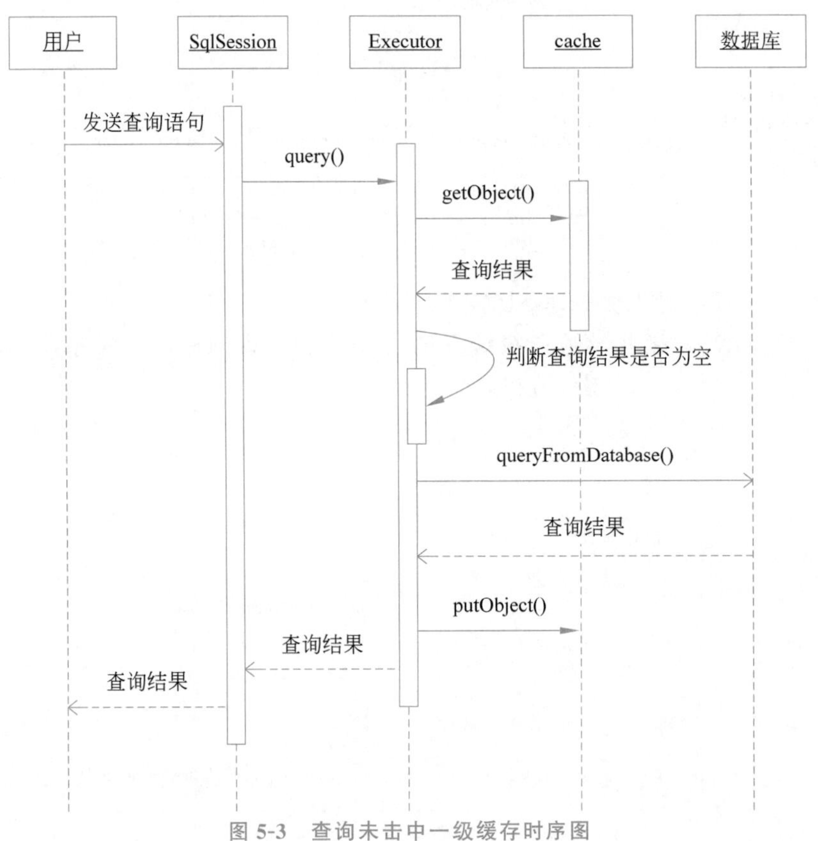

图5-3　查询未击中一级缓存时序图

针对数据的操作包含查询、增加、修改和删除。Executor执行除查询以外的3种操作时，都会转换为更新操作。且执行上述3种操作时，需要提交事务。故当有增加、删除、修改、提交数据，关闭Session操作时，缓存会失效。Session关闭后，PerpetualCache对象不可用，其余情况下则只清空PerpetualCache对象。

在默认情况下，开启的一级缓存为SqlSession级别，即同一SqlSession多次执行同一SQL，直接从内存中取到缓存结果，而无须查询数据库。笔者将通过4个例子讲解。为了便于观察缓存效果，需添加日志配置，通过输出日志确定其是否击中缓存。在XML配置文件mybatis-config.xml中添加日志配置项，代码如下。

```
<settings>
    <setting name="logImpl" value="STDOUT_LOGGING" />
</settings>
```

具体配置信息含义可回顾 2.1.2 节的表 2-1。

5.1.1 相同 SqlSession

输出日志配置完毕，接着编写测试案例。首先是在相同 SqlSession 的前提下，分别从查询和更新两方面测试的案例。

1. 查询

例 5-1：演示根据选课 ID 查询选课记录的方式，其中 SqlSession 和 Mapper 使用的是同一个，连续执行的两条查询语句一致。需要改动的文件为 CcoursesMapper.java、CcoursesMapper.xml 和 TestMB.java。其余文件均沿用例 2-25 的即可。

CcoursesMapper.java 的代码如下。

```
package com.imut.mapper;
import com.imut.pojo.Ccourses;
public interface CcoursesMapper {
    Ccourses selectCcoursesByid(Integer id);
}
```

CcoursesMapper.xml 的代码如下。

```
<?xml version="1.0" encoding="UTF-8"?>
<!DOCTYPE mapper PUBLIC "-//mybatis.org//DTD Mapper 3.0//EN" "http://mybatis.
org/dtd/mybatis-3-mapper.dtd">
<mapper namespace="com.imut.mapper.CcoursesMapper">
    <select id="selectCcoursesByid" resultType="ccourses">
        select * from ccourses where
        <if test="id!=null">
            id = #{id,jdbcType=INTEGER}
        </if>
    </select>
</mapper>
```

TestMB.java 的代码如下。

```
package com.imut.test;
import com.imut.mapper.CcoursesMapper;
import com.imut.pojo.*;
import com.imut.utils.MyBatisUtil;
import org.apache.ibatis.session.SqlSession;
import org.apache.ibatis.session.SqlSessionFactory;
import org.junit.Test;
public class TestMB {
    @Test
    public void testCcoursesMapper(){
        SqlSessionFactory ssf = null;
        SqlSession session = null;
        int id = 6;
        try {
```

```
            ssf = new MyBatisUtil().getSqlSessionFactory();
            session = ssf.openSession();
            CcoursesMapper ccoursesMapper = session.getMapper(CcoursesMapper.class);
            Ccourses ccourses = ccoursesMapper.selectCcoursesByid(id);
            System.out.println(ccourses);
            Ccourses ccourses1 = ccoursesMapper.selectCcoursesByid(id);
            System.out.println(ccourses1);
        } catch (Exception e) {
            e.printStackTrace();
        } finally {
            session.close();
        }
    }
}
```

上述代码实现根据选课 ID 查询选课记录的功能，且查询方法输入 ID 值是通过一个变量传入的，确保一致性。Mapper 的同一个方法连续调用两次，然后输出查询数据。运行结果如图 5-4 所示。

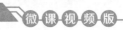

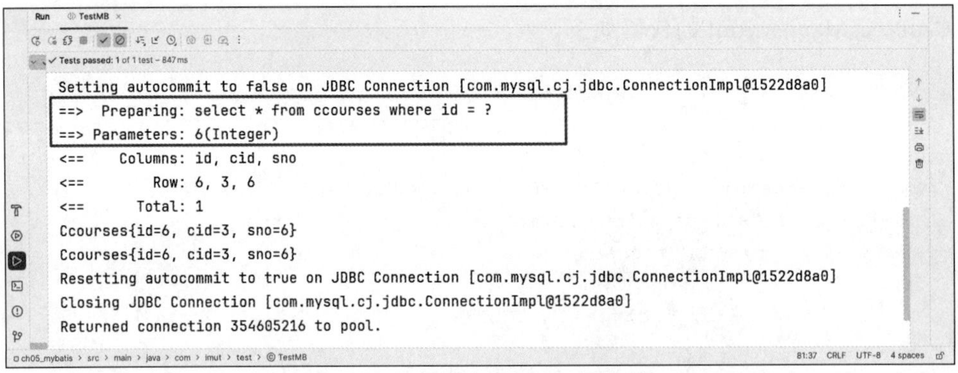
Run
TestMB
Tests passed: 1 of 1 test – 847ms
Setting autocommit to false on JDBC Connection [com.mysql.cj.jdbc.ConnectionImpl@1522d8a0]
==> Preparing: select * from ccourses where id = ?
==> Parameters: 6(Integer)
<== Columns: id, cid, sno
<== Row: 6, 3, 6
<== Total: 1
Ccourses{id=6, cid=3, sno=6}
Ccourses{id=6, cid=3, sno=6}
Resetting autocommit to true on JDBC Connection [com.mysql.cj.jdbc.ConnectionImpl@1522d8a0]
Closing JDBC Connection [com.mysql.cj.jdbc.ConnectionImpl@1522d8a0]
Returned connection 354605216 to pool.
ch05_mybatis > src > main > java > com > imut > test > TestMB
81:37 CRLF UTF-8 4 spaces

图 5-4　一级缓存例 5-1 运行结果

由图 5-4 可知，只向数据库发送了一次 SQL 查询请求，即框住部分。第二次获取的数据并未与数据库交互，而是从缓存中得到。可以看出，两次查询获取的数据一致。

因插入、删除操作最终都会转化为调用 update 方法，即逻辑是一致的，故笔者只以更新操作为例。

2. 更新

例 5-2：演示更新操作令缓存失效的方式，其中 SqlSession 和 Mapper 都使用的是同一个。执行两条相同的查询语句，两次查询之间执行一次更新语句。需要改动的文件为 Ccourses.java、CcoursesMapper.java、CcoursesMapper.xml 和 TestMB.java。其余文件均沿用例 2-25 的即可。

Ccourses.java 的新增代码如下。

```
public Ccourses(Integer id, Integer cid, Integer sno) {
    this.id = id;
    this.cid = cid;
```

```
        this.sno = sno;
    }
```

为了满足不同需求的赋值，为 Ccourses 类添加 3 参构造方法。

CcoursesMapper.java 的新增代码如下。

```
int updateCcoursesByid(Ccourses ccourses);
```

CcoursesMapper.xml 的新增代码如下。

```
<update id="updateCcoursesByid" parameterType="com.imut.pojo.Ccourses">
    update ccourses
    set cid = #{cid,jdbcType=INTEGER},
    sno = #{sno,jdbcType=INTEGER}
    where id = #{id,jdbcType=INTEGER}
</update>
```

TestMB.java 的代码如下。

```
package com.imut.test;
import com.imut.mapper.CcoursesMapper;
import com.imut.pojo.*;
import com.imut.utils.MyBatisUtil;
import org.apache.ibatis.session.SqlSession;
import org.apache.ibatis.session.SqlSessionFactory;
import org.junit.Test;
public class TestMB {
    @Test
    public void testCcoursesMapper(){
        SqlSessionFactory ssf = null;
        SqlSession session = null;
        int id = 2;
        try {
            ssf = new MyBatisUtil().getSqlSessionFactory();
            session = ssf.openSession();
            CcoursesMapper ccoursesMapper = session.getMapper(CcoursesMapper.class);
            Ccourses ccourses = ccoursesMapper.selectCcoursesByid(id);
            System.out.println(ccourses);
            int result = ccoursesMapper.updateCcoursesByid(new Ccourses(1, 2, 3));
            System.out.println(result);
            Ccourses ccourses1 = ccoursesMapper.selectCcoursesByid(id);
            System.out.println(ccourses1);
        } catch (Exception e) {
            e.printStackTrace();
        } finally {
            session.close();
        }
    }
}
```

上述代码实现根据选课 ID 查询选课记录，更新选课记录，再次查询选课记录的功能，且查询方法输入 ID 值是通过一个变量传入的，确保一致性，更新选课记录的 ID 与查询选课记录的 ID 不一致。Mapper 的同一个查询方法调用两次，更新方法在两次查询方法中间调用一次，输出查询数据，数据更新影响条数。运行结果如图 5-5 所示。

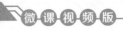

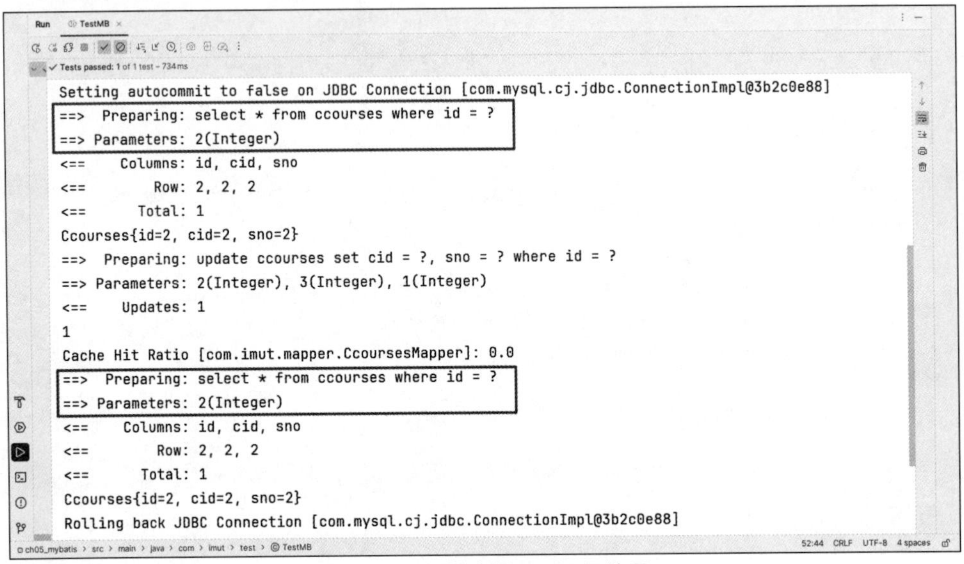
Run TestMB
Tests passed: 1 of 1 test – 734 ms
Setting autocommit to false on JDBC Connection [com.mysql.cj.jdbc.ConnectionImpl@3b2c0e88]
==> Preparing: select * from ccourses where id = ?
==> Parameters: 2(Integer)
<== Columns: id, cid, sno
<== Row: 2, 2, 2
<== Total: 1
Ccourses{id=2, cid=2, sno=2}
==> Preparing: update ccourses set cid = ?, sno = ? where id = ?
==> Parameters: 2(Integer), 3(Integer), 1(Integer)
<== Updates: 1
1
Cache Hit Ratio [com.imut.mapper.CcoursesMapper]: 0.0
==> Preparing: select * from ccourses where id = ?
==> Parameters: 2(Integer)
<== Columns: id, cid, sno
<== Row: 2, 2, 2
<== Total: 1
Ccourses{id=2, cid=2, sno=2}
Rolling back JDBC Connection [com.mysql.cj.jdbc.ConnectionImpl@3b2c0e88]
ch05_mybatis > src > main > java > com > imut > test > TestMB
52:44 CRLF UTF-8 4 spaces

图 5-5　一级缓存例 5-2 运行结果

由图 5-5 可知，共向数据库发送了 3 次 SQL 查询请求，其中向数据库发送了 2 次 SQL 查询请求，即框住部分。也就意味着第 2 次并未击中缓存，而是从数据库中获得的。可以看出，尽管两次查询获取的数据一致，但因为中间执行了更新操作，故缓存被清空，所以第 2 次才需要和数据库交互。这里故意选用更新的 ID 和查询 ID 不一致，避免读者认为是因为更新了查询数据缓存才被清空。且这里并未执行 commit 操作，缓存亦被清空，再次印证无论更新操作是否提交，都将清空缓存。

既然默认开启的一级缓存是基于 SqlSession，接着列举不在同一个 SqlSession 的例子，依然涉及查询和更新两方面。

5.1.2　不同 SqlSession

1. 查询

例 5-3：演示根据选课 ID 查询选课记录的方式，其中 SqlSession 既包含相同的，又包含不同的，连续执行的两条查询语句一致。需要改动的文件为 TestMB.java。CcoursesMapper.java、CcoursesMapper.xml 文件内容详见例 5-2，其余文件均沿用例 2-25 的即可。

TestMB.java 的代码如下。

```
package com.imut.test;
import com.imut.mapper.CcoursesMapper;
import com.imut.pojo.*;
import com.imut.utils.MyBatisUtil;
```

```
import org.apache.ibatis.session.SqlSession;
import org.apache.ibatis.session.SqlSessionFactory;
import org.junit.Test;
public class TestMB {
    @Test
    public void testCcoursesMapper() {
        SqlSessionFactory ssf = null;
        SqlSession session = null;
        SqlSession session1 = null;
        int id = 6;
        try {
            ssf = new MyBatisUtil().getSqlSessionFactory();
            session = ssf.openSession();
            CcoursesMapper ccoursesMapper = session.getMapper(CcoursesMapper.class);
            Ccourses ccourses = ccoursesMapper.selectCcoursesByid(id);
            System.out.println(ccourses);
            Ccourses ccourses1 = ccoursesMapper.selectCcoursesByid(id);
            System.out.println(ccourses1);
            session1 = ssf.openSession();
            CcoursesMapper ccoursesMapper2 = session1.getMapper(CcoursesMapper.class);
            Ccourses ccourses2 = ccoursesMapper2.selectCcoursesByid(id);
            System.out.println(ccourses2);
        } catch (Exception e) {
            e.printStackTrace();
        } finally {
            session.close();
        }
    }
}
```

上述代码分别实现在相同 SqlSession 和不同 SqlSession 前提下，根据同样的选课 ID 查询选课记录的功能，且查询方法输入 ID 值是通过一个变量传入的，确保一致性。即在相同 SqlSession 的情况下连续执行两条一样的查询语句；在不同 SqlSession 的情况下执行与前两次查询相同的 SQL 语句。同样需要输出每次的查询数据。运行结果如图 5-6 所示。

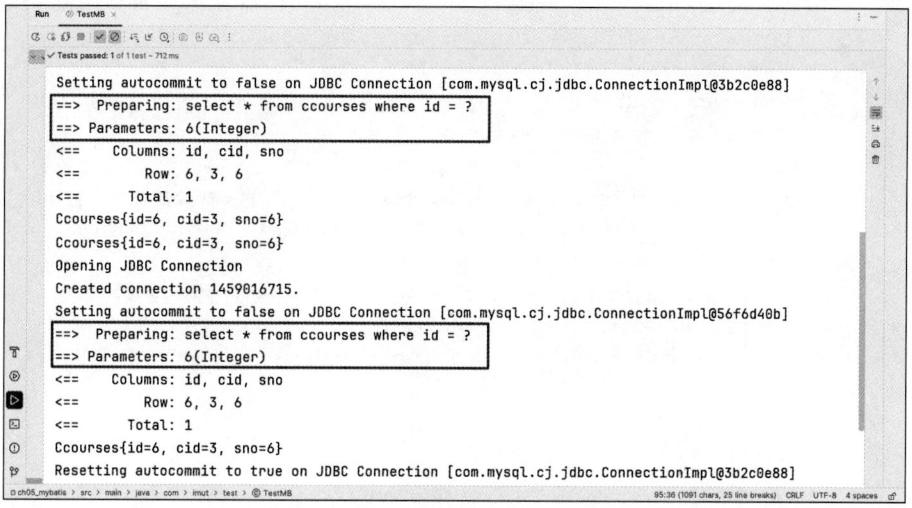

```
Setting autocommit to false on JDBC Connection [com.mysql.cj.jdbc.ConnectionImpl@3b2c0e88]
==>  Preparing: select * from ccourses where id = ?
==> Parameters: 6(Integer)
<==    Columns: id, cid, sno
<==        Row: 6, 3, 6
<==      Total: 1
Ccourses{id=6, cid=3, sno=6}
Ccourses{id=6, cid=3, sno=6}
Opening JDBC Connection
Created connection 1459016715.
Setting autocommit to false on JDBC Connection [com.mysql.cj.jdbc.ConnectionImpl@56f6d40b]
==>  Preparing: select * from ccourses where id = ?
==> Parameters: 6(Integer)
<==    Columns: id, cid, sno
<==        Row: 6, 3, 6
<==      Total: 1
Ccourses{id=6, cid=3, sno=6}
Resetting autocommit to true on JDBC Connection [com.mysql.cj.jdbc.ConnectionImpl@3b2c0e88]
```

图 5-6　一级缓存例 5-3 运行结果

由图 5-6 可知，共向数据库发送了 2 次查询请求，即框住部分。这里共执行了 3 次查询请求，也就意味着某次请求击中缓存，是相同 SqlSession 的两次查询中的第 2 次。可以看出：在相同 SqlSession 的情况下，击中缓存；在不同 SqlSession 的情况下，缓存被清空。本例无论 SqlSession 是否相同，都不会影响查询的结果，3 次查询获取的数据一致。

尽管不同 SqlSession 的查询已经可以证明，一级缓存的作用范围是 SqlSession 内部，但这里依然有必要讲解更新的例子，不仅仅是为了说明一级缓存不同 SqlSession 会失效的问题。

2. 更新

例 5-4：演示不同 SqlSession 缓存失效的方式，其中 SqlSession 既涉及同一个，又涉及不同的，既包含查询操作，又包含更新操作。需要改动的文件为 TestMB.java。CcoursesMapper.java、CcoursesMapper.xml 文件内容详见例 5-2，其余文件均沿用例 2-25 的即可。

TestMB.java 的代码如下。

```
package com.imut.test;
import com.imut.mapper.CcoursesMapper;
import com.imut.pojo.*;
import com.imut.utils.MyBatisUtil;
import org.apache.ibatis.session.SqlSession;
import org.apache.ibatis.session.SqlSessionFactory;
import org.junit.Test;
public class TestMB {
    @Test
    public void testCcoursesMapper(){
        SqlSessionFactory ssf = null;
        SqlSession session = null;
        SqlSession session1 = null;
        int id = 4;
        try {
            ssf = new MyBatisUtil().getSqlSessionFactory();
            session = ssf.openSession();
            CcoursesMapper ccoursesMapper = session.getMapper(CcoursesMapper.class);
            Ccourses ccourses = ccoursesMapper.selectCcoursesByid(id);
            System.out.println(ccourses);
            Ccourses ccourses1 = ccoursesMapper.selectCcoursesByid(id);
            System.out.println(ccourses1);
            session1 = ssf.openSession(true);
            CcoursesMapper ccoursesMapper2 = session1.getMapper(CcoursesMapper.class);
            int result = ccoursesMapper2.updateCcoursesByid(new Ccourses(id, 3, 3));
            System.out.println(result);
            Ccourses ccourses3 = ccoursesMapper.selectCcoursesByid(id);
            System.out.println(ccourses3);
            Ccourses ccourses2 = ccoursesMapper2.selectCcoursesByid(id);
            System.out.println(ccourses2);
        } catch (Exception e) {
            e.printStackTrace();
        } finally {
```

```
            session.close();
        }
    }
}
```

上述代码实现根据选课 ID 查询选课记录,更新选课记录,再次查询选课记录的功能,且查询方法输入 ID 值和更新选课记录的 ID 是通过一个变量传入的,确保一致性。但更新操作是在不同的 Session 中执行的,即先在第 1 个 SqlSession 中连续执行两条相同的查询语句,接着在第 2 个 SqlSession 中执行一条更新语句,然后再使用第 1 个 SqlSession 执行查询语句,最后在第 2 个 SqlSession 中再执行同一条查询语句。即 Mapper 的同一个查询方法调用 4 次,更新方法在两次查询方法后调用 1 次,输出查询数据,数据更新影响条数。运行结果如图 5-7 所示。

```
Setting autocommit to false on JDBC Connection [com.mysql.cj.jdbc.ConnectionImpl@3b2c0e88]
==>  Preparing: select * from ccourses where id = ?
==> Parameters: 4(Integer)
<==    Columns: id, cid, sno
<==        Row: 4, 2, 3
<==      Total: 1
Ccourses{id=4, cid=2, sno=3}
Ccourses{id=4, cid=2, sno=3}
Opening JDBC Connection
Created connection 1459016715.
==>  Preparing: update ccourses set cid = ?, sno = ? where id = ?
==> Parameters: 3(Integer), 3(Integer), 4(Integer)
<==    Updates: 1
1
Ccourses{id=4, cid=2, sno=3}
==>  Preparing: select * from ccourses where id = ?
==> Parameters: 4(Integer)
<==    Columns: id, cid, sno
<==        Row: 4, 3, 3
<==      Total: 1
Ccourses{id=4, cid=3, sno=3}
Resetting autocommit to true on JDBC Connection [com.mysql.cj.jdbc.ConnectionImpl@3b2c0e88]
```

图 5-7　一级缓存例 5-4 运行结果

由图 5-7 可知,共向数据库发送了 3 次 SQL 查询请求,其中向数据库发送了 2 次 SQL 查询请求,而这里共执行了 5 次操作,也就意味着有二次查询击中缓存。由图中框住部分可以看出:两次查询获取的数据不一致,是因为更新操作是在第 2 个 SqlSession 中执行的,而使用第 1 个 SqlSession 的查询语句(在更新语句后面)击中了缓存,故读取的是缓存中的数据,也就是旧数据。第 2 个 SqlSession 的查询操作因为与数据库发生交互,故读取的是新数据。由例 5-4 可知,当使用多个 SqlSession 操作数据库时,可能出现脏数据,故需谨慎使用 Session 级别的一级缓存。

一级缓存支持 statement 级别,每次查询都会清空缓存,故可以在一定程度上解决脏数据问题。这里不再举例说明,感兴趣的读者可以自行尝试。使用前需在 mybatis-config.xml 中添加配置如下。

```
<setting name="localCacheScope" value="STATEMENT"/>
```

5.2 二级缓存

MyBatis 提供二级缓存，同时也可以和其他组件集成，实现自定义缓存。二级缓存有时被称为全局缓存，分为两类开关：全局开关和局部开关。全局开关放置在配置文件 mybatis-config.xml 中，默认是开启状态，即用户无须额外配置。局部开关放置在各个 namespace 中，默认是未开启状态，添加相应元素即可开启。作用域为 namespace，即二级缓存与 Mapper 绑定。二级缓存由同一个 SqlSessionFactory 对象创建的多个 SqlSession 共享。

5.1 节讲过，CachingExecutor 主要用于二级缓存。由图 5-1 可知，CachingExecutor 共有两个属性：Executor 类型的属性用于具体执行任务；剩余 TransactionalCacheManager 类型的属性用于维护二级缓存。维护二级缓存的数据结构依然是 Map，键为 cache，值为 TransactionalCache 对象。TransactionalCache 与 org.apache.ibatis.cache.decorators 包下其余的 9 个类都实现 cache 接口，是装饰器实现类，用于包装装饰器或真实实现类，内部包含 cache 对象。即 cache 使用变体的装饰器模式实现功能的组合，cache 的类关系图如图 5-8 所示。注：图中仅列出部分重点属性、方法和参数。

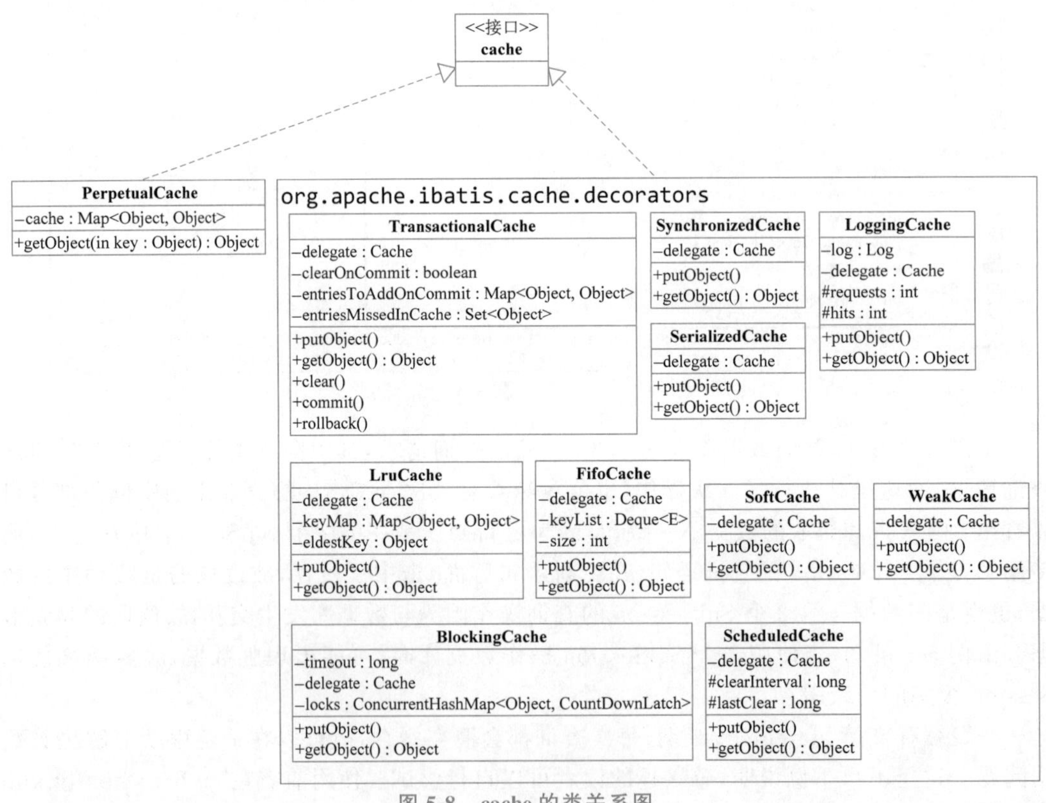

图 5-8 cache 的类关系图

装饰器模式即通过层层包装基础实现类动态地扩展功能。PerpetualCache 是 cache 的基础实现类，TransactionalCache 通过层层包装 PerpetualCache 实现强大的缓存功能。包装顺序由外至内（默认）为：TransactionalCache、SynchronizedCache、LoggingCache、SerializedCache、LruCache、PerpetualCache。其中 LruCache 类可根据用户的具体配置信息替换为其他类，即图 5-8 中与 LruCache 同一行的 FifoCache、SoftCache 或 WeakCache。而 BlockingCache 和 ScheduledCache 是否使用则取决于用户的配置信息。

讲解完 cache 整体结构，接着讲解二级缓存的工作流程。查询有 3 种结果：①击中二级缓存；②未击中二级缓存，但击中一级缓存；③未击中缓存。击中二级缓存的工作流程如图 5-9 所示。

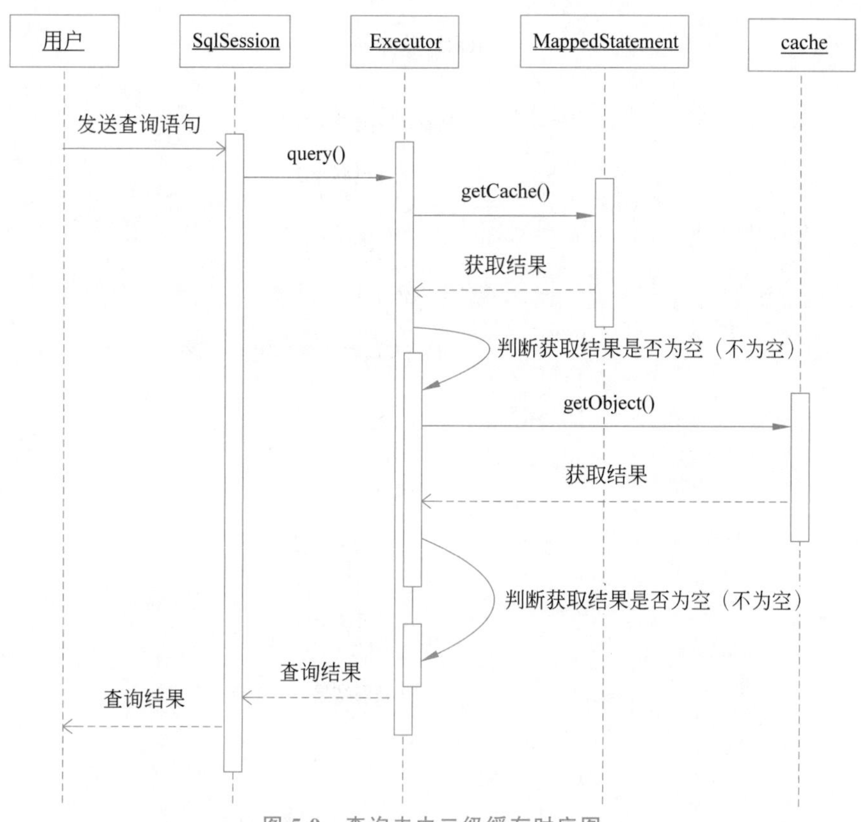

图 5-9 查询击中二级缓存时序图

用户根据自身需求发送查询请求，程序接收到用户的查询请求，判断需要执行的方法，SqlSession 的职责下放至 Executor，即通过调用 Executor 的 query 方法令 Executor 执行具体任务。Executor 则通过调用 MappedStatement 的 getCache 方法获取二级缓存信息，接着判断返回信息是否为空。如不为空，则 Executor 通过调用 cache 的 getObject 方法获取二级缓存中的数据，该方法需要一个 cache 类型的参数 cache(即上一步判断的返回信息)和一个 CacheKey 类型的参数 key。Executor 判断返回数据是否为空，不为空则将获取到的缓存数据返回至 SqlSession，再由 SqlSession 返回给用户。

如果 Executor 判断 cache 返回数据为空，则会调用为一级缓存服务的 Executor 的 query 方法，即走向判断是否击中一级缓存的流程。如果击中一级缓存，则将返回一级缓存的数据，否则调用 queryFromDatabase 方法，从数据库中查询数据。数据库将查询结果返回,用于一级缓存的 Executor 调用 cache 的 putObject 方法将数据放入缓存,接着将查询结果反馈给用于二级缓存的 Executor,再由该 Executor 将结果反馈至 SqlSession,最后由 SqlSession 返回给用户。未命中缓存的工作流程如图 5-10 所示。

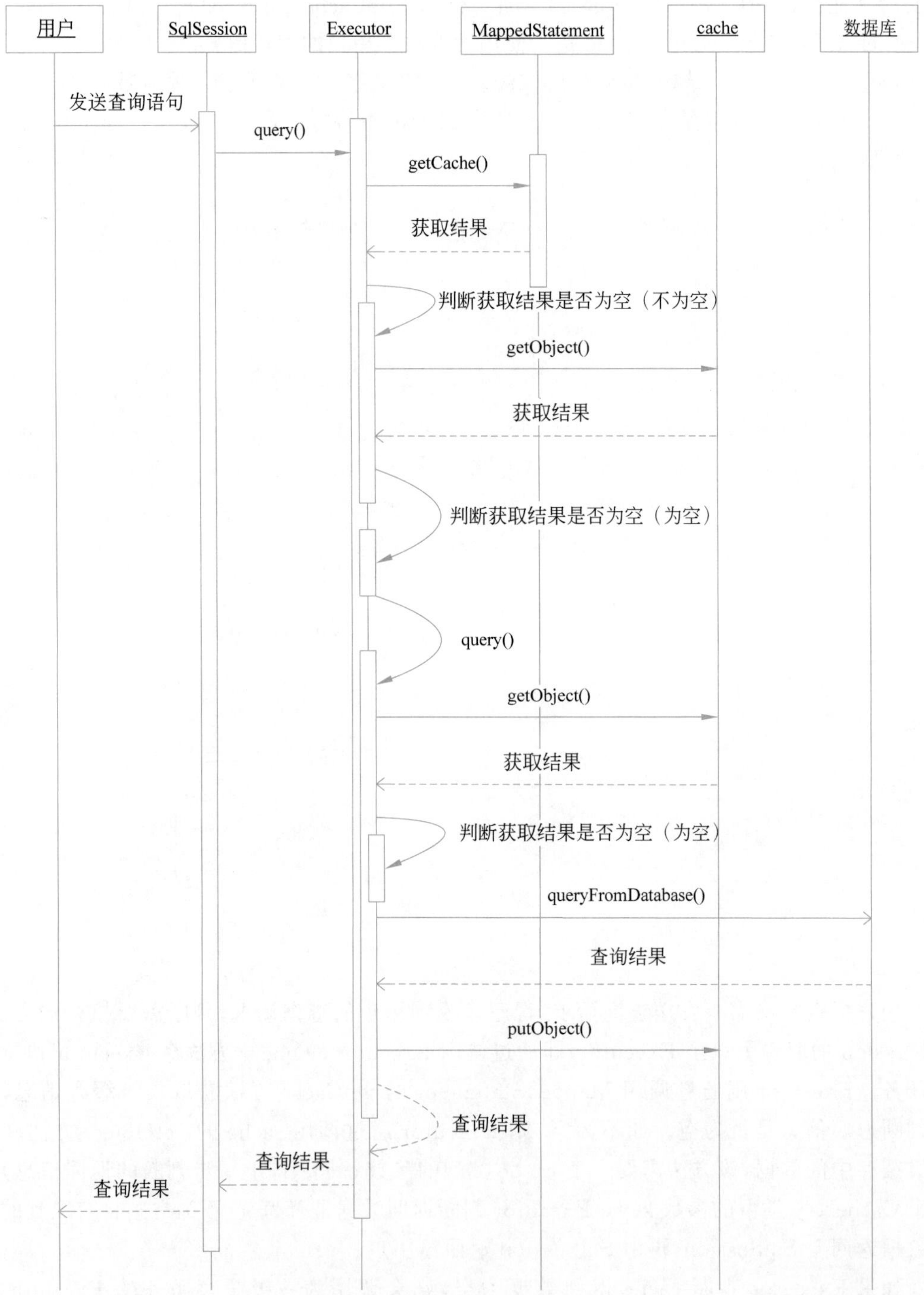

图 5-10　查询未击中缓存时序图

数据查询的顺序(从前至后):二级缓存、一级缓存、数据库。与一级缓存一样,二级缓存的增加、删除、修改操作都会转换为更新操作。且执行上述 3 种操作时，缓存会失效。与一级缓存不同的是：事务提交或 Session 关闭后，二级缓存才会生效。

二级缓存为 Namespace 级别，即不同 SqlSession 多次执行同一 SQL，直接从内存中取到缓存结果，而不需要查询数据库。这里将通过 4 个例子讲解。要使用二级缓存，需要开启开关，因为全局开关默认开启，故只需开启局部开关。开启代码如下所示，即在需要使用二级缓存的 Mapper.xml 中添加 cache 元素。

```
<mapper namespace="com.imut.mapper.CcoursesMapper">
    <cache />
    ...
</mapper>
```

MyBatis 的 cache 使用装饰器模式，SerializedCache 为装饰器实现类。SerializedCache 用于序列化与反序列化缓存数据，故要想使用二级缓存，实体类需要实现 Serializable 接口。代码如下。

```
public class Ccourses implements Serializable {
  ...
}
```

为了便于观测二级缓存效果，接着编写测试案例。在不同 SqlSession 的前提下，分别编写从查询和更新两方面测试的案例。

5.2.1 不同 SqlSession

1. 查询

例 5-5：演示跨 SqlSession 击中二级缓存的方式。需要改动的文件为 TestMB.java。Ccourses.java 和 CcoursesMapper.xml 新增的文件内容详见上页，CcoursesMapper.java 文件内容详见例 5-2，其余文件均沿用例 2-25 的即可。

TestMB.java 的代码如下。

```
package com.imut.test;
import com.imut.mapper.CcoursesMapper;
import com.imut.pojo.*;
import com.imut.utils.MyBatisUtil;
import org.apache.ibatis.session.SqlSession;
import org.apache.ibatis.session.SqlSessionFactory;
import org.junit.Test;
public class TestMB {
    @Test
    public void testCcoursesMapper(){
        SqlSessionFactory ssf = null;
        SqlSession session = null;
        SqlSession session1 = null;
        int id = 6;
        try {
```

```
                ssf = new MyBatisUtil().getSqlSessionFactory();
                session = ssf.openSession();
                CcoursesMapper ccoursesMapper = session.getMapper(CcoursesMapper.class);
                Ccourses ccourses = ccoursesMapper.selectCcoursesByid(id);
                System.out.println(ccourses);
                session.commit();
                session1 = ssf.openSession();
                CcoursesMapper ccoursesMapper2 = session1.getMapper(CcoursesMapper.class);
                Ccourses ccourses2 = ccoursesMapper2.selectCcoursesByid(id);
                System.out.println(ccourses2);
            } catch (Exception e) {
                e.printStackTrace();
            } finally {
                session.close();
            }
        }
    }
```

上述代码实现根据选课 ID 查询选课记录的功能，且查询方法输入 ID 值是通过一个变量传入的，确保一致性。不同 SqlSession 各执行一条查询语句，且两条语句相同。即同一个方法连续调用两次，输出查询数据。运行结果如图 5-11 所示。

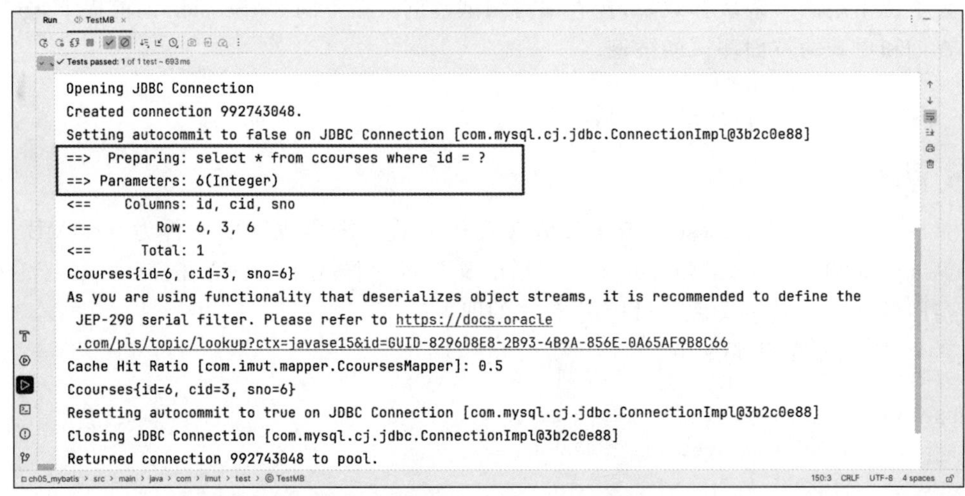

图 5-11　二级缓存例 5-5 运行结果

由图 5-11 可知，只向数据库发送了 1 次 SQL 查询请求，即框住部分。第 2 次获取到的数据并未与数据库交互，而是从二级缓存中得到。可以看出，两次查询获取的数据一致。笔者在两个 SqlSession 中间执行 commit 操作，但缓存并未被清空，再次印证要想二级缓存生效，必须提交事务或关闭 session。

例 5-6：演示根据选课 ID 查询选课记录的方式，其中 SqlSession 既包含同一个的，又包含不同的，执行的查询语句一致。共包含 7 个文件：Ccourses.java、CcoursesMapper.java、CcoursesMapper.xml、mybatis-config.xml、db.properties、MyBatisUtil.java 和 TestMB.java。需要改动的文件为 TestMB.java。Ccourses.java、CcoursesMapper.java 和 CcoursesMapper.xml 文件

沿用例 5-5,其余文件均沿用例 2-25 的即可。

TestMB.java 的代码如下。

```
package com.imut.test;
import com.imut.mapper.CcoursesMapper;
import com.imut.pojo.*;
import com.imut.utils.MyBatisUtil;
import org.apache.ibatis.session.SqlSession;
import org.apache.ibatis.session.SqlSessionFactory;
import org.junit.Test;
public class TestMB {
    @Test
    public void testCcoursesMapper(){
        SqlSessionFactory ssf = null;
        SqlSession session = null;
        SqlSession session1 = null;
        int id = 1;
        try {
            ssf = new MyBatisUtil().getSqlSessionFactory();
            session = ssf.openSession();
            CcoursesMapper ccoursesMapper = session.getMapper(CcoursesMapper.class);
            Ccourses ccourses = ccoursesMapper.selectCcoursesByid(id);
            System.out.println(ccourses);
            Ccourses ccourses1 = ccoursesMapper.selectCcoursesByid(id);
            System.out.println(ccourses1);
            session.commit();
            session1 = ssf.openSession();
            CcoursesMapper ccoursesMapper2 = session1.getMapper(CcoursesMapper.class);
            Ccourses ccourses2 = ccoursesMapper2.selectCcoursesByid(id);
            System.out.println(ccourses2);
        } catch (Exception e) {
            e.printStackTrace();
        } finally {
            session.close();
        }
    }
}
```

上述代码分别实现在相同 SqlSession 和不同 SqlSession 前提下根据同样的选课 ID 查询选课记录的功能,且查询方法输入 ID 值是通过一个变量传入的,确保一致性。在第一个 SqlSession 中,连续执行两条一样的查询语句;在第二个 SqlSession 中,执行与前两次查询相同的 SQL 语句。同样需要输出每次的查询数据。运行结果如图 5-12 所示。

由图 5-12 可知,共向数据库发送了 1 次查询请求,即框住部分。这里共执行了 3 次查询请求,也就意味着 2 次请求击中缓存。分别是第 1 个 SqlSession 的 2 次查询中的第 2 次和第 2 个 SqlSession 中的查询。输出日志包含一条额外信息——命中率,命中率指的是二级缓存的命中率。第 1 次命中率为 0,因为缓存为空;第 2 次命中率为 0,因为缓存虽然不为空,但并未击中二级缓存,而是击中一级缓存;第 3 次命中率为 0.33,因为共执行 3 条查询语

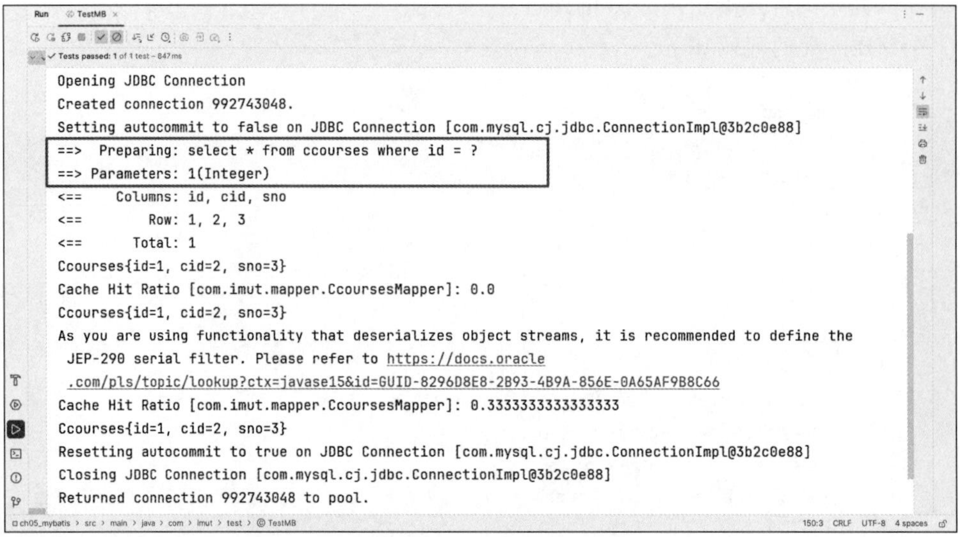
Run TestMB
Tests passed: 1 of 1 test – 847 ms
Opening JDBC Connection
Created connection 992743048.
Setting autocommit to false on JDBC Connection [com.mysql.cj.jdbc.ConnectionImpl@3b2c0e88]
==> Preparing: select * from ccourses where id = ?
==> Parameters: 1(Integer)
<== Columns: id, cid, sno
<== Row: 1, 2, 3
<== Total: 1
Ccourses{id=1, cid=2, sno=3}
Cache Hit Ratio [com.imut.mapper.CcoursesMapper]: 0.0
Ccourses{id=1, cid=2, sno=3}
As you are using functionality that deserializes object streams, it is recommended to define the JEP-290 serial filter. Please refer to https://docs.oracle.com/pls/topic/lookup?ctx=javase15&id=GUID-8296D8E8-2B93-4B9A-856E-0A65AF9B8C66
Cache Hit Ratio [com.imut.mapper.CcoursesMapper]: 0.3333333333333333
Ccourses{id=1, cid=2, sno=3}
Resetting autocommit to true on JDBC Connection [com.mysql.cj.jdbc.ConnectionImpl@3b2c0e88]
Closing JDBC Connection [com.mysql.cj.jdbc.ConnectionImpl@3b2c0e88]
Returned connection 992743048 to pool.
ch05_mybatis › src › main › java › com › imut › test › TestMB
150:3 CRLF UTF-8 4 spaces

图 5-12　二级缓存例 5-6 运行结果

句,只有最后 1 条击中二级缓存,故为 1/3=0.33。

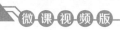

例 5-7:演示根据选课 ID 查询选课记录的方式,其中 SqlSession 既包含同一个的,又包含不同的,执行的查询语句一致。在第一个 SqlSession 中,执行一条查询语句;在第二个 SqlSession 中,连续执行两条与第一次执行的 SQL 相同的查询语句。共包含 7 个文件:Ccourses. java、CcoursesMapper. java、CcoursesMapper. xml、mybatis-config. xml、db. properties、MyBatisUtil.java 和 TestMB.java。需要改动的文件为 TestMB.java。Ccourses. java、CcoursesMapper.java 和 CcoursesMapper.xml 文件沿用例 5-5,其余文件均沿用例 2-25 的即可。

TestMB.java 的代码如下。

```
package com.imut.test;
import com.imut.mapper.CcoursesMapper;
import com.imut.pojo.*;
import com.imut.utils.MyBatisUtil;
import org.apache.ibatis.session.SqlSession;
import org.apache.ibatis.session.SqlSessionFactory;
import org.junit.Test;
public class TestMB {
    @Test
    public void testCcoursesMapper() {
        SqlSessionFactory ssf = null;
        SqlSession session = null;
        SqlSession session1 = null;
        int id = 1;
        try {
            ssf = new MyBatisUtil().getSqlSessionFactory();
            session = ssf.openSession();
            CcoursesMapper ccoursesMapper = session.getMapper(CcoursesMapper.
class);
```

```
            Ccourses ccourses = ccoursesMapper.selectCcoursesByid(id);
            System.out.println(ccourses);
            session.commit();
            session1 = ssf.openSession();
            CcoursesMapper ccoursesMapper2 = session1.getMapper(CcoursesMapper.
class);
            Ccourses ccourses3 = ccoursesMapper2.selectCcoursesByid(id);
            System.out.println(ccourses3);
            Ccourses ccourses2 = ccoursesMapper2.selectCcoursesByid(id);
            System.out.println(ccourses2);
        } catch (Exception e) {
            e.printStackTrace();
        } finally {
            session.close();
        }
    }
}
```

上述代码分别实现在相同 SqlSession 和不同 SqlSession 的前提下根据同样的选课 ID 查询选课记录的功能，且查询方法输入 ID 值是通过一个变量传入的，确保一致性。同样需要输出每次的查询数据。运行结果如图 5-13 所示。

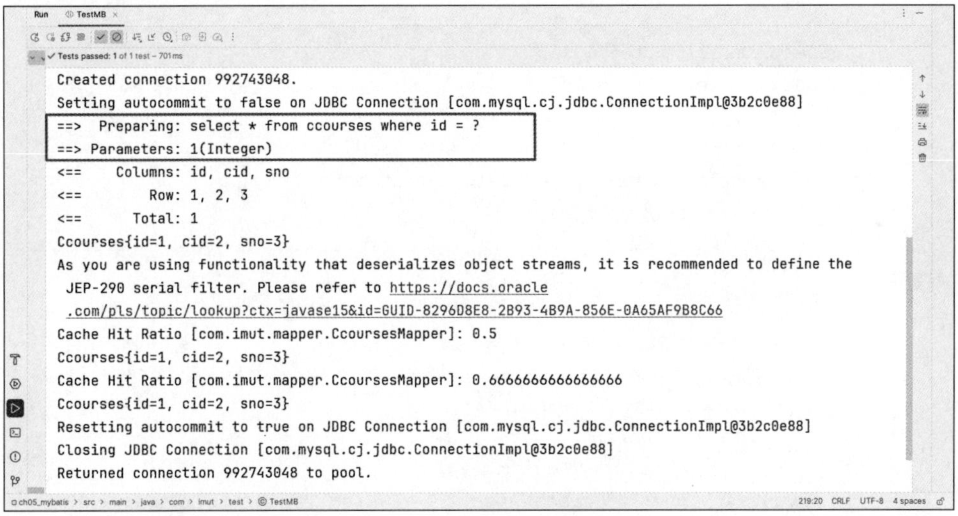
Run TestMB
Tests passed: 1 of 1 test – 701 ms
Created connection 992743048.
Setting autocommit to false on JDBC Connection [com.mysql.cj.jdbc.ConnectionImpl@3b2c0e88]
==> Preparing: select * from ccourses where id = ?
==> Parameters: 1(Integer)
<== Columns: id, cid, sno
<== Row: 1, 2, 3
<== Total: 1
Ccourses{id=1, cid=2, sno=3}
As you are using functionality that deserializes object streams, it is recommended to define the JEP-290 serial filter. Please refer to https://docs.oracle.com/pls/topic/lookup?ctx=javase15&id=GUID-8296D8E8-2B93-4B9A-856E-0A65AF9B8C66
Cache Hit Ratio [com.imut.mapper.CcoursesMapper]: 0.5
Ccourses{id=1, cid=2, sno=3}
Cache Hit Ratio [com.imut.mapper.CcoursesMapper]: 0.6666666666666666
Ccourses{id=1, cid=2, sno=3}
Resetting autocommit to true on JDBC Connection [com.mysql.cj.jdbc.ConnectionImpl@3b2c0e88]
Closing JDBC Connection [com.mysql.cj.jdbc.ConnectionImpl@3b2c0e88]
Returned connection 992743048 to pool.
ch05_mybatis > src > main > java > com > imut > test > TestMB
219:20 CRLF UTF-8 4 spaces

图 5-13 例 5-7 运行结果

由图 5-13 可知，共向数据库发送了一次查询请求，即框住部分。笔者共执行了 3 次查询请求，也就意味着两次请求击中缓存。分别是第 2 个 SqlSession 的两次查询。输出日志的第 1 次命中率为 0，因为缓存为空；第 2 次命中率为 0.5，因为是在不同 SqlSession 中，故击中二级缓存，共执行 2 条查询语句，命中 1 条，即 1/2=0.5；第 3 次命中率为 0.66，因为共执行 3 条查询语句，后 2 条击中二级缓存，故为 2/3≈0.66。

因插入、删除操作最终都会转化为调用 update()方法，即逻辑是一致的，故这里只以更新为例。

2. 更新

例 5-8：演示跨 SqlSession 缓存失效的方式，其中 SqlSession 既涉及同一个的，又涉及不同的，既有查询操作，又包含更新操作。先在第一个 SqlSession 中执行 1 条查询语句、1 条更新语句，然后在第二个 SqlSession 中执行 1 条相同的查询语句。共包含 7 个文件：Ccourses. java、CcoursesMapper. java、CcoursesMapper. xml、mybatis-config. xml、db. properties、MyBatisUtil.java 和 TestMB.java。需要改动的文件为 TestMB.java。Ccourses. java、CcoursesMapper.java 和 CcoursesMapper.xml 文件沿用例 5-5，其余文件均沿用例 2-25 的即可。

TestMB.java 的代码如下。

```
package com.imut.test;
import com.imut.mapper.CcoursesMapper;
import com.imut.pojo.*;
import com.imut.utils.MyBatisUtil;
import org.apache.ibatis.session.SqlSession;
import org.apache.ibatis.session.SqlSessionFactory;
import org.junit.Test;
public class TestMB {
    @Test
    public void testCcoursesMapper() {
        SqlSessionFactory ssf = null;
        SqlSession session = null;
        SqlSession session1 = null;
        int id = 1;
        try {
            ssf = new MyBatisUtil().getSqlSessionFactory();
            session = ssf.openSession();
            CcoursesMapper ccoursesMapper = session.getMapper(CcoursesMapper.
class);
            Ccourses ccourses = ccoursesMapper.selectCcoursesByid(id);
            System.out.println(ccourses);
            int result = ccoursesMapper.updateCcoursesByid(new Ccourses(2, 2, 2));
            System.out.println(result);
            session.commit();
            session1 = ssf.openSession();
            CcoursesMapper ccoursesMapper2 = session1.getMapper(CcoursesMapper.
class);
            Ccourses ccourses2 = ccoursesMapper2.selectCcoursesByid(id);
            System.out.println(ccourses2);
        } catch (Exception e) {
            e.printStackTrace();
        } finally {
            session.close();
        }
    }
}
```

上述代码实现根据选课 ID 查询选课记录，更新选课记录，再次查询选课记录的功能，且查询方法输入 ID 值是通过一个变量传入的，确保一致性。更新选课记录的 ID 是与查询的 ID 不一致。即 Mapper 的同一个查询方法调用 2 次，更新方法在 2 次查询方法中间调用 1 次，输出查询数据，数据更新影响条数。运行结果如图 5-14 所示。

```
Setting autocommit to false on JDBC Connection [com.mysql.cj.jdbc.ConnectionImpl@3b2c0e88]
==>  Preparing: select * from ccourses where id = ?
==> Parameters: 1(Integer)
<==    Columns: id, cid, sno
<==        Row: 1, 2, 3
<==      Total: 1
Ccourses{id=1, cid=2, sno=3}
==>  Preparing: update ccourses set cid = ?, sno = ? where id = ?
==> Parameters: 2(Integer), 2(Integer), 2(Integer)
<==    Updates: 1
1
Committing JDBC Connection [com.mysql.cj.jdbc.ConnectionImpl@3b2c0e88]
As you are using functionality that deserializes object streams, it is recommended to define the
 JEP-290 serial filter. Please refer to https://docs.oracle
 .com/pls/topic/lookup?ctx=javase15&id=GUID-8296D8E8-2B93-4B9A-856E-0A65AF9B8C66
Cache Hit Ratio [com.imut.mapper.CcoursesMapper]: 0.0
Opening JDBC Connection
Created connection 2100572327.
Setting autocommit to false on JDBC Connection [com.mysql.cj.jdbc.ConnectionImpl@7d3430a7]
==>  Preparing: select * from ccourses where id = ?
==> Parameters: 1(Integer)
<==    Columns: id, cid, sno
<==        Row: 1, 2, 3
<==      Total: 1
Ccourses{id=1, cid=2, sno=3}
Resetting autocommit to true on JDBC Connection [com.mysql.cj.jdbc.ConnectionImpl@3b2c0e88]
```

图 5-14　二级缓存例 5-8 运行结果

由图 5-14 可知，共向数据库发送了 3 次 SQL，其中向数据库发送了 2 次 SQL 查询请求，即框住部分。也就意味着第 2 次并未击中缓存，而是从数据库中获得数据。可以看出，尽管 2 次查询获取的数据一致，但因为中间执行了更新操作，故缓存被清空，所以第 2 次才只能和数据库交互。

一级缓存存在脏数据读取的问题，二级缓存亦是如此。二级缓存与 Mapper 对应，Mapper1 开启二级缓存，包含多表(表 1 和表 2)关联查询操作；Mapper2 只包含表 2 的查询、更新操作。当 Mapper1 的二级缓存生效后，接着调用 Mapper2 的语句更新表 2 数据，然后 Mapper1 执行多表(表 1 和表 2)关联查询，则此时读取的数据为脏数据。因为更新操作是在 Mapper2 中执行的，而 Mapper1 的多表关联查询语句击中二级缓存，故读取的是二级缓存中的数据，也就是旧数据。

一级缓存和二级缓存都可能出现脏数据的问题，故可使用自定义缓存，自定义缓存类需实现 cache 接口。如读者能力不足，不建议使用自定义缓存，可使用第三方现有缓存方案，如 Redis、MemCache 等。

5.2.2　cache 元素

cache 元素用于二级缓存，包含 blocking 属性、eviction 属性、flushInterval 属性、

readOnly 属性、size 属性和 type 属性，均为非必填属性。

blocking 属性标识是否开启阻塞缓存，默认值为 false，通常用于多线程场景。设置为 true，cache 将添加 BlockingCache 装饰类，获取缓存数据前会加锁，获取缓存数据后会释放锁。

eviction 属性用于指定缓存清除策略，默认使用最近最少使用（Least Recently Used，LRU）清除策略。共有 4 种：LRU、先进先出（First In First Out，FIFO）、SOFT（软引用）、WEAK（弱引用）。

flushInterval 属性用于指定缓存刷新的时间间隔，单位是 ms，默认无刷新间隔。

readOnly 属性标识缓存是否只读，默认值为 false。设置为 true，cache 将不使用 SerializedCache 装饰。

size 属性用于指定缓存对象引用个数。

type 属性用于指定缓存使用的类，默认值为 PerpetualCache。

例 5-9：演示使用 FIFO 清除策略，每隔 10s 刷新缓存。

```
<cache eviction="FIFO" flushInterval="10000" />
```

例 5-9 所示代码书写位置是需要配置缓存的 Mapper.xml 文件。

5.3 存储过程调用

存储过程是为了满足特定需求的 SQL 语句块，一次编译永久可用，具有运行速度快、稳定、使用便捷的特点。用户通过指定存储过程名称与参数的方式即可调用。MyBatis 支持调用存储过程，为数据库的高效使用添砖加瓦。

MyBatis 调用存储过程与使用 SQL 的流程大致相同：①创建存储过程；②编写接口方法；③编写调用存储过程的 SQL 语句；④编写测试接口的方法。这里利用 2 个例子讲解 MyBatis 是如何调用存储过程的。

例 5-10：演示插入学生信息并返回最新添加学号的方式。共包含 7 个文件：Students.java、StudentsMapper.java、StudentMapper.xml、mybatis-config.xml、db.properties、MyBatisUtil.java 和 TestMB.java。需要改动的文件为 StudentsMapper.java、StudentMapper.xml 和 TestMB.java。Students.java 文件详见例 3-3，其余文件均沿用例 2-25 的即可。

首先创建存储过程，使用的 studnets 表为 3.1 节所创建，这里不再赘述建表语句，创建存储过程的 SQL 代码如下。

```
delimiter $$
    CREATE PROCEDURE insert_students(OUT s_sno INTEGER, IN s_sname VARCHAR(20),
IN s_ssex VARCHAR(20), IN s_sage INTEGER, IN s_sclaid INTEGER)
    BEGIN
        INSERT INTO students (sname, ssex, sage, sclaid) VALUES (s_sname, s_ssex,
s_sage, s_sclaid);
        SET s_sno=LAST_INSERT_ID();
```

```
    END
$$
delimiter ;
```

注意：因存储过程包含 SQL 语句，SQL 语句的结束标识符为分号，故存储过程创建语句会被分割为不完整的多条 SQL。需要首先使用 DELIMITER 关键字修改结束符，再创建存储过程，最后将结束符修改回分号。

将调用存储过程的接口方法添加至 StudentsMapper.java，StudentsMapper.java 的代码如下。

```
package com.imut.mapper;
import com.imut.pojo.Students;
public interface StudentsMapper {
    int addStudent(Students students);
}
```

具体的 SQL 语句需要在 StudentsMapper.xml 中编写，StudentsMapper.xml 的代码如下。

```
<?xml version="1.0" encoding="UTF-8" ?>
<!DOCTYPE mapper PUBLIC "-//mybatis.org//DTD Mapper 3.0//EN" "http://mybatis.
org/dtd/mybatis-3-mapper.dtd">
<mapper namespace="com.imut.mapper.StudentsMapper">
    <insert id="addStudent" parameterType="com.imut.pojo.Students"
statementType="CALLABLE">
        {call insert_students(#{sno,mode=OUT,jdbcType=INTEGER}, #{sname,mode
=IN}, #{ssex,mode=IN}, #{sage,mode=IN}, #{sclaid,mode=IN})}
    </insert>
</mapper>
```

insert 元素的 statementType 属性用于指定 SQL 语句执行的类型，值为 STATEMENT，PREPARED(默认)或 CALLABLE。statementType 的值为 CALLABLE，即标识其为存储过程语句。insert 元素包裹部分为调用存储过程的 SQL，包含存储过程名称和参数。#{}包裹部分为参数，其中 mode 代表参数模式，包含 IN、OUT、INOUT 3 种。当 mode 为 OUT 或 INOUT 时，必须列出 jdbcType。

TestMB.java 的代码如下。

```
package com.imut.test;
import com.imut.mapper.CcoursesMapper;
import com.imut.mapper.StudentsMapper;
import com.imut.pojo.*;
import com.imut.utils.MyBatisUtil;
import org.apache.ibatis.session.SqlSession;
import org.apache.ibatis.session.SqlSessionFactory;
import org.junit.Test;
public class TestMB {
    @Test
```

```
    public void testStudentsMapper() {
        SqlSessionFactory ssf = null;
        SqlSession session = null;
        try {
            ssf = new MyBatisUtil().getSqlSessionFactory();
            session = ssf.openSession(true);
            StudentsMapper studentsMapper = session.getMapper(StudentsMapper.
class);
            Students students = new Students();
            students.setSname("Bob");
            students.setSsex("男");
            students.setSage(23);
            students.setSclaid(3);
            int result = studentsMapper.addStudent(students);
            if (result==1) {
                System.out.println("数据插入成功");
                System.out.println("插入数据的学号:" + students.getSno());
            } else {
                System.out.println("数据插入失败");
            }
        } catch (Exception e) {
            e.printStackTrace();
        } finally {
            session.close();
        }
    }
}
```

例 5-10 的运行结果如图 5-15 所示。

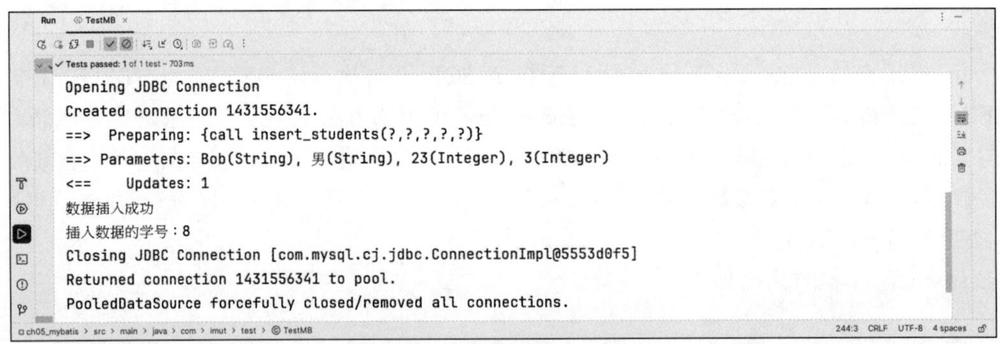

图 5-15　例 5-10 的运行结果

例 5-11：演示根据传入参数查询相应性别的学生人数的方式。共包含 7 个文件：Students. java、StudentsMapper. java、StudentsMapper. xml、mybatis-config. xml、db. properties、MyBatisUtil.java 和 TestMB.java。需要改动的文件为 StudentsMapper.java、StudentsMapper.xml 和 TestMB.java。Students.java 文件详见例 3-3，其余文件均沿用例 2-25 的即可。

首先创建存储过程，要求当传入参数为 1 时，输出男学生数量，否则输出女学生数量，创建存储过程代码如下。

```
delimiter $$
CREATE PROCEDURE select_student(OUT student_count INTEGER, IN s_ssex INTEGER)
BEGIN
    IF s_ssex = 1 THEN
        SELECT COUNT(sno) FROM students WHERE ssex = "男" INTO student_count;
    ELSE
        SELECT COUNT(sno) FROM students WHERE ssex = "女" INTO student_count;
    END IF;
END
$$
delimiter ;
```

存储过程创建完毕,接着编写接口方法。StudentMapper.java 的新增代码如下。

```
Integer selectStuCountBySex(@Param("map") Map<String, Integer> map);
```

注:返回值类型应为封装数据类型,否则返回值为 Null 时会报错。传入参数类型为 Map,故需使用@Param 注解。

接口方法编写完毕,接着完成具体的 SQL。parameterMap 属性用于多参数传值(不建议使用),与 parameterMap 元素共同使用。parameterMap 的属性值与 parameterMap 元素的 id 属性值一致,即使用相应的 parameterMap 元素定义。parameterMap 元素中定义具体的参数,使用的是 parameter 元素。StudentMapper.xml 的新增代码如下。

```
<parameterMap id="getStudentCount" type="java.util.Map">
    <parameter property="student_count" jdbcType="INTEGER" mode="OUT" />
    <parameter property="ssex" jdbcType="VARCHAR" mode="IN" />
</parameterMap>
<select id="selectStuCountBySex" parameterMap="getStudentCount"statementType=
"CALLABLE">
    {call select_student(#{map.student_count, mode=OUT, jdbcType=INTEGER},
#{map.ssex, mode=IN})}
</select>
```

功能编写完毕,接着编写测试用例。TestMB.java 的代码如下。

```
package com.imut.test;
import com.imut.mapper.CcoursesMapper;
import com.imut.mapper.StudentsMapper;
import com.imut.pojo.*;
import com.imut.utils.MyBatisUtil;
import org.apache.ibatis.session.SqlSession;
import org.apache.ibatis.session.SqlSessionFactory;
import org.junit.Test;
import java.util.HashMap;
import java.util.Map;
public class TestMB {
```

```
    @Test
    public void testProcedure() {
        SqlSessionFactory ssf = null;
        SqlSession session = null;
        try {
            ssf = new MyBatisUtil().getSqlSessionFactory();
            session = ssf.openSession(true);
            StudentsMapper studentMapper = session.getMapper(StudentsMapper.
class);
            Map<String, Integer> map = new HashMap<String, Integer>();
            map.put("student_count", 0);
            map.put("ssex", 1);
            studentMapper.selectStuCountBySex(map);
            int stu_count = map.get("student_count");
            System.out.println("学生中男生有:" + stu_count + " 人");
        } catch (Exception e) {
            e.printStackTrace();
        } finally {
            session.close();
        }
    }
}
```

注：获取返回结果时，使用 map 的 get 方法。无须赋值给对象，因为最终返回结果会赋值给传出的参数。

例 5-11 的运行结果如图 5-16 所示。

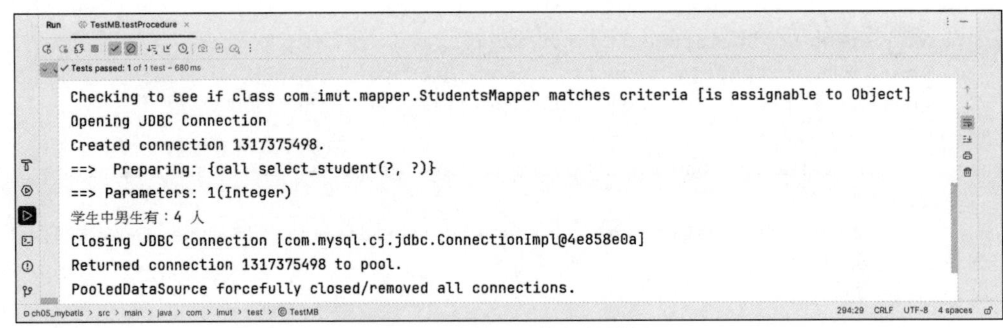

图 5-16 例 5-11 的运行结果

5.4 综合案例

本章案例主要实现模拟演示一级缓存和二级缓存的功能。

5.4.1 案例设计

本章案例划分为持久层和数据访问层。持久层由实体类 Student.java 组成。数据访问层由接口和映射文件组成，接口的名称统一以 Mapper 结尾，映射文件名称与接口文件名称相同。本章案例使用的主要文件如表 5-1 所示。

表 5-1　本章案例使用的文件

文　　件	所在包/路径	功　　能
log4j.properties	src/	日志输出配置文件
db.properties	src/	数据库连接的属性文件
mybatis-config.xml	src/	MyBatis 配置文件
Ccourse.java	com.imut.pojo	封装选课信息的类
Student.java	com.imut.pojo	封装学生信息的类
StudentMapper.java	com.imut.mapper	定义操作学生表方法的接口
StudentMapper.xml	com.imut.mapper	Student 类的映射文件
MybatisUtil.java	com.imut.utils	获取 SqlSession 对象的工具类
StudentTest.java	com.imut.test	测试操作学生信息方法的类

5.4.2　案例演示

使用 Junit4 测试执行 StudentTest.java 中的方法。执行 listStudentsCache1Test()方法测试一级缓存,在控制台输出的结果如图 5-17 所示。

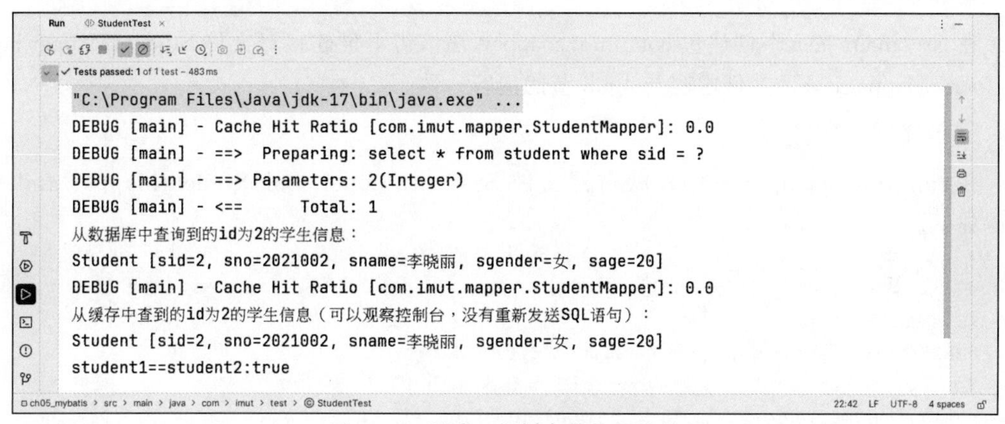

图 5-17　测试一级缓存的执行结果

可以观察到第一次查询发送了 SQL 语句,是从数据库中查到的信息,而第二次查询没有发送 SQL 语句,是从缓存中查到的数据,且 student1 与 student2 是同一个对象。

执行 listStudentsCache2Test()方法测试二级缓存,在控制台输出的结果如图 5-18 所示。

可以观察到第一次查询发送了 SQL 语句,是从数据库中查询到的信息,而第二次查询没有发送 SQL 语句,且缓存命中率为 0.5,即第二次查询是从二级缓存中取到的数据。

5.4.3　代码实现

1. 创建项目，添加依赖

创建名为 ch05_mybatis 的 Maven 工程,编写 pom.xml 文件,内容与 3.4.3 节中的一致。

```
DEBUG [main] - ==>  Preparing: select * from student where sid = ?
DEBUG [main] - ==> Parameters: 2(Integer)
DEBUG [main] - <==      Total: 1
从数据库中查询到的id为2的学生信息：
Student [sid=2, sno=2021002, sname=李晓丽, sgender=女, sage=20]
DEBUG [main] - Cache Hit Ratio [com.imut.mapper.StudentMapper]: 0.5
从二级缓存中查到的id为2的学生信息（可以观察控制台，没有重新发送SQL语句）：
Student [sid=2, sno=2021002, sname=李晓丽, sgender=女, sage=20]
```

图 5-18　测试二级缓存的执行结果

2. 创建配置文件

创建配置文件 log4j.properties、db.properties 与 mybatis-config.xml，内容与 3.4.3 节中的一致。

3. 创建实体类

创建实体类文件 Course.java、Ccourse.java、Student.java 以及 MybatisUtil.java，内容与 3.4.3 节中对应的文件内容一致。

4. 创建接口文件

在/src/main/java 下创建包 com.imut.mapper，在该包下创建接口文件 StudentMapper.java 和 CourseMapper.java。文件内容与 4.6.3 节的内容一致。

5. 创建映射文件

在 com.imut.mapper 下创建映射器文件 StudentMapper.xml，StudentMapper.xml 的代码如下。

```
<?xml version="1.0" encoding="UTF-8" ?>
<!DOCTYPE mapper
PUBLIC "-//mybatis.org//DTD Mapper 3.0//EN"
"http://mybatis.org/dtd/mybatis-3-mapper.dtd">
<mapper namespace="com.imut.mapper.StudentMapper">
    <cache></cache>
    <select id="getStudentById" resultType="com.imut.pojo.Student">
        select * from student where sid = #{sid}
    </select>
    <select id="listStudents" resultType="com.imut.pojo.Student">
        select * from student
    </select>
    <select id="getCcoursesById" resultType="com.imut.pojo.Ccourse" >
         select cc.id, s.sid, s.sname, c.cid, c.cname from ccourse cc, student s,
course c
            where cc.sid=s.sid and c.cid=cc.cid and s.sid=#{sid}
    </select>
    <insert id="addStudent">
        insert into student (sno,sname,sgender,sage)
```

```
            values(#{sno},#{sname},#{sgender},#{sage})
        </insert>
        <update id="updateStudent">
            update student set sno=#{sno},sname=#{sname},sgender=#{sgender},sage
=#{sage}
            where sid=#{sid}
        </update>
        <delete id="deleteStudent">
            delete from student where sid=#{sid}
        </delete>
</mapper>
```

6. 创建测试类

在 com.imut 目录下新建 test 目录,在 com.imut.test 目录创建测试类 StudentTest.java,StudentTest.java 的代码如下。

```
package com.imut.test;
import java.io.InputStream;
import java.util.List;
import org.apache.ibatis.io.Resources;
import org.apache.ibatis.session.SqlSession;
import org.apache.ibatis.session.SqlSessionFactory;
import org.apache.ibatis.session.SqlSessionFactoryBuilder;
import org.junit.Test;
import com.imut.mapper.StudentMapper;
import com.imut.pojo.Ccourse;
import com.imut.pojo.Student;
import com.imut.utils.MybatisUtil;
public class StudentTest {
    @Test
    public void getStudentByIdTest() {
        String resource = "mybatis-config.xml";
        SqlSession sqlSession = null;
        try {
            InputStream inputStream = Resources.getResourceAsStream(resource);
             SqlSessionFactory sqlSessionFactory = new SqlSessionFactoryBuilder().
build(inputStream);
            sqlSession = sqlSessionFactory.openSession();
            StudentMapper mapper = sqlSession.getMapper(StudentMapper.class);
            Student student = mapper.getStudentById(2);
            System.out.println("查询出的编号为 2 的学生的学号、姓名、性别、年龄为:");
            System.out.println(student);
        } catch (Exception e) {
            e.printStackTrace();
        } finally {
            sqlSession.close();
        }
    }
```

```
@Test
public void getCcoursesByIdTest() {
    SqlSession sqlSession = MybatisUtil.getSqlSession();
    StudentMapper mapper = sqlSession.getMapper(StudentMapper.class);
    List<Ccourse> list = mapper.getCcoursesById(1);
    System.out.println("所有学生的学号、姓名、性别、年龄为:");
    for (Ccourse cc : list) {
        System.out.println(cc);
    }
    sqlSession.close();
}
@Test
public void listStudentsTest() {
    SqlSession sqlSession = MybatisUtil.getSqlSession();
    StudentMapper mapper = sqlSession.getMapper(StudentMapper.class);
    List<Student> list = mapper.listStudents();
    System.out.println("所有学生的学号、姓名、性别、年龄为:");
    for (Student s : list) {
        System.out.println(s);
    }
    sqlSession.close();
}
@Test
public void addStudentTest() {
    SqlSession sqlSession = MybatisUtil.getSqlSession();
    StudentMapper mapper = sqlSession.getMapper(StudentMapper.class);
    Student student = new Student();
    student.setSno("20210106");
    student.setSname("李林");
    student.setSgender("男");
    student.setSage(21);
    mapper.addStudent(student);
    sqlSession.commit();
    sqlSession.close();
}
@Test
public void updateStudentTest() {
    SqlSession sqlSession = MybatisUtil.getSqlSession();
    StudentMapper mapper = sqlSession.getMapper(StudentMapper.class);
    Student student = mapper.getStudentById(2);
    student.setSname("李晓丽");
    mapper.updateStudent(student);
    sqlSession.commit();
    sqlSession.close();
}
@Test
public void deleteStudentTest() {
    SqlSession sqlSession = MybatisUtil.getSqlSession();
    StudentMapper mapper = sqlSession.getMapper(StudentMapper.class);
    mapper.deleteStudent(6);
```

```
        sqlSession.commit();
        sqlSession.close();
    }
    @Test
    public void listStudentsCache1Test() {
        SqlSession sqlSession = MybatisUtil.getSqlSession();
        StudentMapper mapper = sqlSession.getMapper(StudentMapper.class);
        Student student1 = mapper.getStudentById(2);
        System.out.println("从数据库中查询到的 id 为 2 的学生信息:");
        System.out.println(student1);
        Student student2 = mapper.getStudentById(2);
        System.out.println("从缓存中查到的 id 为 2 的学生信息(可以观察控制台,没有重
新发送 SQL 语句):");
        System.out.println(student2);
        System.out.println("student1==student2:"+(student1==student2));
        sqlSession.close();
    }
    @Test
    public void listStudentsCache2Test() {
        SqlSession sqlSession1 = MybatisUtil.getSqlSession();
        SqlSession sqlSession2 = MybatisUtil.getSqlSession();
        StudentMapper mapper1 = sqlSession1.getMapper(StudentMapper.class);
        StudentMapper mapper2 = sqlSession2.getMapper(StudentMapper.class);
        Student student1 = mapper1.getStudentById(2);
        System.out.println("从数据库中查询到的 id 为 2 的学生信息:");
        System.out.println(student1);
        sqlSession1.close();
        Student student2 = mapper2.getStudentById(2);
        System.out.println("从二级缓存中查到的 id 为 2 的学生信息(可以观察控制台,没
有重新发送 SQL 语句):");
        System.out.println(student2);
        sqlSession2.close();
    }
}
```

5.5 习题

1. 单选题

(1) 以下(　　)语句不会清空缓存。

A. insert　　B. delete　　C. update　　D. select

(2) cache 元素的属性不包括(　　)。

A. flushInterval　　B. size　　C. type　　D. name

2. 填空题

(1) select 元素的 useCache 属性值默认是________。

（2）MyBatis 缓存包括________、________。

（3）存储过程的参数类型有________、________、________。

3. 简答题

（1）一级缓存和二级缓存的作用域分别是什么？

（2）二级缓存的开启方式及二级缓存的触发机制是什么？

（3）一级缓存失效的情况有哪些？

4. 编程题

编写存储过程并调用，查询新闻标题中包含“爱国”的新闻信息的条数。其中“爱国”一词可由参数控制。新闻表结构如表 5-2 所示。

表 5-2　新闻表结构

字段名	字段类型	字段含义
newsid	int	新闻 id
newstitle	varchar(40)	新闻标题
newsauthor	varchar(10)	新闻作者

第2部分

Spring 篇

第6章 Spring基础

本章详细讲解轻量级 Java 企业级应用开发框架 Spring，它旨在简化开发流程，通过优化底层服务设计让开发者更专注于业务逻辑。Spring 具有优良的设计和分层架构，解决了传统框架臃肿低效的问题。Spring 的核心是基于控制反转（Inversion of Control，IoC）的轻量级容器和面向切面编程（Aspect Oriented Programming，AOP）。

6.1 Spring 框架概述

Spring 兴起之前，Java 企业级应用开发主要通过 EJB（Enterprise Java Beans）完成。EJB 是服务器端的组件模型，由于它过于依靠 EJB 容器，并存在缓慢、复杂等问题，因此逐渐被 Spring 取代。讲解 Spring 框架前，首先讲解企业级应用开发。

6.1.1 企业级应用开发与 Spring

企业级应用是为商业组织和大型企业设计的解决方案和应用。这些应用结构复杂，涉及大量外部资源、事务、数据和用户，同时功能多样，对安全性和性能有较高要求。因此，企业级应用通常部署多个交互应用，并与其他企业应用连接，形成复杂的分布式企业应用集群。除了强大的功能，企业级应用还需易于扩展和维护，以适应未来业务需求的变化。然而，EJB 作为解决企业级应用问题的架构体系，存在开发效率低、难度大和性能上的不足。

Spring 自 2004 年首个版本发布，至今已有 20 年。其诞生源于 SUN 公司 EJB 的失败，特别是 EJB2 烦琐的配置和抽象概念而使发展受到很大限制。尽管 EJB3 减少了配置冗余，但仍依赖 EJB 容器，限制了其灵活性。这种依赖导致开发缓慢，测试复杂，限制了快速开发和测试的可能性。

EJB 的衰落催生了 Spring 的崛起。Spring 是一个轻量级框架，其核心理念是将所有 Java 类视为资源，即 Bean，并由 Spring 提供的 IoC 容器管理。这种基于 Bean 的编程方式解决了 EJB 的问题。

此外，使用 EJB 需要测试人员不断部署和配置，增加了测试复杂性。而 Spring 框架的非侵入性或低侵入性特点，使得 Java 应用即使离开 Spring 也能正常运行，提高了灵活性。Spring 并不是要取代现有技术，而是整合这些技术，提供一个支持它们开发的模板。

6.1.2 Spring 框架简介

Spring 框架是一个轻量级、分层的开源框架，以 IoC 和 AOP 为核心，使用 JavaBean 替代 EJB，简化开发过程。Spring 提供以下策略。

(1) 对于简单的 Java 对象(POJO)，提供轻量级和低侵入的编程，通过配置扩展功能强调接口编程和 OOD 理念，降低耦合度，提高可读性和可扩展性。

(2) 提供切面编程，简化企业核心应用，如数据库开发的代码结构，使开发者更专注于业务开发，避免滥用 try…catch…finally 语句。

(3) Spring 框架通过模板类整合各种框架和技术，如 Hibernate、MyBatis、Redis 等，简化技术使用。

Spring 框架贯穿表现层、业务逻辑层和持久层，提供与 Spring MVC、Struts 的整合功能，管理事务、记录日志等，并整合 MyBatis、Hibernate、JdbcTemplate 等技术。虽然贯穿于各层，但不取代已有框架，以高开放性与其无缝整合。

Spring 框架简化 Java 企业级应用开发，提供强大、稳定的功能，无额外负担，具有得体和优雅的气质，使现有技术更易于使用，促进良好的编程习惯。Spring 坚持不重复实现已有解决方案，如对象持久化和 ORM，而是提供支持，使现有技术更易用。

6.1.3 Spring 框架的优势

Spring 框架的主要优势如下。

(1) 降低耦合度，方便开发：通过 IoC 容器，Spring 管理对象生命周期和控制依赖关系，避免硬编码导致的过度耦合。

(2) 支持 AOP 编程：Spring 使用 AOP 进行权限拦截和安全监控，减少传统 OOP 方法的代码冗余和复杂性。

(3) 支持声明式事务：在 Spring 中，可以通过配置文件直接管理数据库事务，提高开发效率。

(4) 方便程序测试：Spring 集成了 Junit，便于进行单元测试。

(5) 方便集成优秀框架：Spring 提供广泛的基础平台，支持多种优秀开源框架，如 MyBatis、Hibernate、Quartz 等。

(6) 降低 JavaEE API 难度：Spring 封装了 JavaEE 中难以使用的 API，使其更易于理解和调用。

6.1.4 Spring 框架的体系结构

Spring 框架采用分层和模块化架构，包括核心容器、数据访问及集成、Web、AOP 等功能，如图 6-1 所示。该框架包含约 20 个模块，由 1300 多个文件构成，分为核心容器、AOP、设备支持、数据访问及集成、Web、报文发送和 Test 等 6 个模块集合。开发者可根据业务需求选择所需模块，并集成其他第三方框架，提高开发效率和针对性。

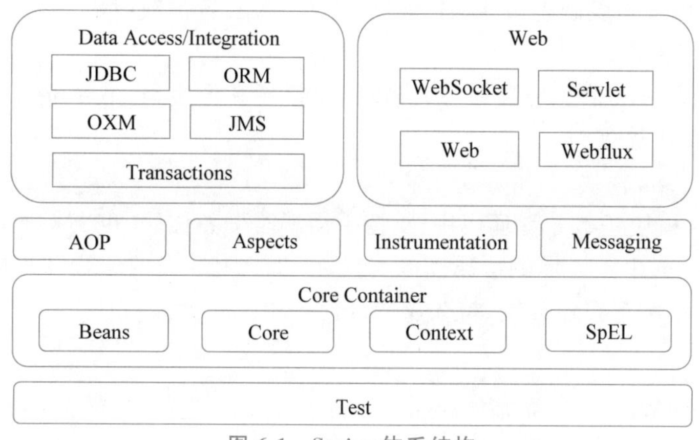

图 6-1 Spring 体系结构

本书仅讲解主要模块。

1. 核心容器（Core Container）

核心容器是 Spring 的基石，包括 Beans、Core、Context 和 SpEL 模块。

(1) Beans 模块。

Beans 模块定义了 BeanFactory 接口，实现了工厂模式。它管理 Bean 的创建、配置和管理，并使用 IoC 分离配置和代码。BeanFactory 在 Bean 被使用时才进行实例化和装配。

(2) Core 模块。

Core 模块提供了 IoC 和 DI 功能，是 Spring 框架的基本组成部分。

(3) Context 模块。

Context 模块扩展了 BeanFactory，增加了 Bean 生命周期控制、事件体系和资源加载透明化等功能。它还提供了企业级支持，如邮件访问、远程访问和任务调度。ApplicationContext 是 Context 的核心接口，它自动实例化单实例 Bean，并装配依赖。

(4) SpEL 模块。

SpEL 模块是 Spring 3.0 版本后新增的模块，提供强大的表达式语言支持。它可以查询和管理运行中的对象、调用方法、操作数组和集合等。语法类似于传统 EL，但增加了更多功能，尤其是函数调用和字符串模板函数。SpEL 与 Spring IoC 容器无缝交互。

2. 数据访问及集成（Data Access/Integration）

数据访问及集成用于访问和操作数据，包括 JDBC、ORM、OXM、JMS 和 Transactions 模块。

(1) JDBC 模块。

JDBC 模块是 Spring 的 JDBC 抽象框架实现，它简化数据库操作编码，解析数据库错误代码。提供 JDBC 模板、关系数据库对象化、Simple JDBC 和事务管理。核心类包括 JdbcTemplate、Simple JdbcTemplate 和 Named Parameter JdbcTemplate。

(2) ORM 模块。

ORM 模块集成 MyBatis、Hibernate、JPA 和 JDO，用于资源管理、DAO 实现和事务策略。可将对象关系映射框架与 Spring 特性组合使用。

(3) OXM 模块。

OXM 模块提供抽象层，支持对象/XML 映射，如 JAXB、Castor、XML Beans、JiBX 和

XStream 等。将 Java 对象映射成 XML 数据，或将 XML 数据映射成 Java 对象。

(4) JMS 模块。

JMS 模块提供 Java 消息传递服务，支持信息使用和产生，自 Spring 4.1 版本后支持与 Spring-message 模块集成。

(5) Transactions 模块。

Transactions 模块是事务管理模块，支持手动编写事务代码，也支持通过注解或配置实现自动处理的声明式事务。提供特殊接口实现，以及 POJO 类的编程和声明式事务。通过 AOP 配置可以灵活配置在业务层、持久层等任何一层。

3. Web

Spring 框架的 Web 层功能基于 ApplicationContext，提供了 WebSocket、Servlet、Web 和 Webflux 等模块。

(1) WebSocket 模块

WebSocket 模块是 Spring 4.0 版本后新增的模块，支持全双工通信协议，适配不同的 WebSocket 引擎，提供额外服务。

(2) Servlet 模块。

Servlet 模块也称为 Webmvc 模块，包含 Spring MVC 和 REST Web 服务实现。

(3) Web 模块。

Web 模块提供 Web 集成功能，如多文件上传、使用 Servlet 监听器的 IoC 容器初始化和 Web 应用上下文。还包括 HTTP 客户端和远程处理支持。

(4) Webflux 模块。

Webflux 模块是一个新的非阻塞函数式 Reactive Web 框架，用于建立异步、事件驱动的服务，扩展性良好。

4. 其他模块

Spring 框架除了核心功能外，还包含了多个模块，包括 AOP、Aspects、Instrumentation、Messaging 和 Test 模块，下面分别介绍这些模块的功能。

(1) AOP 模块。

AOP 模块是 Spring 的核心模块之一，实现了 AOP 的思想。AOP 通过动态代理技术提供了一系列横切实现，如前置通知、返回通知、异常通知等。同时，还通过 Pointcut 接口匹配切入点，实现方法拦截和切入点匹配，从而扩展相关方法。AOP 模块将代码按照功能进行分离，降低了耦合性。

(2) Aspects 模块。

Aspects 模块提供了与 AspectJ 的集成功能。AspectJ 是一个功能强大的 AOP 框架，Aspects 模块为 Spring AOP 提供了多种实现方法，增强了 Spring 的 AOP 功能。

(3) Instrumentation 模块。

Instrumentation 模块提供了类工具的支持和类加载器的实现，适用于特定的应用服务器。它基于 JAVA SE 中的“java.lang.instrument”设计，主要是在 JVM 启用时生成代理类，允许程序员运行时修改类的字节，从而改变类的功能，实现 AOP 的功能。

（4）Messaging 模块。

Messaging 模块是 Spring 4.0 版本后新增的模块，提供了对消息传递体系结构和协议的支持，使得 Spring 能够更好地与其他消息传递系统集成。

（5）Test 模块。

Test 模块支持使用 JUnit 或 TestNG 对 Spring 组件进行单元测试和集成测试。它提供了一致加载 ApplicationContext 和这些 ApplicationContext 的缓存。此外，Test 模块还提供了可用于独立测试代码的模拟对象，帮助开发人员更方便地进行测试工作。

6.1.5 Spring 框架的下载及目录结构

编写本书时，Spring 5.3.22 为最新版本，本书的项目代码是以 Spring 5.2.5 版本为例。Spring 框架开发所需的 jar 包分为两部分，即 Spring 框架包和第三方依赖包。

1. Spring 框架包

Spring 5.2.5 版本的框架压缩包名称为 spring-5.2.5.RELEASE-dist.zip，可以在地址 https://repo.spring.io/artifactory/libs-release-local/org/springframework/spring/5.2.5.RELEASE 下载。下载完成后，将压缩包解压到自定义的文件夹中，解压后的文件目录结构如图 6-2 所示。

名称	修改日期	类型	大小
docs	2020/3/24 11:32	文件夹	
libs	2020/3/24 11:32	文件夹	
schema	2020/3/24 11:32	文件夹	
license.txt	2020/3/24 11:23	文本文档	16 KB
notice.txt	2020/3/24 11:23	文本文档	1 KB
readme.txt	2020/3/24 11:23	文本文档	1 KB

6 个项目

图 6-2　Spring 的目录结构

其中，docs 文件夹包含 Spring 框架的相关文档，包括 API 参考文档、开发手册；libs 文件夹包含 Spring 框架各个模块的 JAR 文件和源码；schema 文件夹包含开发所需的 schema 文件，这些文件中定义了 Spring 框架相关配置文件的约束。

打开 libs 目录，可以看到 63 个 JAR 文件，如图 6-3 所示。

名称	修改日期	类型	大小
spring-aop-5.2.5.RELEASE.jar	2020/3/24 11:25	Executable Jar File	364 KB
spring-aop-5.2.5.RELEASE-javadoc.jar	2020/3/24 11:31	Executable Jar File	1,250 KB
spring-aop-5.2.5.RELEASE-sources.jar	2020/3/24 11:31	Executable Jar File	353 KB
spring-aspects-5.2.5.RELEASE.jar	2020/3/24 11:25	Executable Jar File	47 KB
spring-aspects-5.2.5.RELEASE-javadoc.jar	2020/3/24 11:31	Executable Jar File	79 KB
spring-aspects-5.2.5.RELEASE-sources.jar	2020/3/24 11:31	Executable Jar File	31 KB
spring-beans-5.2.5.RELEASE.jar	2020/3/24 11:25	Executable Jar File	670 KB
spring-beans-5.2.5.RELEASE-javadoc.jar	2020/3/24 11:31	Executable Jar File	2,035 KB
spring-beans-5.2.5.RELEASE-sources.jar	2020/3/24 11:31	Executable Jar File	631 KB

63 个项目

图 6-3　Spring 框架中的 JAR 包

可以看出,libs 目录中的 jar 包分为 3 类。以 RELEASE.jar 结尾的是 Spring 框架字节码文件.class 的压缩包;以 RELEASE-javadoc.jar 结尾的是 Spring 框架 API 文档的压缩包;以 RELEASE-sources.jar 结尾的是 Spring 框架源文件的压缩包。整个 Spring 框架由 21 个模块组成,每个模块都提供了这 3 类压缩包。

libs 目录的 21 个模块中有 4 个 Spring 的基础包,它们分别对应 Spring 核心容器的 4 个模块。

(1) spring-core-5.2.5.RELEASE.jar。

spring-core-5.2.5.RELEASE.jar 包含 Spring 框架基本的核心工具类。Spring 框架的其他组件都要用到这个包里的类,是其他组件的基本核心。

(2) spring-beans-5.2.5.RELEASE.jar。

spring-beans-5.2.5.RELEASE.jar 包含所有应用都要用到的 jar 包,即访问配置文件、创建和管理 Bean,以及进行 IoC 或者依赖注入(Dependency Injection,DI)操作相关的所有类。

(3) spring-context-5.2.5.RELEASE.jar。

spring-context-5.2.5.RELEASE.jar 为 Spring 核心提供了大量扩展,包含了使用 Spring ApplicationContext 特性时所需的全部类、JDNI 所需的全部类、UI 方面的用来与模板引擎如 Velocity、FreeMarker、JasperReports 集成的类,以及校验 Valid 方面的相关类。

(4) spring-expression-5.2.5.RELEASE.jar。

spring-expression-5.2.5.RELEASE.jar 定义了 Spring 框架的表达式语言。

2. 第三方依赖包

使用 Sping 框架开发时,除了使用自带的 jar 包,其核心容器还需要依赖 commons.logging 的 JAR 包。该 JAR 包可以在网址 https://commons.apache.org/proper/commons-logging/download_logging.cgi 下载。下载完成后,会得到一个名为 commons-logging-1.2-bin.zip 的压缩包。将压缩包解压到自定义目录,即可找到 commons-logging-1.2.jar。

Spring 框架作为开源框架,提供了相关的源文件,在学习和开发过程中,可以通过阅读源文件,了解 Spring 框架的底层实现。这不仅有利于正确理解和运用 Spring 框架,也有助于开拓思路,提升自身的编程水平。

6.2 Spring 的容器机制

6.2.1 容器机制简介

容器是 Spring 框架的核心,类似超级工厂,启动时管理所有配置过的类,称为 Bean。Bean 无须遵循特定规范,只要是 Java 类并配置在容器中,Spring 就能管理。Spring 通过 XML 或注解获取配置信息,管理 Bean 的创建和生命周期。Bean 在容器中运行,专注于自身功能,无须关注容器情况。Spring 提供容器 API 中最常用的接口是 BeanFactory 和 ApplicaitonContext 接口,其中 ApplicaitonContext 是 BeanFactory 的子接口,如图 6-4 所示。下面详细介绍这两个接口。

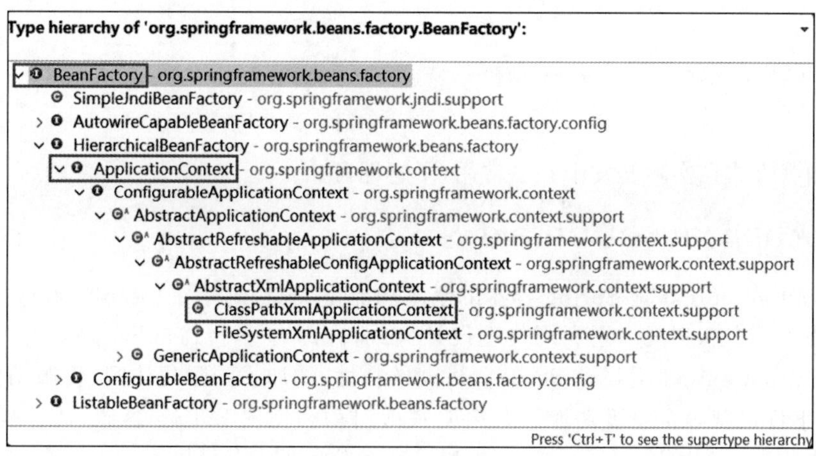

图 6-4 BeanFactory 接口与 ApplicationContext 接口

6.2.2 BeanFactory 接口

BeanFactory 是 Spring 容器的核心设置，是 Spring 控制反转和依赖注入最底层的实现，负责管理 Bean 的注册、注入和依赖等。Spring 容器通过依赖注入管理 Bean，实现了 Bean 的配置和与实际应用代码的解耦。因为 Spring 的核心模型是 Bean 模型，所以 BeanFactory 是所有 Spring 应用的核心，必须有效管理 Bean，以支持应用正常运行。BeanFactory 由 org.springframework.beans.facytory.BeanFactory 接口定义，提供了完整的依赖注入服务支持。它是一个管理 Bean 的工厂，主要负责初始化 Bean，调用管理 Bean 生命周期方法，为其他容器提供最基本的规范。BeanFactory 接口提供了一系列操作 Bean 的方法，具体如表 6-1 所示。

表 6-1 BeanFactory 接口的方法

方法名称	描述
ObjectgetBean(String name)	根据名称获取 Bean
<T>TgetBean(String name,Class<T> requiredType)	根据名称、类型获取 Bean
Object getBean(String name, Object... args)	根据名称获取 Bean
<T>T getBean(Class<T> requiredType)	根据类型获取 Bean
booleanisSingleton(String name)	是否为单实例
booleanisPrototype(String name)	是否为多实例
boolean isTypeMatch(String name,ResolvableType typeToMatch)	名称、类型是否匹配
Class<?> getType(String name)	根据名称获取类型

表 6-1 列举了 BeanFactory 接口的方法，开发者调用这些 API 即可完成对 Bean 的操作，无须关注 Bean 的实例化过程。

Spring 中提供了几种 BeanFactary 的实现类，其中最常用的是 XmlBeanFactory，它可以读取 XML 文件，并根据 XML 文件中的配置信息来管理 Bean，其加载配置信息的语法如下。

```
BeanFactory beanFactory=new XmlBeanFactory(new FilesystemResources(
"applicationContext.xml"));
```

这种加载方式在实际开发中并不常用，了解即可。

6.2.3 ApplicationContext 接口

ApplicaitonContext 是 Spring 容器的基本功能定义类，继承了 BeanFactory 接口，是面向开发者的工厂多态实现。它利用代理设计模式，内部有一个 BeanFactory 实例执行其功能。ApplicaitonContext 增强了 BeanFactory 的特性，增加了国际化访问、资源访问、多配置文件加载、事件机制等企业级功能。它声明式地启动和创建 Spring 容器，适合构建复杂的企业级应用程序。ApplicaitonContext 可以预初始化单例 Bean，执行 setter 方法，提高程序获取 Bean 实例的性能。Spring 提供了多种 ApplicaitonContext 实现类供开发者使用，具体如表 6-2 所示。

表 6-2 ApplicaitonContext 接口的实现类

类 名 称	描 述
ClassPathXmlApplicationContext	具体的容器的实现类，从类路径加载配置文件，创建 ApplicaitonContext 实例
FileSystemXmlApplicationContext	具体的容器的实现类，从文件系统加载配置文件，创建 ApplicaitonContext 实例
AnnotationConfigApplicationContext	从注解中加载配置文件，创建 ApplicaitonContext 实例
WebApplicationContext	在 Web 应用中使用，从相对于 Web 根目录的路径中加配置文件，创建 ApplicaitonContext 实例
ConfigurableWebApplicationContext	扩展了 WebApplicationContext，它允许通过配置的方式实例化 WebApplicationContext，扩展出了 refresh()方法和 close()方法，支持手动刷新、关闭容器

创建 ApplicationContext 接口实例，通常采用 ClassPathXmlApplicationContext 和 FileSystemXmlApplicationContext 两种方法实现。

ClassPathXmlApplicationContext 类会从类路径 classPath 中寻找指定的 XML 配置文件，找到并装载完成 ApplicationContext 的实例化工作，其使用语法如下。

```
ApplicationContext applicationContext = new ClassPathXmlApplicationContext
(String configLocation);
```

参数 configLocation 用于指定 Spring 框架配置文件的名称和位置。如果其值为 applicationContext.xml，则可去类路径中查找名称为 applicationContext.xml 的配置文件。

FileSystemXmlApplicationContext 会从指定的文件系统路径(绝对路径)中寻找指定的 XML 配置文件，并装载完成 ApplicationContext 的实例化工作，其使用语法如下。

```
ApplicationContext applicationContext = new FileSystemXmlApplicationContext
(String configLocation);
```

FileSystemXmlApplicationContext 是独立的 XML 应用上下文，通过程序在初始化时导入 Bean 配置文件，并创建 Bean 实例。读取 Spring 框架的配置文件时，FileSystemXmlApplicationContext 不再从类路径中读取配置文件，而是通过参数指定配置文件的位置，从文件系统或 URL 得到上下文定义文件，例如 D:\workspaces\application Context.xml。文件位置或 URL 可以是绝对或相对路径。如果在参数中写的不是带有盘符的绝对路径，方法调用时就会默认使用项目的工作路径。

使用 Spring 框架时，可以通过实例化其中任何一个类来创建 ApplicationContext 容器。通常在 Java 项目中，采用通过 ClassPathXmlApplicationContext 类来实例化 ApplicationContext 容器的方式。而在 Web 项目中，ApplicationContext 容器的实例化工作会交由 Web 服务器来完成。Web 服务器实例化 ApplicationContext 容器时，通常会使用基于 ContextLoaderListener 实现的方式，此种方式只需要在 web.xml 中添加代码如下。

```
<!-指定 Spring 配置文件的位置有多个配置文件时，以逗号分隔-->
<context-param>
<param-name>contextConfigLocat<ion</param-name>
<--Spring 将加载 Spring 目录下的 applicationContext.xml 文件-->
<param-value>classpath:spring/applicationContext.xml</param-value>
</context-param>
<!-指定以 ContextLoaderListener 方式启动 Spring 容器-->
<listener>
<listener-class>org.springframework.web.context.ContextLoaderListener
</listener-class>
</1istener>
```

在三大框架整合时，将采用基于 ContextLoaderListener 的方式由 Web 服务器实例化 ApplicationContext 容器。

BeanFactory 和 ApplicationContext 两种容器都是通过 XML 配置文件加载 Bean 的。二者的主要区别在于对 Bean 属性的注入处理。如果 Bean 的属性没有正确注入，使用 BeanFacotry 加载后，第一次调用 getBean()方法时会抛出异常，而 ApplicationContext 则在初始化时自检，这样有利于提前检查依赖属性是否正确注入。因此，在实际开发中，通常都优先选择使用 ApplicationContext，只有系统资源较少时才考虑使用 BeanFactory。

创建 Spring 框架容器后就可以获取其中的 Bean，通常可以调用以下几种方法。

(1) Object getBean(String name)：根据容器中 Bean 的 id 或 name 来获取指定的 Bean，获取后需要进行强制类型转换，是较常用的方法。

(2) <T>T getBean(Class<T> requiredType)：根据类的类型来获取 Bean 的实例。由于此方法为泛型方法，因此获取 Bean 之后并不需要进行强制类型转换。但是当容器中存在多个 Class 相同的 Bean 时，使用该方法获取对象时，就会由于找不到唯一的 Bean 而抛出异常。

(3) <T>T getBean(String name, Class<T> requiredType)：通过 Bean 的 id 与 Class 来获取 Bean 对象，结合了前面两种方法。

6.2.4 容器的启动过程

Spring 容器启动包括 3 个基本步骤：定位、载入与注册 BeanDefinition。

(1) 定位：Spring 容器通过 ResourceLoader 找到具体的 Resource，可以是 XML 文件或注解。

(2) 载入：读取配置信息，将<bean>配置转换为 Spring 容器内部的数据结构 BeanDefinition，便于管理 Bean。

(3) 注册：将 BeanDefinition 注册到 Spring 容器中，通过 HashMap 对象持有数据。

完成这三个步骤后，Bean 在 Spring 容器中完成定义。默认情况下，Bean 实例化在容器启动过程中完成。可通过配置 Bean 的 lazy-init 属性控制实例化时机，其属性有 3 个可选值：default、false、true。其中，default 和 false 在容器启动时实例化，true 则在第一次请求时实例化。

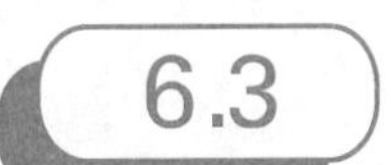

6.3 依赖注入与控制反转

6.3.1 控制反转

控制反转(IoC)是一个抽象概念。以制作热橙汁为例，无饮品店时，需自己准备果汁机、橙子和开水，即主动创造。有饮品店时，只需下订单，饮品店会制作好橙汁，并送到手上，这是 IoC 的思想，即将创建和管理对象的控制权从个人转移到饮品店。在软件开发中，控制反转意味着将对象的创建和依赖关系的管理委托给框架或容器，使代码更简洁和可维护。不精通某领域时，可把创建对象的主动权转交给别人。在 Spring 中，实现控制反转的是 IoC 容器，通过依赖注入(DI)实现。

现实中的开发者通常由多人组成团队。以开发电商网站为例，团队中有人熟悉商品交易流程，有人熟悉财务处理。在交易过程中，商品交易流程需调用财务接口。财务开发人员需开发简单逻辑的接口，其他开发者只需调用接口即可。

财务接口开发完成后，可发布到 Spring IoC 容器中，其他开发者通过容器调用财务接口完成操作，无须了解财务模块的内部细节。同样，交易接口也可以发布到容器中，供财务开发人员调用，如图 6-5 所示，Spring IoC 容器为开发带来便利。

图 6-5 Spring IoC 容器的便利性

IoC 降低了对象间耦合度。有些类不需要了解实现细节，只需要知道用途。对象产生依赖 IoC 容器，而非开发者主动行为。主动创建责任在开发者，被动模式下责任归 IoC 容器。这种方式降低了对象间的耦合度。

6.3.2 依赖注入

依赖注入与控制反转的含义相同，只不过是从两个角度描述的同一个概念。

当某个 Java 对象（调用者）需要调用另一个 Java 对象（被调用者，即被依赖对象）时，在传统模式下，调用者通常会采用“new 被调用者”的代码方式来创建对象，如图 6-6 所示。这种方式会导致调用者与被调用者之间的耦合性增加，不利于后期项目的升级和维护。

使用 Spring 框架之后，对象的实例不再由调用者来创建，而是由 Spring 框架的容器来创建，它会负责控制程序之间的关系，而不是由调用者的程序代码直接控制。这样，控制权由应用代码转移到 Spring 框架容器，控制权发生了反转，这就是 Spring 框架的控制反转。

从 Spring 框架容器的角度来看，它负责将被依赖对象赋值给调用者的成员变量，相当于为调用者注入了其依赖实例，这就是 Spring 框架的依赖注入，如图 6-7 所示。

图 6-6 调用者创建被调用者对象　　图 6-7 将被调用者对象注入调用者对象

相对于“控制反转”，“依赖注入”的说法也许更容易理解一些，即由容器（如 Spring 框架）负责把组件所“依赖”的具体对象“注入”（赋值）给组件，从而避免组件之间以硬编码的方式结合在一起。

6.4 综合案例

本章案例主要实现由 Spring 容器管理，并创建 Student 类的实例化对象的功能。

6.4.1 案例设计

案例使用 Spring 的控制反转，通过 Spring 容器构建类的实例。本章案例使用的主要文件如表 6-3 所示。

表 6-3 本章案例使用的主要文件

文　件	所在包/路径	功　能
log4j.properties	src/	日志输出配置文件
applicationContext.xml	src/	Spring 的配置文件
Student.java	com.imut.pojo	封装学生信息的类
StudentTest.java	com.imut.test	测试类

6.4.2 案例演示

使用 Junit4 测试执行 StudentTest.java 中的方法。运行 test()方法，在控制台输出的结果如图 6-8 所示。

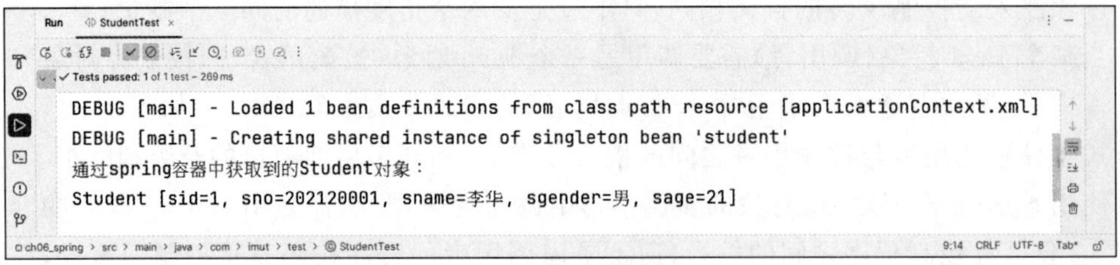

```
Run   StudentTest
✓ Tests passed: 1 of 1 test – 269 ms
DEBUG [main] - Loaded 1 bean definitions from class path resource [applicationContext.xml]
DEBUG [main] - Creating shared instance of singleton bean 'student'
通过spring容器中获取到的Student对象：
Student [sid=1, sno=202120001, sname=李华, sgender=男, sage=21]
ch06_spring > src > main > java > com > imut > test > StudentTest        9:14  CRLF  UTF-8  Tab*
```

图 6-8 运行结果

6.4.3 代码实现

1. 创建项目，添加依赖

创建名为 ch06_spring 的 Maven 工程，添加所需依赖，代码如下。

```
<dependencies>
    <dependency>
        <groupId>org.springframework</groupId>
        <artifactId>spring-core</artifactId>
        <version>6.0.11</version>
    </dependency>
    <dependency>
        <groupId>org.springframework</groupId>
        <artifactId>spring-beans</artifactId>
        <version>6.0.11</version>
    </dependency>
    <dependency>
        <groupId>org.springframework</groupId>
        <artifactId>spring-expression</artifactId>
        <version>6.0.11</version>
    </dependency>
    <dependency>
        <groupId>org.springframework</groupId>
        <artifactId>spring-context</artifactId>
        <version>6.0.10</version>
    </dependency>
    <dependency>
        <groupId>junit</groupId>
        <artifactId>junit</artifactId>
        <version>4.12</version>
    </dependency>
    <dependency>
        <groupId>log4j</groupId>
        <artifactId>log4j</artifactId>
```

```
        <version>1.2.17</version>
    </dependency>
</dependencies>
```

2. 编写配置文件

在/src/main/resources 下分别创建日志输出配置文件 log4j.properties 和 Spring 配置文件 applicationContext.xml。

配置文件 log4j.properties 的代码如下。

```
# Global logging configuration
log4j.rootLogger=DEBUG, stdout
# MyBatis logging configuration...
# log4j.logger.com.imut=DEBUG
# Console output...
log4j.appender.stdout=org.apache.log4j.ConsoleAppender
log4j.appender.stdout.layout=org.apache.log4j.PatternLayout
log4j.appender.stdout.layout.ConversionPattern=%5p [%t] - %m%n
```

配置文件 applicationContext.xml 的代码如下。

```
<?xml version="1.0" encoding="UTF-8"?>
<beans xmlns="http://www.springframework.org/schema/beans"
      xmlns:xsi="http://www.w3.org/2001/XMLSchema-instance"
      xsi:schemaLocation="http://www.springframework.org/schema/beans
      http://www.springframework.org/schema/beans/spring-beans.xsd">
    <bean id="student" class="com.imut.pojo.Student">
        <property name="sid" value="1"></property>
        <property name="sno" value="202120001"></property>
        <property name="sname" value="李华"></property>
        <property name="sgender" value="男"></property>
        <property name="sage" value="21"></property>
    </bean>
</beans>
```

3. 创建实体类

在/src/main/java 下新建包 com.imut.pojo,在该包下创建实体类 Student.java。

程序 Student.java 的代码如下。

```
public class Student implements Serializable{
    private int sid;
    private String sno;
    private String sname;
    private String sgender;
    private int sage;
    //此处省略 getXxx()、setXxx()和 toString()方法。
}
```

4. 编写测试类

在 src 下新建包 com.imut.test，在该包下新建测试类 StudentTest.java，测试通过 Spring 容器创建 Student 对象的功能。

程序 StudentTest.java 的代码如下。

```
public class StudentTest {
    @Test
    public void test() {
        ApplicationContext context=new
            ClassPathXmlApplicationContext("applicationContext.xml");
        Student student=(Student)context.getBean("student");
        System.out.println("通过 Spring 容器中获取到的 Student 对象:");
        System.out.println(student);
    }
}
```

6.5 习题

1. 选择题

(1) 下面关于 Spring 框架的说法，错误的是(　　)。

A. Spring 框架是一个轻量级框架

B. Spring 框架颠覆了已经有较好解决方案的领域，如 MyBatis 框架

C. Spring 框架可以实现与多种框架的无缝集成

D. Spring 框架的核心机制是“依赖注入”

(2) 下面关于依赖注入的说法，正确的是(　　)。

A. 依赖注入的目标是在代码之外管理程序组件间的依赖关系

B. 依赖注入即“面向接口”编程

C. 依赖注入是面向对象技术的替代品

D. 依赖注入的使用会增大程序的规模

(3) 若 Spring 配置文件中有如下代码片段，则下面说法正确的是(　　)。

```
<bean id="userInfo" class="cn.user.UserInfo">
<property name="userName" value="john" />
<property name="userAge" value="26"/>
</bean>
```

A. Userlnfo 中一定声明了属性：private String userName；

B. Userlnfo 中一定声明了属性：private Integer user Age；

C. Userlnfo 中一定有 public void setUserName(String username)方法

D. Userlnfo 中一定有 public void setUserAge(Integer userAge)方法

2. 简答题

(1) 简述 Spring 框架的优点。

(2) 简述 Spring 的 IoC 和 DI。

3. 操作题

在控制台上使用 IoC 输出,并通过 Spring 实现依赖注入,输出内容如下。

```
张三说:"好好学习,天天向上!"
TOM 说:"Study hard. Improve every day!"
```

第7章 使用Spring管理Bean

第 6 章介绍了 Spring 框架的控制反转思想、原理、容器机制，通过案例展示了其基本使用方法，使我们对控制反转和依赖注入有了初步了解。Spring 的容器功能是其基础，它管理 Bean 的生命周期和依赖关系，降低类间耦合，简化开发，提高效率。因此，深入理解 Spring 管理 Bean 的技术细节至关重要。本章将详细讲解使用 Spring 管理 Bean 的知识，学习将 Bean 注入 Spring IoC 容器中的方法。

7.1 Bean 的配置

Spring 框架是一个大型工厂，其基础是容器，用于生产和管理 Bean。在 Java 项目中使用此工厂，需配置 XML 或 Properties 格式的配置文件。XML 格式最常用，通过文件注册并管理 Bean 依赖关系。本书将用 XML 文件讲解 Bean 属性和定义。Spring XML 配置文件根元素是<beans>，包含多个<bean>子元素，每个子元素定义了一个 Bean 及其装配方式。<bean>元素有多个属性及其子元素描述 Bean 信息，如表 7-1 所示。

表 7-1 <bean>元素的常用属性及其子元素

属性或子元素名称	描　述
id	表示一个 Bean 的唯一标识符，Spring 容器对 Bean 的配置、管理都通过该属性来完成
name	Spring 容器同样可以通过此属性对容器中的 Bean 进行配置和管理，name 属性中可以为 Bean 指定多个名称，每个名称之间用逗号或分号隔开
class	该属性指定了 Bean 的具体实现类，它必须是一个完整的类名，即使用类的全限定名
scope	用来设定 Bean 实例的作用域，其属性值有 singleton（单例）、prototype（原型）、request、session、global Sessions application 和 websocket()，默认值为 singleton

续表

属性或子元素名称	描　述
constructor-arg	<bean>元素的子元素,可以使用此元素传入构造参数进行实例化。该元素的 index 属性指定构造参数的序号(从 0 开始),type 属性指定构造参数的类型,参数值可以通过 ref 属性或 value 属性直接指定,也可以通过 ref 或 value 子元素指定
setter.property	<bean>元素的子元素,用于调用 Bean 实例中的 setter 方法完成属性赋值,从而完成依赖注入。该元素的 name 属性指定 Bean 实例中的相应属性名,ref 属性或 value 属性用于指定参数值
Autowiring mode	自动装配协作对象
ref	<property>、<constructor-arg>等元素的属性或子元素,可以用于指定对 Bean 工厂中某个 Bean 实例的引用
value	<property>、<constructor-arg>等元素的属性或子元素,可以用于直接指定一个常量值
list	用于封装 list 或数组类型的依赖注入
set	用于封装 set 类型属性的依赖注入
map	用于封装 map 类型属性的依赖注入

表 7-1 概括了<bean>元素的常用属性和子元素,但还有更多属性和子元素可通过网上资料获取。

在配置文件中,普通 Bean 通常只需定义 id(或 name)和 class 属性。若未指定 id 和 name,Spring 将使用 class 值作为 id。除了列出的属性,<bean>元素还包括配置依赖信息的子元素,如<property>用于属性注入,<constructor-arg>用于自动寻找并注入 Bean 的构造函数参数。

7.2 Bean 的实例化

在面向对象的程序中,要使用某个类的对象,就需要先实例化这个对象。同样,在 Spring 框架中,要使用容器中的 Bean,也需要实例化 Bean。实例化 Bean 有 3 种方式,分别为构造器实例化、静态工厂方式实例化和实例工厂方式实例化,其中最常用的是构造器实例化。下面将分别对这 3 种实例化 Bean 的方式进行详细讲解。

7.2.1 构造器实例化

构造器实例化是指 Spring 框架的容器通过 Bean 对应类中默认的无参构造方法来实例化 Bean。下面通过一个案例来演示 Spring 框架的容器通过构造器来实例化 Bean。

(1) 在 IDEA 中,创建一个名为 ch07_spring_01 的 Java 项目,在该项目的 lib 目录中加入 Spring 框架支持和依赖的 JAR 包。

(2) 在项目的 src/main/java 目录下创建一个 com.imut.instance.constructor 包,并在该包中创建 Beanl 类,如例 7-1 所示。

例 7-1：Bean1.java。

```
package com.imut.instance.constructor;
public class Bean1 {
}
```

(3) resources 目录下创建 Spring 的配置文件 instanceConstructor.xml，在配置文件中定义一个 id 为 bean1 的 Bean，并通过 class 属性指定其对应的实现类为 Bean1，如例 7-2 所示。

例 7-2：instanceConstructor.xml。

```
<? xml version="1.0" encoding="UTF-8"? >
<beans xmlns="http://www.springframework.org/schema/beans"
    xmlns:xsi="http://www.w3.org/2001/XMLSchema-instance"
    xsi:schemaLocation="http://www.springframework.org/schema/beans
    http://www.springframework.org/schema/beans/spring-beans.xsd">
    <bean id="bean1" class="com.imut.instance.constructor"/>
</beans>
```

(4) 在 com.imut.instance.constructor 包中创建测试类 instanceTest1，来测试构造器是否能实例化 Bean，代码如例 7-3 所示。

例 7-3：instanceTestl.java。

```
package com.imut.instance.constructor;
import org.junit.Test;
import org.springframework.context.ApplicationContext;
import org.springframework.context.support.ClassPathXmlApplicationContext;
public class instanceTest1 {
@Test
    public void testBean1() {
    //定义配置文件路径
    String xmlPath = "applicationContext.xml";
    //ApplicationContext 加载配置文件时,对 Bean 进行实例化
    ApplicationContext applicationContext = new ClassPathXmlApplicationContext
(xmlPath);
    Bean1 bean = (Bean1) applicationContext.getBean ("bean1");
    System.out.println (bean);
    }
}
```

在构造器实例化案例代码中，首先定义了配置文件的路径，然后 Spring 框架的容器 ApplicationContext 会加载配置文件。加载时，Spring 框架的容器会通过 id 为 Bean1 的实现类 bean1 中默认的无参构造方法，对 Bean 进行实例化。执行程序后，控制台的输出结果如图 7-1 所示。

可以看出，Spring 框架的容器已经成功实例化 Bean1，并输出了结果。

```
"C:\Program Files\Java\jdk-17\bin\java.exe" ...
DEBUG [main] - Refreshing org.springframework.context.support.ClassPathXmlApplicationContext@2781e022
DEBUG [main] - Loaded 1 bean definitions from class path resource [instanceConstructor.xml]
DEBUG [main] - Creating shared instance of singleton bean 'bean1'
com.imut.instance.constructor.Bean1@55b0dcab
```

图 7-1　构造器实例化案例执行结果

7.2.2　静态工厂方式实例化

使用静态工厂是实例化 Bean 的另一种方式。这种方式要求开发者利用一个静态工厂的方法来创建 Bean 的实例，其 Bean 配置中的 class 属性指定的不再是 Bean 实例的实现类，而是静态工厂类。同时还需要使用 factory-method 属性来指定所创建的静态工厂方法。下面通过一个案例演示使用静态工厂方式实例化 Bean。

(1) 在 ch07_spring_01 项目的 src/main/java 目录下创建一个 com.imut.instance.static_factory 包，在该包中创建 Bean2 类，该类与 Bean1 一样，不需要添加任何方法。

(2) 在 com.imut.instance.static_factory 包中创建一个 Bean2Factory 类，并在类中创建一个静态方法 createBean()来返回 Bean2 实例，如例 7-4 所示。

例 7-4：Bean2Factory.java。

```
package com.imut.instace.static_factory;
public class Bean2Factory {
    public static Bean2 createBean() {
        return new Bean2();
    }
}
```

(3) 在 resources 目录中创建 Spring 框架配置文件 instance-staticFactory.xml，代码如例 7-5 所示。

例 7-5：instance-staticFactory.xml。

```
<?xml version="1.0" encoding="UTF-8"?>
<beans xmlns="http://www.springframework.org/schema/beans"
    xmlns:xsi="http://www.w3.org/2001/XMLSchema-instance"
    xsi:schemaLocation="http://www.springframework.org/schema/beans
    http://www.springframework.org/schema/beans/spring-beans.xsd">
    <bean id="bean2" class="com.imut.instace.static_factory.Bean2Factory"
        factory-method="createBean"/>
</beans>
```

在上述配置文件中，首先通过<bean>元素的 id 属性定义了一个名称为 bean2 的 Bean。由于使用的是静态工厂方法，所以需要通过 class 属性指定其对应的工厂实现类为 Bean2Factory()。由于使用这种方式配置 Bean 后，Spring 容器仍不知道哪个是所需要的工厂方法，因此需要使用 factory-method 属性，以告诉 Spring 容器其方法名称为 createBean。

(4) 在 com.imut.instance.static_factory 包中创建一个测试类 instanceTest2，来测试使用静态工厂方式是否能实例化 Bean，编辑后的代码如例 7-6 所示。

例 7-6：instanceTest2.java。

```
package com.imut.instace.static_factory;
import org.junit.Test;
import org.springframework.context.ApplicationContext;
import org.springframework.context.support.ClassPathXmlApplicationContext;
public class instanceTest2 {
@Test
    public void testBean1() {
    //定义配置文件路径
    String xmlPath = "applicationContext.xml";
    //ApplicationContext 加载配置文件时，对 Bean 进行实例化
    ApplicationContext applicationContext = new
        ClassPathXmlApplicationContext(xmlPath);
    Bean2 bean = (Bean2) applicationContext.getBean ("bean2");
    System.out.println (bean);
    }
}
```

执行程序后，控制台的输出结果如图 7-2 所示。

```
"C:\Program Files\Java\jdk-17\bin\java.exe" ...
DEBUG [main] - Refreshing org.springframework.context.support.ClassPathXmlApplicationContext@27a8c74e
DEBUG [main] - Loaded 1 bean definitions from class path resource [instance-staticFactory.xml]
DEBUG [main] - Creating shared instance of singleton bean 'bean2'
com.imut.instance.static_factory.Bean2@2002fc1d
```

图 7-2　静态工厂方式实例化的案例执行结果

可以看到使用自定义的静态工厂方法已成功实例化了 Bean2。

7.2.3　实例工厂方式实例化

还有一种实例化 Bean 的方式，就是采用实例工厂。在此种方式的工厂类中，不再使用静态方法创建 Bean 实例，而是采用直接创建 Bean 实例的方式。同时，在配置文件中，需要实例化的 Bean 也不是通过 class 属性直接指向的实例化类，而是通过 factory-bean 属性指向配置的实例工厂，然后使用 factory-method 属性确定使用工厂中的哪个方法。下面通过一个案例来演示实例工厂方式的使用。

(1) 在 ch07_spring_01 项目的 src/main/java 目录下创建一个 com.imut.instance.factory 包，在该包中创建 Bean3 类，该类与 Bean1 一样，不需要添加任何方法。

(2) 在 com.imut.instance.factory 包中创建工厂类 Bean3Factory，在类中使用默认无参构造方法输出“bean3 工厂实例化中”语句，并使用 createBean()方法创建 Bean3 对象，如例 7-7 所示。

例 7-7：Bean3Factory.java。

```
package com.imut.instace.factory;
public class Bean3Factory {
    public Bean3 createBean() {
        return new Bean3();
    }
}
```

(3) 在 resources 目录中创建 Spring 配置文件 instanceFactory.xml，设置相关配置，如例 7-8 所示。

例 7-8：instanceFactory.xml。

```
<?xml version="1.0" encoding="UTF-8"?>
<beans xmlns="http://www.springframework.org/schema/beans"
    xmlns:xsi="http://www.w3.org/2001/XMLSchema-instance"
    xsi:schemaLocation="http://www.springframework.org/schema/beans
    http://www.springframework.org/schema/beans/spring-beans.xsd">
    <bean id="myBean3Factory" class="com.imut.instance.factory.Bean3Factory"/>
    <!-- 通过 factory-bean 属性指向配置的实例工厂,并通过 factory-method=属性确定
使用工厂中的哪个方法-->
    <bean id="bean3" factory-bean="myBean3Factory" factory-method=
"createBean"/>
</beans>
```

在上述配置文件中，先配置一个工厂 Bean，然后配置需要实例化的 Bean。在 id 为 bean3 的 Bean 中，使用 factory-bean 属性指向配置的实例工厂，该属性值就是工厂 Bean 的 id。使用 factory-method 属性来确定使用工厂中的 createBean()方法。

(4) 在 com.imut.instance.factory 包中创建测试类 instanceTest3，来测试实例工厂方式能否实例化 Bean，代码如例 7-9 所示。

例 7-9：instanceTest3.java。

```
package com.imut.instance.factory;
import org.junit.Test;
import org.springframework.context.ApplicationContext;
import org.springframework.context.support.ClassPathXmlApplicationContext;
public class instanceTest3 {
@Test
    public void testBean1() {
    //定义配置文件路径
    String xmlPath = "applicationContext.xml";
    //ApplicationContext 加载配置文件时,对 Bean 进行实例化
    ApplicationContext applicationContext = new ClassPathXmlApplicationContext
                                            (xmlPath);
    Bean3 bean = (Bean3) applicationContext.getBean ("bean3");
    System.out.println (bean);
    }
}
```

执行程序后,控制台的输出结果如图 7-3 所示。

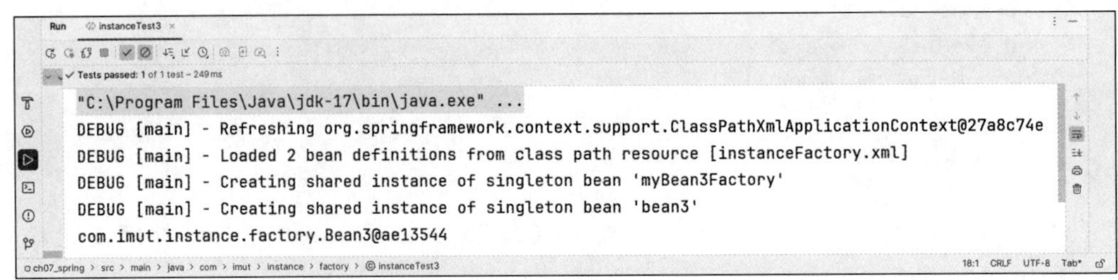

图 7-3　实例工厂方式实例化的案例执行结果

从上述实例化方法中可以看到,Spring 中提供了两种类型的 Bean。一种是普通 Bean,另一种是 FactoryBean。使用 FactoryBean 可以参与到 Bean 对象的创建过程中。

7.3　基于 XML 的 Bean 装配方式

通过前面的讲解,读者已经对 Spring IoC 的理念和设计有了基本认识。Bean 的装配可以理解为依赖关系注入,即如何将 Bean 装配到 Spring 容器中。Spring 容器支持多种形式的 Bean 装配方式,包括基于 XML 的装配、基于注解(Annotation)的装配和自动装配等。首先介绍基于 XML 的装配实现依赖注入的方式。

7.3.1　常用的依赖注入方式

在 Spring 中实现 IoC 容器的方法是依赖注入。依赖注入是一个处理过程,在对象被构造后或者从工厂返回后,仅通过构造函数参数、工厂方法的参数或者属性赋值的方式来定义它们的依赖。依赖注入的作用是在使用 Spring 框架创建对象时动态地将其所依赖的对象(如属性值)注入 Bean 组件中。Spring 框架的依赖注入通常有两种实现方式,一种是通过 setter 访问器对属性的赋值,称为设值注入(Setter Injection),另一种是通过构造方法赋值的能力,称为构造注入(Constructor Injection)。

1. 使用属性的 setter 方法注入

使用属性的 setter 方法注入是 Spring 主流的依赖注入方式。它利用 Java Bean 规范所定义的 setter 方法来完成注入,灵活且可读性高。属性 setter 注入是在被注入的类中声明一个 setter 方法,并通过该 setter 方法的参数注入对应的值。这种方式消除了使用构造方法注入时出现多个参数的可能性。Spring 框架其实是使用 Java 的反射机制实现的 setter 方法注入。在实例化 Bean 的过程中,首先会调用无参构造器或无参静态工厂方法实例化 Bean,然后通过反射的方式调用 setter 方法来注入属性值。

因此,设值注入要求一个 Bean 必须满足以下两点要求。

(1) 提供一个默认的无参构造方法。

(2) 为需要注入的属性提供对应的 setter 方法。

使用设值注入时,在 Spring 框架配置文件中,需要使用＜bean＞元素的子元素＜property＞来为每个属性注入值。使用属性的 setter 方法在 Spring 框架的容器中实现依赖注入的步骤具体如下。

(1) 在 Java 工程 ch07_spring_01 的 src/main/java 目录中创建包 com.imut.pojo。

(2) 在 src/main/java 目录的包 com.imut.pojo 下新建 Course1 类,如例 7-10 所示。

例 7-10: Course1.java。

```
package com.imut.pojo;
public class Course1 {
    private int cid;
    private String cno;
    private String cname;
    //此处省略 getter()、setter()和 toString()方法
}
```

在以上代码中,Course1 类提供了成员变量、setter()方法和 toString()方法。当 Spring 通过属性注入相应的值以后,通过 toString()方法可以获取这些注入的值。

(3) 在工程 ch07_spring_01 的 resources 目录下新建配置文件 xmlBean.xml,Spring 将通过该文件获取 Bean 的配置信息。这里 Spring 通过调用属性对应的 setter 方法给对象的属性进行值的注入。在配置文件中使用标签<property name="setter 方法的属性名" value="注入字面值" ref="引用其他的 bean">进行值的注入,如例 7-11 所示。

例 7-11: xmlBean.xml。

```
<?xml version="1.0" encoding="UTF-8"?>
<beans xmlns="http://www.springframework.org/schema/beans"
      xmlns:xsi="http://www.w3.org/2001/XMLSchema-instance"
      xsi:schemaLocation="http://www.springframework.org/schema/beans
      http://www.springframework.org/schema/beans/spring-beans.xsd">
    <!--使用属性的 setter 方法进行依赖注入-->
    <bean id="course1" class="com.imut.pojo.Course1">
        <property name="cid" value="1"></property>
        <property name="cno" value="350225001"></property>
        <property name="cname" value="软件工程"></property>
    </bean>
</beans>
```

在以上代码中,<property>元素的 name 属性指定该类的成员变量名称,value 属性给对应的成员变量注入值。在第 6 章的综合案例中,Student 对象就是以属性的 setter 方法注入的,具体见例 6-3。

(4) 在项目 ch07_spring_01 的 src/main/java 目录下的 com.imut.test 包下新建测试类 CourseTest,如例 7-12 所示。

例 7-12: CourseTest.java。

```
package com.imut.test;
import org.junit.Test;
import org.springframework.beans.factory.BeanFactory;
import org.springframework.context.ApplicationContext;
```

```
import org.springframework.context.support.ClassPathXmlApplicationContext;
import com.imut.pojo.Course1;
public class CourseTest {
    @Test
    public void test() {
        ApplicationContext context = new ClassPathXmlApplicationContext
("xmlBean.xml");
        Course1 course = (Course1) context.getBean("course1");
        System.out.println(course);
    }
}
```

在上述代码中，程序通过 Spring 获取 Course1 对象，进而输出 Course1 对象的信息，执行测试类 CourseTest，执行结果如图 7-4 所示。

```
DEBUG [main] - Creating shared instance of singleton bean 'teacherList'
DEBUG [main] - Creating shared instance of singleton bean 'singleUtilBean'
DEBUG [main] - Creating shared instance of singleton bean 'teacherMap'
DEBUG [main] - Creating shared instance of singleton bean 'teacherListP'
DEBUG [main] - Creating shared instance of singleton bean 'teacherSpEL'
Course [cid=1, cno=350225001, cname=软件工程]
```

图 7-4　使用属性的 setter()方法注入的结果

从以上执行结果可以看出，控制台输出 Course1 对象的信息，其中成员变量的值和配置文件<property>元素的 value 值相同。由此可见，Spring 实现了属性注入。

2. 使用构造方法注入

构造方法注入指 IoC 容器使用构造方法注入被依赖的实例。基于构造方法的依赖注入通过调用带参数的构造方法实现，每个参数代表一个依赖。调用带参的静态工厂方法构造 Bean 与此十分类似。

使用构造方法注入，需要在被注入的类中声明一个构造方法，该构造方法可以是有参或无参的。Spring 读取配置信息后，会通过反射方式调用构造方法来实例化对象。如果是有参构造方法，需要在构造方法中传入所需的参数值，进而创建类对象。具体做法是在配置文件里使用<bean>元素的子元素<constructor-arg>来定义构造方法的参数，使用其 value 属性(或子元素)来设置该参数的值。使用构造方法在 Spring 框架的容器中实现依赖注入的步骤具体如下。

(1) 在 Java 工程 ch07_spring_01 的 src/main/java 目录的包 com.imut.pojo 下新建 Course2 类，如例 7-13 所示。

例 7-13：Course2.java。

```
package com.imut.pojo;
public class Course2{
    private int cid;
    private String cno;
```

```
    private String cname;
    public Course2(int cid, String cno, String cname) {
        super();
        this.cid = cid;
        this.cno = cno;
        this.cname = cname;
    }
    //此处省略 getter()、setter()和 toString()方法
}
```

在以上代码中，Course2 类提供了有参构造方法和 toString()方法，当 Spring 通过构造器注入相应的值后，通过 toString()方法可以获取这些注入的值。

(2) 修改工程 ch07_spring_01 的 resources 目录下的配置文件 xmlBean.xml，在 applicationContext.xml 文件的 bean 元素中添加配置信息，具体代码如例 7-14 所示。

例 7-14：xmlBean.xml。

```
<?xml version="1.0" encoding="UTF-8"?>
<beans xmlns="http://www.springframework.org/schema/beans"
     xmlns:xsi="http://www.w3.org/2001/XMLSchema-instance"
     xsi:schemaLocation="http://www.springframework.org/schema/beans
     http://www.springframework.org/schema/beans/spring-beans.xsd">
   <!-- 通过构造方法注入为 course2 的属性赋值 -->
   <bean id="course2" class="com.imut.pojo.Course2">
      <constructor-arg name="cid" value="2"></constructor-arg>
      <constructor-arg name="cno" value="350225001"></constructor-arg>
      <constructor-arg name="cname" value="专业英语"></constructor-arg>
   </bean>
</beans>
```

在以上代码中，<constructor-arg>元素用于给类构造方法的参数注入值，Spring 通过构造方法注入获取这些值，最终这些值将会赋给 Spring 创建的 Course2 对象。

(3) 修改工程 ch07_spring_01 的 src/main/java 目录下包 com.imut.test 的测试类 CourseTest，如例 7-15 所示。

例 7-15：CourseTest.java。

```
package com.imut.test;
import org.junit.Test;
import org.springframework.beans.factory.BeanFactory;
import org.springframework.context.ApplicationContext;
import org.springframework.context.support.ClassPathXmlApplicationContext;
import com.imut.pojo.Course2;
public class CourseTest {
    @Test
    public void test() {
        ApplicationContext context = new
ClassPathXmlApplicationContext("xmlBean.xml");
        Course2 course = (Course2) context.getBean("course2");
```

```
        System.out.println(course);
    }
}
```

在上述代码中,程序通过 Spring 获取 Course2 对象,进而输出 Course2 对象的信息,执行测试类 CourseTest,执行结果如图 7-5 所示。

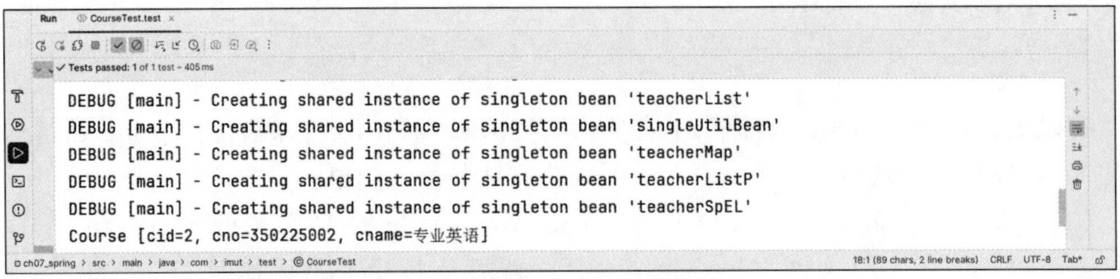

```
Run  CourseTest.test
✓ Tests passed: 1 of 1 test – 405 ms
DEBUG [main] - Creating shared instance of singleton bean 'teacherList'
DEBUG [main] - Creating shared instance of singleton bean 'singleUtilBean'
DEBUG [main] - Creating shared instance of singleton bean 'teacherMap'
DEBUG [main] - Creating shared instance of singleton bean 'teacherListP'
DEBUG [main] - Creating shared instance of singleton bean 'teacherSpEL'
Course [cid=2, cno=350225002, cname=专业英语]
ch07_spring > src > main > java > com > imut > test > CourseTest    18:1 (89 chars, 2 line breaks)  CRLF  UTF-8  Tab*
```

图 7-5　使用构造方法注入的结果

从以上执行结果可以看出,控制台输出 Course2 对象的信息,其中成员变量的值和配置文件<constructor-arg>元素的 value 值相同。由此可见,Spring 实现了构造方法注入。

<constructor-arg>的 name 属性可以使用指定构造参数名称赋值,消除二义性。此外,也可以使用 index 属性指定构造函数的参数,代码如下。

```
<bean id="course2" class="com.imut.pojo.Course2">
        <constructor-arg index="0" value="2"></constructor-arg>
        <constructor-arg index="1" value="350225002"></constructor-arg>
        <constructor-arg index="2" value="专业英语"></constructor-arg>
</bean>
```

也可以使用 type 属性给参数指定类型,容器通过类型匹配给参数赋值,代码如下。

```
<bean id="course2" class="com.imut.pojo.Course2">
        <constructor-arg type="int" value="2"></constructor-arg>
        <constructor-arg type=""java.lang.String"" value="350225002">
</constructor-arg>
</bean>
```

但是当参数有多个相同类型时,多个简单类型参数的赋值存在二义性,代码如下。

```
<bean id="course2" class="com.imut.pojo.Course2">
        <constructor-arg type="int" value="2"></constructor-arg>
        <constructor-arg type=""java.lang.String"" value="350225002">
</constructor-arg>
        <constructor-arg type=""java.lang.String" value="专业英语">
</constructor-arg>
</bean>
```

7.3.2　注入不同数据类型

Spring 的配置文件提供不同的标签来实现各种不同类型参数的注入,这些标签对于设值注入和构造注入都适用。下面将以设值注入为例介绍。对于构造注入,只需将所介绍的标签添加到<constructor-arg>与</constructor-arg>中间即可。

1. 设值注入数据(基本数据类型、String 等类型)

字面值数据是指可用字符串表示的值,可以通过<value>元素标签或 value 属性进行注入。基本数据类型及其封装类、String 等类型都可以采取设值注入的方式。

元素<property/>的属性 value 可以为指定的参数赋值。Spring 的转换服务将这些值从 String 转换为合适的类型。代码如下。

```
<bean id="myDataSource" class="org.apache.commons.dbcp.BasicDataSource"
destroy-method="close">
    <property name="driverClassName" value="com.mysql.jdbc.Driver"/>
    <property name="url" value="jdbc:mysql://localhost:3306/mydb"/>
    <property name="username" value="root"/>
    <property name="password" value="masterkaoli"/>
</bean>
```

对于基本数据类型及其包装类、字符串,除了可以使用 value 属性,还可以通过<value>子元素进行注入,代码如下。

```
<bean id="myDataSource" class="org.apache.commons.dbcp.BasicDataSource"
destroy-method="close">
    <property name="driverClassName">
        <value>com.mysql.jdbc.Driver</value>
    </property>
    <property name="url">
        <value>jdbc:mysql://localhost:3306/mydb</value>
    </property>
    <property name="username">
        <value>root</value>
    </property>
    <property name="password" >
        <value>masterkaoli</value>
    </property>
</bean>
```

注入字面值时,如遇到特殊字符(&、<、>、"、'),通常可以采用两种办法处理,即使用<![CDATA[...]]>方式,或者把特殊字符替换为实体引用。需要注意的是,<![CDATA[...]]>只能用在<value>子标签中。代码如下。

```
<bean id="course3" class="com.imut.pojo.Course2">
    <constructor-arg name="cid" value="3"></constructor-arg>
    <constructor-arg name="cno" value="350225003"></constructor-arg>
    <!--<constructor-arg name="cname"
        value="&lt;&lt;Java 程序设计 &gt;&gt;"></constructor-arg>-->
    <constructor-arg name="cname">
        <value><![CDATA[<<Java 程序设计>>]]></value>
```

```
    </constructor-arg>
</bean>
```

XML 中有 5 个预定义的实体引用，如表 7-2 所示。

表 7-2　XML 预定义的实体引用

符　　号	实体引用	符　　号	实体引用
<	<.	'	'
>	>	"	"
&	&		

严格地讲，XML 中仅有字符<和 & 是非法的，其他 3 个符号>、'、"是合法的，但是把它们都替换为实体引用是个好习惯。

2. 注入其他 Bean 组件

组成应用程序的 Bean 经常需要相互协作，以完成应用程序的功能。要使 Bean 能够相互访问，就必须在 Bean 配置文件中指定对 Bean 的引用。在 Bean 的配置文件中，可以通过<ref>元素或 ref 属性为 Bean 的属性或构造器参数指定对 Bean 的引用。也可以在属性或构造器里包含 Bean 的声明，这样的 Bean 称为内部 Bean。注入其他 Bean 组件的代码如例 7-16 所示。

例 7-16：xmlBean.xml。

```
<!--定义 Teacher 对象,并指定 id 为 teacher-->
<bean id="teacher" class="com.imut.pojo.Teacher">
        <property name="tid" value="1"></property>
        <property name="tname" value="软件工程"></property>
        <!--<property name="tcourse" ref="course1"></property>-->
        <property name="tcourse">
            <ref bean="course1"/>
        </property>
</bean>
```

在工程 ch07_spring_01 的 src/main/java 目录下包 com.imut.pojo 的新建类 Teacher 如例 7-17 所示。

例 7-17：Teacher.java。

```
package com.imut.pojo;
public class Teacher {
    private String tid;
    private String tname;
    private Course1 tcourse;
    //此处省略 getter()、setter()和 toString()方法
}
```

在工程 ch07_spring_01 的 src/main/java 目录下包 com.imut.test 的新建测试类

TeacherTest 如例 7-18 所示。

例 7-18：TeacherTest.java。

```
package com.imut.test;
import org.junit.Test;
import org.springframework.context.ApplicationContext;
import org.springframework.context.support.ClassPathXmlApplicationContext;
import com.imut.pojo.Teacher;
public class teacherTest {
    @Test
    public void test() {
        ApplicationContext context = new
            ClassPathXmlApplicationContext("xmlBean.xml");
        Teacher teacher = (Teacher) context.getBean("teacher");
        System.out.println(teacher);
    }
}
```

执行测试类 TeacherTest，执行结果如图 7-6 所示。

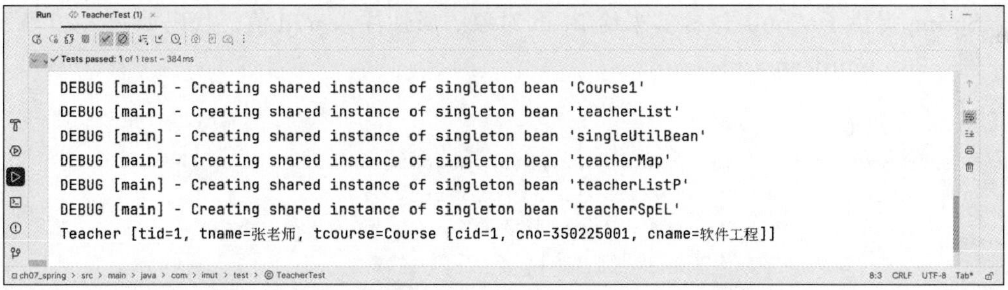

```
Run    TeacherTest (1)
✓ Tests passed: 1 of 1 test – 384ms
DEBUG [main] - Creating shared instance of singleton bean 'Course1'
DEBUG [main] - Creating shared instance of singleton bean 'teacherList'
DEBUG [main] - Creating shared instance of singleton bean 'singleUtilBean'
DEBUG [main] - Creating shared instance of singleton bean 'teacherMap'
DEBUG [main] - Creating shared instance of singleton bean 'teacherListP'
DEBUG [main] - Creating shared instance of singleton bean 'teacherSpEL'
Teacher [tid=1, tname=张老师, tcourse=Course [cid=1, cno=350225001, cname=软件工程]]
ch07_spring > src > main > java > com > imut > test > TeacherTest        8:3  CRLF  UTF-8  Tab*
```

图 7-6 执行测试类 TeacherTest 注入其他 Bean 组件的结果

3. 使用内部 Bean

当 Bean 实例仅仅被一个特定的属性使用时，可以将其声明为内部 Bean。内部 Bean 声明直接包含在<property>或者<constructor-arg>元素里，不需要设置任何 id 或者 name 属性。内部 Bean 只能在当前 bean 的内部使用，不能在其他任何地方使用。代码如例 7-19 所示。

例 7-19：xmlBean.xml。

```
<!--使用内部 Bean-->
    <!--定义 Teacher 对象，并指定 id 为 teacher-->
    <bean id="teacher1" class="com.imut.pojo.Teacher">
        <property name="tid" value="2"></property>
        <property name="tname" value="王老师"></property>
        <property name="tcourse">
        <!--内部 Bean-->
            <bean class="com.imut.pojo.Course1">
                <property name="cid" value="4"></property>
                <property name="cno" value="350225004"></property>
```

```
            <property name="cname" value="人工智能"></property>
        </bean>
    </property>
</bean>
```

这样，这个 Course1 类型的 Bean 就只能被 teacher1 使用，无法被其他的 Bean 引用。执行测试类 TeacherTest，执行结果如图 7-7 所示。

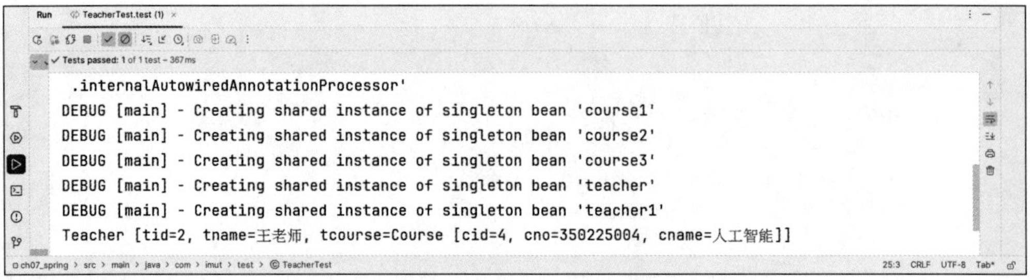

```
Run    TeacherTest.test (1)
✓ Tests passed: 1 of 1 test – 367ms
 .internalAutowiredAnnotationProcessor'
DEBUG [main] - Creating shared instance of singleton bean 'course1'
DEBUG [main] - Creating shared instance of singleton bean 'course2'
DEBUG [main] - Creating shared instance of singleton bean 'course3'
DEBUG [main] - Creating shared instance of singleton bean 'teacher'
DEBUG [main] - Creating shared instance of singleton bean 'teacher1'
Teacher [tid=2, tname=王老师, tcourse=Course [cid=4, cno=350225004, cname=人工智能]]
ch07_spring > src > main > java > com > imut > test > TeacherTest        25:3  CRLF  UTF-8  Tab*
```

图 7-7　执行 TeacherTest 类注入内部 bean 的结果

4. null 值、空字符串值和级联属性

Spring 提供了<null/>标签来给 Bean 对象的属性注入 null 值。代码如例 7-20 所示。

例 7-20：xmlBean.xml。

```
<!-- null 值-->
<bean id="teacherNull" class="com.imut.pojo.Teacher">
    <property name="tid" value="3"></property>
    <property name="tname" value="赵老师"></property>
    <property name="tcourse"><null/></property>
</bean>
```

执行测试类 TeacherTest，执行结果如图 7-8 所示。

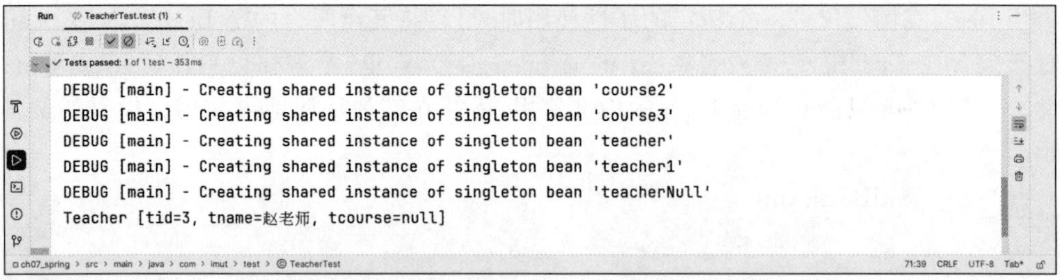

```
Run    TeacherTest.test (1)
✓ Tests passed: 1 of 1 test – 353ms
DEBUG [main] - Creating shared instance of singleton bean 'course2'
DEBUG [main] - Creating shared instance of singleton bean 'course3'
DEBUG [main] - Creating shared instance of singleton bean 'teacher'
DEBUG [main] - Creating shared instance of singleton bean 'teacher1'
DEBUG [main] - Creating shared instance of singleton bean 'teacherNull'
Teacher [tid=3, tname=赵老师, tcourse=null]
ch07_spring > src > main > java > com > imut > test > TeacherTest        71:39  CRLF  UTF-8  Tab*
```

图 7-8　执行 TeacherTest 类注入 null 值的结果

也可以使用<value></value>注入空字符串值，为 Teacher 类对象的属性 tcourse 注入空字符串值<property name="tcourse"><value></value></property>。

Spring 支持通过级联的方式给对象的属性进行值的注入。级联的方式给对象的属性注入值，会改变原本的对象的属性值，因为引用的是同一个对象。级联方式注入值的代码如例 7-21 所示。

例 7-21：xmlBean.xml。

```
<!--级联属性 -->
    <bean id="teacherCascade" class="com.imut.pojo.Teacher">
        <property name="tid" value="3"></property>
        <property name="tname" value="赵老师"></property>
        <property name="tcourse" ref="course1"></property>
        <!-- 级联属性 : 会改变本来对象的属性的值. -->
        <property name="tcourse.cname" value="项目管理"></property>
    </bean>
```

执行测试类 TeacherTest，执行结果如图 7-9 所示。

```
DEBUG [main] - Creating shared instance of singleton bean 'teacher1'
DEBUG [main] - Creating shared instance of singleton bean 'teacherNull'
DEBUG [main] - Creating shared instance of singleton bean 'teacherCascade'
Teacher [tid=3, tname=赵老师, tcourse=Course [cid=1, cno=350225001, cname=项目管理]]

Process finished with exit code 0
```

图 7-9 执行 TeacherTest 类注入级联属性的结果

用级联的方式给 tcourse 对象的 cname 属性注入值后，course1 的 cname 属性就由原来的"软件工程"变为了"项目管理。"

5. 注入集合类型属性

在实际开发中，根据业务需求，有时需要将集合注入 Bean，例如，List、Set、Map、Array 等。此时，Spring 配置文件可以分别通过内置的 XML 标签<list>、<set>、<map>、<array>等来配置集合属性。

(1) List Set 集合类型与 Array 数组类型的属性注入。

如果对象中包含 java.util.List 类型的属性，则配置该属性需要使用<list>标签。在<list>标签包含一些元素，可以使用<value>指定简单的常量值，使用<ref>指定对其他 Bean 的引用，使用<bean>指定内置 Bean 定义，使用<null/>指定空元素，也可以使用另一个<list>标签来嵌套配置。

修改 Teacher 类，把属性 tcourse 的类型改为集合 List<Course2>，得到 TeacherList 类，将 TeacherList 类放到 com.imut.pojo 包下。核心代码如例 7-22 所示。

例 7-22：TeacherList.java。

```
public class TeacherList {
    private String tid;
    private String tname;
    private List<Course2> tcourse;
}
```

在配置文件 xmlBean.xml 中的配置代码如例 7-23 所示。

例 7-23：xmlBean.xml。

```
<!--List 集合类型的属性注入-->
```

```
<bean id="teacherList" class="com.imut.pojo.TeacherList">
    <property name="tid" value="4"></property>
    <property name="tname" value="康老师"></property>
    <property name="tcourses">
        <!-- 使用 list 标签构造集合 -->
        <list>
            <ref bean="course2"/>
            <ref bean="course3"/>
        </list>
    </property>
</bean>
```

其中，course2 和 course3 为引用已有的 Course 类对象。测试类 TeacherListTest 的核心代码如例 7-24 所示。

例 7-24：TeacherListTest.java。

```
public class TeacherListTest {
    @Test
    public void test() {
        ApplicationContext context = new
            ClassPathXmlApplicationContext("xmlBean.xml");
        TeacherList teacherList = (TeacherList)
                        context.getBean("teacherList",TeacherList.class);
        System.out.println(teacherList);
    }
}
```

测试类 TeacherListTest 的执行结果如图 7-10 所示。

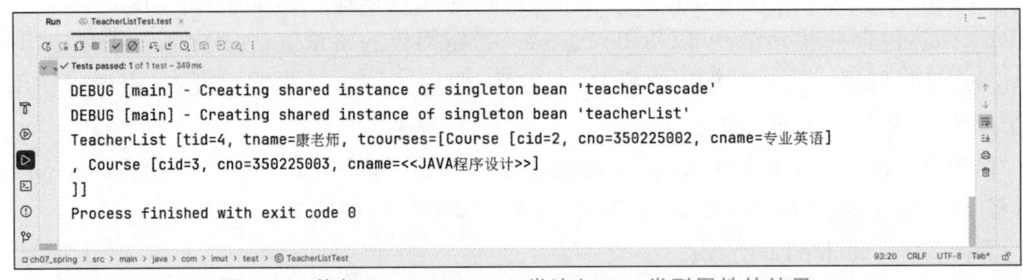

图 7-10　执行 TeacherListTest 类注入 List 类型属性的结果

数组 Array 类型属性的配置，可以使用 Spring 的内置 XML 标签<array>，也可以使用<list>标签进行值的注入，因为 List 类型可以视为数组。也就是说，可以像配置 List 一样来配置数组类型的属性，只需要把<list>标签替换成<array>，对于 Array 类型的属性，推荐使用<list></list>标签进行值的注入。

配置 java.util.Set 类型的属性，需要使用<set>标签，定义元素的方法与 List 类似。<set>标签中间也可以使用<value><ref>等标签注入集合元素。

(2) Map 集合类型的属性注入。

配置 java.util.Map 类型的属性，需要使用<map>标签。在<map>标签里可以使用多个<entry>作为子标签，每个<entry>条目包含一个键<key>和一个值<value>，必须在<key>标签里定义键。因为键和值的类型没有限制，所以可以自由地为它们指定<value>、<ref>、<bean>或<null>元素。还可以将 Map 的键和值作为<entry>的属性定义，简单常量、字面值可以使用 key 和 value 来定义，Bean 引用则需要通过 key-ref 和 value-ref 属性定义。

在 com.imut.pojo 包下修改 Teacher 类，把属性 tcourses 的类型改为 Map<String, Course2>，得到新的类 TeacherMap，核心代码如例 7-25 所示。

例 7-25：TeacherMap.java。

```
public class TeacherMap {
    private String tid;
    private String tname;
    private Map<String,Course2> tcourses;
}
```

在 applicationContext 中的配置代码如例 7-26 所示。

例 7-26：xmlBean.xml。

```
<!-- Map 集合类型的属性注入 -->
    <bean id="teacherMap" class="com.imut.pojo.TeacherMap">
        <property name="tid" value="5"></property>
        <property name="tname" value="田老师"></property>
        <property name="tcourses">
            <map>
                <entry key="授课 1" value-ref="course2"></entry>
                <entry key="授课 2" value-ref="course3"></entry>
            </map>
        </property>
    </bean>
```

其中，course2 和 course3 为引用已有的 Course 类对象，需要通过 value-ref 属性定义。测试类 TeacherMapTest 的核心代码如例 7-27 所示。

例 7-27：TeacherMapTest.java。

```
public class TeacherMapTest {
    @Test
    public void test() {
        ApplicationContext context = new
            ClassPathXmlApplicationContext("xmlBean.xml");
        TeacherMap teacherMap = (TeacherMap) context.getBean("teacherMap",
                    TeacherMap.class);
        System.out.println(teacherMap);
    }
}
```

测试类 TeacherMapTest 的执行结果如图 7-11 所示。

```
DEBUG [main] - Creating shared instance of singleton bean 'teacherList'
DEBUG [main] - Creating shared instance of singleton bean 'teacherMap'
TeacherList [tid=4, tname=康老师, tcourses=[Course [cid=2, cno=350225002, cname=专业英语]
, Course [cid=3, cno=350225003, cname=<<JAVA程序设计>>]
]]
Process finished with exit code 0
```

图 7-11　执行 TeacheMapTest 类注入 Map 类型属性的结果

（3）独立的集合 Bean。

使用基本的集合标签定义集合时，通常不能将集合作为独立的 Bean 定义，这会导致其他 Bean 无法引用该集合，所以无法在不同 Bean 之间共享集合。为了解决这个问题，Spring 提供了<util:list>、<util:set>和<util:map>标签来定义独立的集合 Bean，方便对集合的重复使用。也可以使用 util schema 里的集合标签定义独立的集合 Bean，但使用 util schema 里的集合标签必须在<beans>根元素里添加 util schema 定义。

使用之前需要导入新的命名空间。在配置文件的声明信息中加入如下代码。

```
xmlns:util=http://www.springframework.org/schema/util
xsi="http://www.springframework.org/schema/util
http://www.springframework.org/schema/util/spring-util-4.0.xsd"
```

导入新的命名空间后，在 XML 配置文件的声明信息如图 7-12 所示。

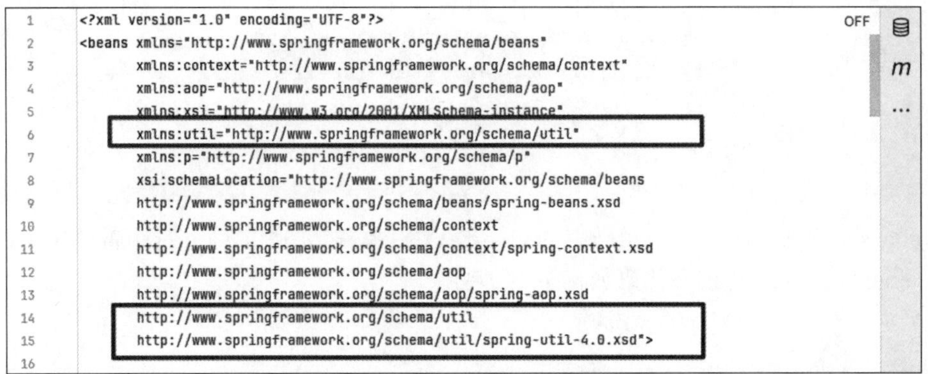

```
<?xml version="1.0" encoding="UTF-8"?>
<beans xmlns="http://www.springframework.org/schema/beans"
       xmlns:context="http://www.springframework.org/schema/context"
       xmlns:aop="http://www.springframework.org/schema/aop"
       xmlns:xsi="http://www.w3.org/2001/XMLSchema-instance"
       xmlns:util="http://www.springframework.org/schema/util"
       xmlns:p="http://www.springframework.org/schema/p"
       xsi:schemaLocation="http://www.springframework.org/schema/beans
       http://www.springframework.org/schema/beans/spring-beans.xsd
       http://www.springframework.org/schema/context
       http://www.springframework.org/schema/context/spring-context.xsd
       http://www.springframework.org/schema/aop
       http://www.springframework.org/schema/aop/spring-aop.xsd
       http://www.springframework.org/schema/util
       http://www.springframework.org/schema/util/spring-util-4.0.xsd">
```

图 7-12　XML 文件中的声明信息

为对象 teacherList 中的 List 类型属性 tcourses 注入值时，引用独立的集合 Bean。首先在 applicationContext 中使用标签<util:list></util:list>为独立的集合 Bean 定义。然后在 Bean 对象 teacherList 中引用独立集合 Bean 的 id 为 tcourses 属性注入值。核心配置代码如例 7-28 所示。

例 7-28：xmlBean.xml。

```
<!-- 使用 util 来定义独立的集合 Bean -->
    <util:list id="singleUtilBean">
        <ref bean="course2"/>
```

```
            <ref bean="course3"/>
        </util:list>
    <!--List 集合类型的属性注入-->
        <bean id="teacherList" class="com.imut.pojo.TeacherList">
            <property name="tid" value="4"></property>
            <property name="tname" value="康老师"></property>
            <property name="tcourses" ref="singleUtilBean"></property>
        </bean>
```

重新执行 teacherListTest 类,执行结果如图 7-13 所示。

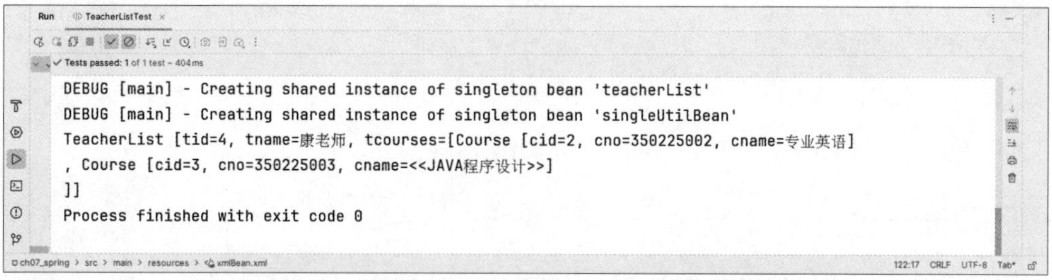

```
DEBUG [main] - Creating shared instance of singleton bean 'teacherList'
DEBUG [main] - Creating shared instance of singleton bean 'singleUtilBean'
TeacherList [tid=4, tname=康老师, tcourses=[Course [cid=2, cno=350225002, cname=专业英语]
, Course [cid=3, cno=350225003, cname=<<JAVA程序设计>>]
]]
Process finished with exit code 0
```

图 7-13　执行 TeacheListTest 类为属性注入独立的集合 Bean

7.3.3　使用 P: 命名空间注入

Spring 的配置文件一般基于<property>元素来配置 Bean 的属性,但是当 Bean 实例的属性足够多时,使用大量的<property>元素配置属性就显得冗余。为了简化 XML 文件的配置,越来越多的 XML 文件采用属性而非子元素配置信息。Spring 从 2.5 版本开始引入了一个新的 P: 命名空间,可以通过<bean>元素属性的方式配置 Bean 的属性。

Spring 引入的 P: 命名空间是为了简化 XML 配置文件,特别是在 Bean 具有众多属性时。它允许使用<bean>元素的属性方式配置 Bean 的属性,提高了配置文件的可读性和维护性。这种紧凑的配置方式可以像设置普通 XML 属性一样来设置 Bean 的属性,从而减少了烦琐的<property>元素配置。P: 命名空间在处理大型配置文件时尤其有用,使 XML 文件更易于管理。

引入 P: 命名空间,首先修改配置文件 applicationContext.xml,导入新的命名空间。首先单击 XML 文件左下角的 Namespaces 选项卡,然后选中要导入的 P: 命名空间。导入新的命名空间后,在 XML 文件的声明信息中会加入对应的声明,如下所示。

```
<beans xmlns="http://www.springframework.org/schema/beans"
    xmlns:xsi="http://www.w3.org/2001/XMLSchema-instance"
    xmlns:util="http://www.springframework.org/schema/util"
    xmlns:p=http://www.springframework.org/schema/p
...
</beans>
```

以上代码的第 4 行用于引入 P: 命名空间,使用 P: 名空间时,要首先添加此行代码。然后在 applicationContext 文件中添加 TeacherList 类型的 Bean 对象,id 为 teacherListP,使用 P: 命名空间为其属性注入值。核心代码如例 7-29 所示。

例 7-29：xmlBean.xml。

```
<bean id="teacherListP" class="com.imut.pojo.TeacherList"
        p:tid="6" p:tname="小花老师"  p:tcourses-ref="singleUtilBean">
</bean>
```

在以上代码中，p：tid、p：tname、p：tcourses-ref 用于设置 TeacherList 类的属性，它们可以实现和<property>元素同样的功能。修改在 com.imut.test 包下的测试类 TeacherListTest，添加代码如下。

```
TeacherList teacherListP = (TeacherList) context.getBean("teacherListP",
TeacherList.class);
        System.out.println(teacherListP);
```

执行测试类 TeacherListTest，执行结果如图 7-14 所示。

```
DEBUG [main] - Creating shared instance of singleton bean 'teacherMap'
DEBUG [main] - Creating shared instance of singleton bean 'singleUtilBean'
DEBUG [main] - Creating shared instance of singleton bean 'teacherListP'
TeacherList [tid=6, tname=小花老师, tcourses=[Course [cid=2, cno=350225002, cname=专业英语]
, Course [cid=3, cno=350225003, cname=<<Java程序设计>>]
]]
```

图 7-14　使用 P：命名空间注入

从以上执行结果可以看出，控制台输出了 TeacherList 类的 tid 值、tname 值和 tcourses 对象的信息。经过分析发现，程序输出的信息和配置文件中 P：属性设置的值是匹配的，因此，程序实现了使用 P：命名空间注入值。

7.3.4　使用 SpEL 注入

SpEL，即 Spring Expression Language（Spring 表达式语言），是一种强大的表达式语言，用于查询和操作运行时对象图。SpEL 的主要作用是简化开发工作，减少代码逻辑和配置信息的编写。

SpEL 的语法与 EL 表达式语言类似，它以 #{...} 为定界符，所有在花括号中的字符都被认为是 SpEL 表达式。SpEL 的功能包括为 Bean 属性动态赋值、通过 Bean 的 ID 引用其他 Bean、调用对象的方法、引用对象的属性、计算表达式的值、匹配正则表达式等。

1. 用 SpEL 注入字面值

使用 SpEL 为 Bean 对象的属性注入字面值，可以通过设置 value 属性完成。如 value ="#{整数、小数、科学记数、布尔型、'字符串'}"。具体如例 7-30 所示。

例 7-30：注入不同类型的字面值。

```
整数:<property name="count" value="#{5}"/>
小数:<property name="frequency" value="#{89.7}"/>
```

```
科学记数:<property name="capacity" value="#{1e4}"/>
字符串:<property name="name" value="#{'Chuck'}"/>
或<property name='name' value='#{"Chuck"}'/>
布尔型:<property name="enabled" value="#{false}"/>
```

2. SpEL 引用其他的 Bean

使用 SpEL 引用其他的 Bean，即可以通过 SpEL 和 value 属性配置 Bean 之间的引用关系，如 value＝"＃{Bean 的 id 值}"。修改例 7-16 的 applicationContext 配置文件，为 Teacher 类对象的 tcourses 注入值时，用 SpEL 方式完成，核心代码如例 7-31 所示。

例 **7-31**：xmlBean.xml。

```
<!-- 使用 SpEL 注入 -->
    <bean id="teacherSpEL" class="com.imut.pojo.Teacher">
        <property name="tid" value="7"></property>
        <property name="tname" value="小美老师"></property>
        <!-- 引用其他对象 -->
        <property name="tcourses" value="#{course1}"></property>
   </bean>
```

修改测试类 TeacherTest，如例 7-32 所示。

例 **7-32**：TeacherTest.java。

```
public class TeacherTest {
    @Test
    public void test() {
        ApplicationContext context = new
            ClassPathXmlApplicationContext("xmlBean.xml");
        Teacher teacher3 = (Teacher) context.getBean("teacherSpEL");
        System.out.println(teacher3);
    }
}
```

执行测试类 TeacherTest 得到的结果如图 7-15 所示。

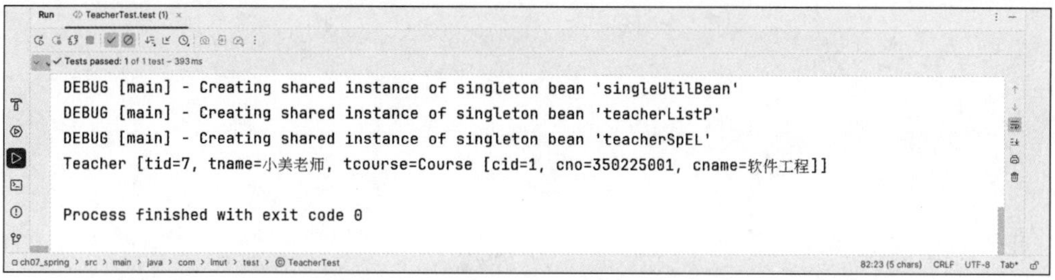

```
Run   TeacherTest.test (1)
Tests passed: 1 of 1 test – 393ms
DEBUG [main] - Creating shared instance of singleton bean 'singleUtilBean'
DEBUG [main] - Creating shared instance of singleton bean 'teacherListP'
DEBUG [main] - Creating shared instance of singleton bean 'teacherSpEL'
Teacher [tid=7, tname=小美老师, tcourse=Course [cid=1, cno=350225001, cname=软件工程]]

Process finished with exit code 0
ch07_spring > src > main > java > com > imut > test > TeacherTest      82:23 (5 chars)  CRLF  UTF-8  Tab*
```

图 7-15　使用 SpEL 注入的执行结果

3. SpEL 引用对象的属性及调用对象的方法

使用 SpEl 还可以引用对象的属性或调用对象的方法，通过设置 value 属性完成，如 value＝"＃{对象.属性名}"　value＝"＃{对象.方法}"。

4. SpEL 调用静态属性及静态方法

通过 T()调用一个类的静态方法，它将返回一个 Class Object，然后再调用相应的方法或属性，如 value="#{T(java.lang.Math).PI}" value="#{T(java.lang.Math).random()}"。

5. SpEL 进行运算及判断

SpEL 支持算数运算、比较运算、逻辑运算，条件运算符、字符串连接和正则表达式等操作。如 value="#{+ - * / % ^.... > < >= <= ==.... }"。具体如例 7-33 所示。

例 7-33：SpEL 进行运算及判断。

算数运算符：+、-、*、/、%、^。

```
<property name="adjusteAmount" value="#{counter.total + 42}"/>
<property name="adjusteAmount" value="#{counter.total - 20}"/>
<property name="circumference" value="#{2 * T(java.lang.Math).PI * circle.
radius}"/>
<property name="average" value="#{counter.total / counter.count}"/>
<property name="remainder" value="#{counter.total % counter.count}"/>
<property name="area" value="#{T(java.lang.Math).PI * circle.radius ^ 2}"/>
```

字符串连接符+。

```
<property name="name" value="#{performer.firstName + ' ' + performer.
lastName}"/>
```

比较运算符：<、>、==、<=、>=、lt、gt、eq、le、ge。

```
<property name="equal" value="#{counter.total == 100}"/>
<property name="hasCapacity" value="#{counter.total le 100000}"/>
```

逻辑运算符号：and、or、not、|。

```
<property name="largeCircle" value="#{shap.kind == 'circle' and shape.
perimeter gt 10000}"/>
<property name="outOfStack" value="#{!product.available}"/>
<property name="outOfStack" value="#{not product.available}"/>
```

条件运算符：?。

```
<property name="ifelse" value="#{songSelector.selectSong() == 'piano' ?
'Jingle Bells' : 'Silent Night'}"/>
```

正则表达式 matches。

```
<property name="regular" value="#{admin.email matches '[a-zA-Z0-9._%+-]+
@[a-zA-Z0-9.-]+\\.[a-zA-Z]{2,4}'}"/>
```

7.4 基于注解的 Bean 装配方式

Spring 2.0 版本开始引入注解的配置方式，将 Bean 的配置信息和 Bean 实现类结合在一起，直接写入代码中，使用注解可以减少 XML 文件的配置内容，简化配置工作，并且注解能够实现自动装配，提供的功能也更为强大。

在 Spring 中使用注解定义 Bean，并通过组件扫描的方式来让 Spring IoC 容器发现 Bean。组件扫描通过定义资源的方式，让 Spring IoC 容器扫描对应的包，从而把 Bean 装配进来。

使用注解配置的方式在许多项目中已经成为主流，能够有效减少 XML 配置的复杂性。然而，也有一些情况使用 XML 配置文件更为合适，比如系统存在多个公共的配置文件（如多个 properties 和 XML 文件），需要统一管理公共资源配置，或者涉及第三方类时，用 XML 的方式来完成会更加明确一些。因此，目前常见的做法是主要采用注解配置，辅以 XML 配置，以便更好地满足项目的需求。

7.4.1 常用的注解及使用注解定义 Bean

Spring 支持的注解相对较多，其中常用的注解如表 7-3 所示。

表 7-3 常用的注解

注 解	描 述
@Component	指定一个普通的 Bean，是一个泛化概念，仅仅表示一个组件（Bean），并且可以作用在任何层次，使用时只需将该注解标注在相应类上即可
@Controller	指定一个控制器组件 Bean，通常作用在控制层（如 Spring MVC 的 Controller），用于将控制层的类标识为 Spring 中的 Bean，其功能与@Component 相同
@Service	指定一个业务逻辑组件 Bean，通常作用在业务层（Service 层），用于将业务层的类标识为 Spring 中的 Bean，功能上等同于@Component
@Repository	指定一个 DAO 组件 Bean，将数据访问层（DAO 层）的类标识为 Spring 中的 Bean，其功能与@Component 相同
@Scope	指定 Bean 实例的作用域
@Value	指定 Bean 实例的注入值
@Autowired	指定要自动装配的对象
@Qualifier	指定要自动装配的对象名称，通常与@Autowired 联合使用
@Resource	指定要注入的对象
@PostConstruct	指定 Bean 实例完成初始化后调用的方法
@PreDestroy	指定 Bean 实例销毁之前调用的方法

表 7-3 中的前 4 个注解用于定义 Bean，虽然@Repository、@Service 与@Controller 的功能与@Component 注解的功能相同，但为了使标注类本身用途更加清晰，一般在实际开发中分别使用@Repository、@Service 与@Controller 对实现类进行标注。例 7-34 是使用

@Component 注解实现 Bean 组件的定义。

在 Java 项目工程 ch07_spring_01 的 src/main/java 目录下新建包 com.imut.annotation.pojo。在包 com.imut.annotation.pojo 下创建 Student 类，并使用@Component 注解。

例 7-34：@Component 注解实现 Bean 组件的定义。

```
package com.imut.annotation.pojo;
import org.springframework.stereotype.Component;
@Component
public class Student{
    private int sid;
    private String sno;
    private String sname;
    private String sgender;
    private int sage;

    public void printStudent() {
        System.out.println("Student printStudent()...");
    }
}
```

例 7-35、例 7-36、例 7-37 分别使用注解@Repository、@Service 与@Controller 定义 Bean 组件。

在工程 ch07_spring_01 的 src/main/java 目录下新建包 com.imut.annotation.dao，在该包中创建接口 StudentDao 和实现类 StudentDaoJdbcImpl。在 src/main/java 目录下新建包 com.imut.annotation.service，在该包中创建 StudentService 类。在 src/main/java 目录下新建包 com.imut.annotation.controller，在该包中创建 StudentController 类。

例 7-35：使用注解@Repository 定义 Bean 组件。

```
//StudentDao.java
package com.imut.annotation.dao;
public interface StudentDao {
}
//StudentDaoJdbcImpl.java
package com.imut.annotation.dao;
import org.springframework.stereotype.Repository;
//使用注解@Repository 实现 DAO 层 Bean 组件的定义
@Repository
public class StudentDaoJdbcImpl implements StudentDao{
}
```

以上代码通过注解定义了一个名为 studentDao 的 Bean，使用@Repository 注解将 StudentDaoImpl 类标识为 Spring 框架中的 Bean，其写法相当于配置文件中<bean id="studentDao" class="com.imut.annotation.dao.StudentDaoImpl"/>的编写。在 addStudent()方法中输出打印一句话，用于验证是否成功调用了该方法。

例 7-36：使用注解@Service 定义 Bean 组件。

```
package com.imut.annotation.service;
import org.springframework.beans.factory.annotation.Autowired;
import org.springframework.beans.factory.annotation.Qualifier;
import org.springframework.stereotype.Service;
import com.imut.annotation.dao.StudentDao;
@Service
public class StudentService {}
```

例 7-37：使用注解@Controller 定义 Bean 组件。

```
package com.imut.annotation.controller;
import org.springframework.beans.factory.annotation.Autowired;
import org.springframework.stereotype.Component;
import org.springframework.stereotype.Controller;
import com.imut.annotation.service.StudentService;
@Controller
public class StudentController {
}
```

@Controller 注解的作用就相当于在 Spring 的配置中做如下配置。

```
< bean  id =" studentController "  class =" com. imut. annotation. controller.
StudentController">
    <property name="studentService" ref="studentService"></property>
</bean>
```

其中，默认的 id 是类名首字母小写，也可以指定 id：@Controller("id 值")或者@Controller(value="id 值")。

从上述例子中可以看到，@Componet 注解可以指定一个普通的 Bean，是一个泛化概念，@Repository、@Service 与@Controller 是特定的注解，可以指定特定的 Bean。特定的注解可使组件的用途更加清晰，并且 Spring 框架在以后的版本中可能会为它们添加特殊的功能，所以推荐使用特定的注解来标注特定的实现类。

7.4.2 加载注解定义的 Bean

Spring 框架的注解方式虽然简化了 XML 文件中 Bean 的配置，但仍需要在 Spring 配置文件中配置那些使用了特定注解的组件。为此，Spring 框架还提供了一种更高效的注解配置方式，即通过扫描指定的包路径下的所有 Bean 类，无须再配置烦琐的 XML 元素(如<property>)，具体配置方式如下。

```
<context: component-scan  base-package="Bean 所在的包路径"/>
```

该配置表示使用 context 命名空间，通知 Spring 扫描指定包下的所有 Bean 类，进行注解解析。其中，base-package 属性指定一个需要扫描的基类包，Spring 容器将会扫描这个基类包及其子包中的所有类，将加了注解的类管理到 IoC 容器中。当需要扫描多个包时，可以使用逗号分隔。对于扫描到的组件，Spring 有默认的命名策略。使用非限定类名，并且默认将类名的第一个字母小写，也可以在注解中通过 value 属性值标识组件的名称。

使用注解定义完 Bean 组件后，就可以使用注解的配置信息启动 Spring 框架的容器，其关键代码如例 7-38 所示。需要注意的是，配置注解前，要为工程导入 AOP 的包，同时还需要在配置文件 annotationBean.xml 中配置 context 命名空间。

例 7-38：annotationBean.xml。

```
<?xml version="1.0" encoding="UTF-8"?>
<beans xmlns="http://www.springframework.org/schema/beans"
    xmlns:xsi="http://www.w3.org/2001/XMLSchema-instance"
    xmlns:context="http://www.springframework.org/schema/context"
    xsi:schemaLocation="http://www.springframework.org/schema/beans
        http://www.springframework.org/schema/beans/spring-beans.xsd
        http://www.springframework.org/schema/context
        http://www.springframework.org/schema/context/spring-context-4.0.xsd">
    <!-- 配置组件扫描 -->
        <context:component-scan base-package="com.imut.annotation" />
...
</beans>
```

以上代码中，首先在 Spring 框架配置文件中添加对 context 命名空间的声明，然后使用 context 命名空间的 component-scan 标签扫描注解标注的类。base-package 属性指定了需要扫描的基准包(多个包名可用逗号隔开)。Spring 会扫描这些包中所有的类，以获取 Bean 的定义信息。

加载了注解定义的 Bean 之后，可以运行测试类 StudentTest，测试例 7-34、例 7-35、例 7-36、例 7-37 的执行效果。测试类 StudentTest 的代码如例 7-39 所示。

在工程 ch07_spring_01 的 src/main/java 目录下新建包 com.imut.annotation.test，在该包中创建测试类 StudentTest。

例 7-39：StudentTest.java。

```
public class StudentTest {
    @Test
    public void test() {
        ApplicationContext ctx = new
                ClassPathXmlApplicationContext("annotationBean.xml");
        Student student= ctx.getBean("student",Student.class);
        student.printStudent();
        //StudentController
        StudentController studentController= ctx.getBean("studentController",
StudentController.class);
        System.out.println(studentController);
        //StudentService
        StudentService studentService = ctx.getBean("studentService",
StudentService.class);
        System.out.println(studentService);
        //StudentDao
        StudentDao studentDao = ctx.getBean("studentDaoJdbcImpl",
StudentDaoJdbcImpl.class);
```

```
        System.out.println(studentDao);
    }
}
```

测试类 StudentTest 的执行效果如图 7-16 所示。

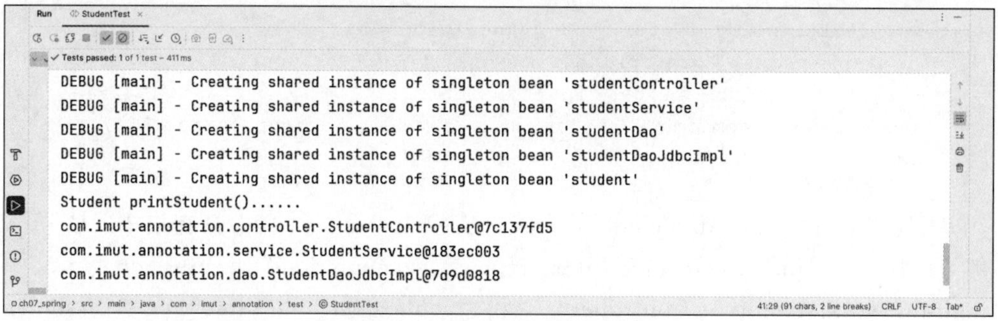
Run StudentTest
Tests passed: 1 of 1 test – 411 ms
DEBUG [main] - Creating shared instance of singleton bean 'studentController'
DEBUG [main] - Creating shared instance of singleton bean 'studentService'
DEBUG [main] - Creating shared instance of singleton bean 'studentDao'
DEBUG [main] - Creating shared instance of singleton bean 'studentDaoJdbcImpl'
DEBUG [main] - Creating shared instance of singleton bean 'student'
Student printStudent()......
com.imut.annotation.controller.StudentController@7c137fd5
com.imut.annotation.service.StudentService@183ec003
com.imut.annotation.dao.StudentDaoJdbcImpl@7d9d0818
ch07_spring › src › main › java › com › imut › annotation › test › StudentTest
41:29 (91 chars, 2 line breaks) CRLF UTF-8 Tab*

图 7-16　加载注解定义的类

如果仅希望扫描特定的类而非基包下的所有类,可使用 resource-pattern 属性过滤特定的类,核心代码如例 7-40 所示。

例 7-40:resource-pattern 属性。

```
<!-- 配置组件扫描  只扫描以 dent 结尾的类 -->
<context:component-scan  base-package="com.imut.annotation"
Resource-pattern=" **/ * dent.class"   />
```

<context:component-scan>还可以包含子节点。其中,<context:include-filter>子节点表示要包含的目标类。<context:exclude-filter>子节点表示要排除在外的目标类。<context:component-scan>下可以拥有若干个<context:include-filter>和<context:exclude-filter>子节点。<context:include-filter>和<context:exclude-filter>子节点支持多种类型的过滤表达式,如表 7-4 所示。

表 7-4　<context:include-filter>和<context:exclude-filter>子节点的过滤表达式

类别	示　例	说　明
annotation	com.imut.XxxAnnotation	所有标注了 XxxAnnotation 的类。该类型采用目标类是否标注了某个注解进行过滤
assinable	com.imut.XxxService	所有继承或扩展 XxxService 的类。该类型采用目标类是否继承或扩展某个特定类进行过滤
aspectj	com.imut.. * Service+	所有类名以 Service 结束的类及继承或扩展它们的类。该类型采用 AspectJ 表达式进行过滤
regex	com.\imut\.anno\.. *	所有 com.imut.anno 包下的类。该类型采用正则表达式,根据类的类名进行过滤
custom	com.imut.XxxTypeFilter	采用 XxxTypeFilter 通过代码的方式定义过滤规则。该类必须实现 prg.springframework.core.type.TypeFilter 接口

使用<context:include-filter> 或者<context:exclude-filter>时,需要将<context:

component-scan>的 use-default-filters 属性设置为 false，表示不使用默认的过滤方式，而是通过 annotation 或是 assignable 执行要具体扫描或是不扫描的类。核心代码如下所示。

例 7-41：annotationBean.xml。

```
<context:component-scan base-package="com.imut.annotation" use-default-
filters="false">
    <context:include-filter type="annotation"
        expression = "org.springframework.stereotype.Controller" />
<!--<context:include-filter type="assignable"
    expression = "com.imut.annotation.controller.StudentController"/> -->
</context:component-scan>
```

以上代码通过<context:include-filter>过滤@Controller 注解，只有扫描@Controller 注解的类。<context:include-filter type = " annotation " expression = " org.springframework.stereotype.Controller"/>和<context:include-filter type="assignable" expression = "com.imut.annotation.controller.StudentController"/>都可以实现过滤@Controller 注解的目的。annotation 采用指定注解的方式，而 assignable 采用指定具体类的方式。

7.4.3 使用注解完成 Bean 组件装配

<context:component-scan>元素还会自动注册 AutowiredAnnotationBeanPostProcessor 实例，该实例可以自动装配具有@Autowired、@Resource 和@Inject 注解的属性。其中@Autowired 和@Resource 都是用于对 Bean 的属性值进行装配的。@Autowired 默认按照 Bean 类型装配，而@Resource 默认按照 Bean 实例名称进行装配。

使用@Autowired 进行组件的装配时，首先会使用 byType 的方式匹配兼容类型的 Bean。如果能确定唯一匹配的 Bean，则装配成功。否则就尝试使用 byName 的方式进行匹配，如果能唯一确定一个 bean，则装配成功。如果不能唯一确定一个 Bean，则装配失败，抛出异常。使用@Autowired 标注的属性，默认情况下必须被装配，但可以使用 required=false 来设置为非必需。@Autowired 注解可标注在定义的成员变量上，也可标注在对应的 set 方法上。还可以使用@Qualifier("指定具体要装配的 Bean 的 id 值")进一步指定具体要装配的 Bean，特别是当 IoC 容器里存在多个类型兼容的 Bean 时，通过类型的自动装配将无法工作。此时可以在@Qualifier 注解里提供 Bean 的名称。此外，Spring 还允许对方法的参数标注@Qualifiter 以指定要注入 Bean 的名称。

修改例 7-35 的 StudentDao 类，为 StudentDao 添加方法 addStudent()，在 addStudent()方法中输出打印一句话，用于验证是否成功调用了该方法。代码如例 7-42 所示。

例 7-42：使用注解@Repository 定义 Bean 组件。

```
//StudentDao.java
package com.imut.annotation.dao;
public interface StudentDao {
    public void addStudent();
}
//StudentDaoJdbcImpl.java
```

```
package com.imut.annotation.dao;
import org.springframework.stereotype.Repository;
//使用注解@Repository实现 DAO层 Bean组件的定义
@Repository
public class StudentDaoJdbcImpl implements StudentDao{
    @Override
    public void addStudent() {
        System.out.println("StudentDaoJdbcImpl addStudent()...");
    }
}
```

修改例 7-36 中的 StudentService 类，添加成员变量 studentDao，并采用@Autowired 自动装配。创建 toAddStudent()方法，在其中调用 studentDao.addStudent()，用于验证是否成功装配了 studentDao，代码如例 7-43 所示。

例 7-43：使用@Autowired 自动装配@service 定义的 Bean 组件的成员变量。

```
@Service
public class StudentService {
    @Autowired
    //也可以用@Qualifier指定具体要装配的 Bean 的 id值，进一步指定要装配的 Bean
    //@Qualifier("studentDao")
    private StudentDao studentDao;
    public void  toAddStudent() {
        System.out.println("StudentService AddStudent()...");
        studentDao.addStudent();
    }
}
```

同样地，修改例 7-37 中的 StudentController 类，添加成员变量 studentService，并采用@Autowired 自动装配。在 regist()方法中调用 studentService.toAddStudent()，用于验证是否成功装配了 studentService。代码如例 7-44 所示。

例 7-44：使用@Autowired 自动装配@Controller 定义 Bean 组件的成员变量。

```
@Controller
public class StudentController {
    @Autowired
    private StudentService studentService ;
    public void regist() {
        System.out.println("StudentController regist()...");
        studentService.toAddStudent();
    }
}
```

修改 StudentTest 类，通过调用 studentController.regist()方法验证 studentDao 和 studentService 的自动装配是否成功。代码如例 7-45 所示。

例 7-45：StudentTest.java。

```
public class StudentTest {
    @Test
    public void test() {
        ApplicationContext ctx = new
                ClassPathXmlApplicationContext("annotationBean.xml");
        //StudentController
        StudentController studentController= ctx.getBean("studentController",
StudentController.class);
        studentController.regist();
    }
}
```

测试类 studentTest 的执行结果如图 7-17 所示，可以看到由 Controller 层调用 Service 层，再调用 DAO 层的结果，说明 Controller 类和 Service 类的成员变量由@Autowired 自动装配成功。

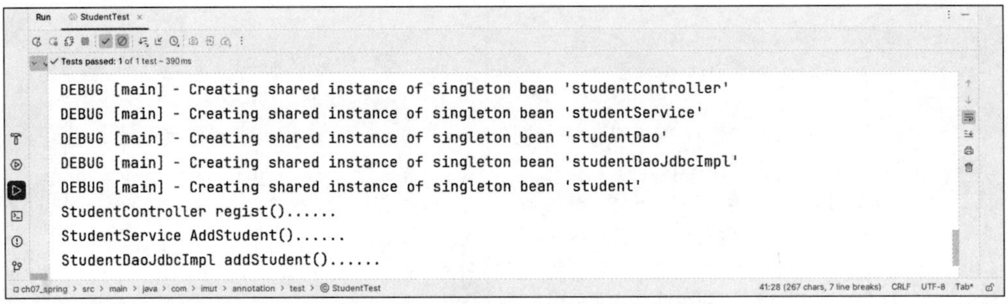

```
Run    StudentTest
Tests passed: 1 of 1 test – 390 ms
DEBUG [main] - Creating shared instance of singleton bean 'studentController'
DEBUG [main] - Creating shared instance of singleton bean 'studentService'
DEBUG [main] - Creating shared instance of singleton bean 'studentDao'
DEBUG [main] - Creating shared instance of singleton bean 'studentDaoJdbcImpl'
DEBUG [main] - Creating shared instance of singleton bean 'student'
StudentController regist()......
StudentService AddStudent()......
StudentDaoJdbcImpl addStudent()......
ch07_spring > src > main > java > com > imut > annotation > test > StudentTest
41:28 (267 chars, 7 line breaks)  CRLF  UTF-8  Tab*
```

图 7-17　@Autowired 自动装配的执行结果

@Autowired 注解自动装配具有兼容类型的单个 Bean 属性。类的构造器、普通字段（即使是非 public）、一切具有参数的方法都可以应用@Authwired 注解。

@Authwired 注解也可以应用在数组类型的属性上，此时 Spring 将会把所有匹配的 Bean 进行自动装配。@Authwired 注解也可以应用在集合属性上，此时 Spring 读取该集合的类型信息，然后自动装配所有与之兼容的 Bean。

@Authwired 注解用在 java.util.Map 上时，若该 Map 的键值为 String，Spring 将自动装配与之 Map 值类型兼容的 Bean，此时 Bean 的名称作为键值。

Spring 还提供了@Resource 和@Inject 注解。这两个注解与@Autowired 注解的功用类似。@Resource 注解要求提供一个用于指定 Bean 名称的属性，若该属性为空，则自动采用标注处的变量或方法名作为 Bean 的名称。@Inject 和@Autowired 注解一样也是按类型匹配注入的 Bean，但@Inject 没有 required 属性。通常来说，推荐使用@Autowired 注解。

7.4.4　自动装配

Spring 框架的自动装配功能不仅可以通过注解实现依赖注入时使用自动装配，还可以基于 XML 的配置使用自动装配简化配置。

Spring 的<bean>元素中包含一个 autowire 属性，可以通过设置 autowire 的属性值来自动装配 Bean。所谓自动装配，就是将一个 Bean 自动注入其他 Bean 的 Property 中。

autowire 属性有 5 个值，其值及说明如表 7-5 所示。

表 7-5 <bean>元素的 autowire 属性值及说明

属性值	说明
default(默认值)	由<bean>上级标签<beans>的 default-autowire 属性值确定。如<beans default-autowire="byName">,则该<bean>元素中的 autowire 属性对应的属性值就为 byName
byName	根据属性名自动装配。BeanFactory 查找容器中的全部 Bean,找出 id 与属性的 setter 方法匹配的 Bean。找到即自动注入,否则什么都不做
byType	根据属性类型自动装配。BeanFactory 查找容器中的全部 Bean,如果正好有一个与依赖属性类型相同的 Bean,就自动装配这个属性;如果有多个这样的 Bean,Spring 无法决定注入哪个 Bean,则抛出一个致命异常;如果没有匹配的 Bean,则什么都不会发生,即属性不会被设置
constructor	与 byType 的方式类似,不同之处在于它应用于构造器参数。如果在容器中没有找到与构造器参数类型一致的 Bean,则抛出异常
no	在默认情况下,不使用自动装配,Bean 依赖就必须通过 ref 元素定义

对于 byType 属性值(根据类型自动装配),若 IoC 容器中有多个与目标 Bean 类型一致的 Bean,Spring 将无法判定哪个 bean 最合适该属性,所以不能执行自动装配。对于 byName 属性值(根据名称自动装配),必须将目标 bean 的名称和属性名设置的完全相同。autowire 属性要么根据类型自动装配,要么根据名称自动装配,不能两者兼而有之。

对于 constructor 属性值(通过构造器自动装配),当 Bean 中存在多个构造器时,此种自动装配方式将会很复杂,不推荐使用。

在 Bean 配置文件里设置 autowire 属性进行自动装配将会装配 Bean 的所有属性。然而,若只希望装配个别属性时,autowire 属性就不够灵活了。一般情况下,在实际的项目中很少使用自动装配功能,因为和自动装配功能所带来的好处比起来,明确清晰的配置文档更有说服力一些。基于 XML 方式配置 Bean,一般不建议使用自动装配。基于注解方式配置 Bean,推荐使用自动装配。

基于 XML 配置的自动装配如例 7-46 所示。

例 7-46:annotationBean.xml。

```
<bean id="studentDao" class="com.imut.annotation.dao.StudentDaoJdbcImpl" />
<bean id="studentService" class="com.imut.annotation.service.StudentService"
    autowire="byName" />
```

在上述配置文件中,用于配置 studentService 的<bean>元素中除了 id 和 class 属性,还增加了 autowire 属性,并将其属性值设置为 byName。在默认情况下,配置文件需要通过 ref 来装配 Bean,但设置了 autowire="byName"后,Spring 会自动寻找 studentService Bean 的属性,并将其属性名称与配置文件中定义的 Bean 做匹配。由于 StudentService 中定义了 studentDao 属性及其 setter 方法,这与配置文件中 id 为 studentDao 的 Bean 相匹配,所以 Spring 会自动地将 id 为 userDao 的 Bean 装配到 id 为 studentService 的 Bean 中。

执行测试类 StudentTest 类,用 System.out.println(studentService)输出 student Service 对象的信息,控制台的输出结果与预期相同,使用自动装配同样完成了依赖注入,结果如图 7-18 所示。

```
DEBUG [main] - Creating shared instance of singleton bean 'studentController'
DEBUG [main] - Creating shared instance of singleton bean 'studentService'
DEBUG [main] - Creating shared instance of singleton bean 'studentDao'
DEBUG [main] - Creating shared instance of singleton bean 'studentDaoJdbcImpl'
DEBUG [main] - Creating shared instance of singleton bean 'student'
com.imut.annotation.service.StudentService@7c137fd5
```

图 7-18　基于 XML 的自动装配

7.5　Bean 与 Bean 之间的关系

Bean 与 Bean 之间的关系有继承关系和依赖关系。本节介绍 Spring 的 IoC 容器中如何配置 Bean 的继承关系和依赖关系。

7.5.1　Bean 与 Bean 之间的继承关系

Spring 允许 Bean 配置继承，被继承的 Bean 为父 Bean，继承的 Bean 为子 Bean。子 Bean 可继承父 Bean 的配置，包括属性配置，并可覆盖继承的配置。Spring 支持在<bean>中使用 parent 指定父 Bean，实现 Bean 之间的继承关系。父 Bean 可作为配置模板或 Bean 实例。若仅作为模板，可将<bean>的 abstract 属性设为 true，这样 Spring 不会实例化这个抽象 Bean，且可省略 class 配置。需要注意的是，不是所有<bean>属性都会被继承，如 autowire 和 abstract 属性。也可以忽略父 Bean 的 class 属性，让子 Bean 指定自己的类，共享相同的属性配置，但此时 abstract 必须设为 true。

在 Java 工程项目 ch07_spring_01 中新建包 com.imut.relation.pojo，包 com.imut.relation.pojo 下新建 Course1 类(与例 7-10 相同)和 Teacher 类(与例 7-17 相同)，此处不再赘述。在 Spring 容器中配置 Teacher 的 Bean 的继承关系，核心代码如例 7-47 所示。

例 7-47：relationBean.xml。

```
<!--  父 Bean -->
<bean id="teacher1" class="com.imut.relation.pojo.Teacher" >
    <property name="tid" value="1"></property>
    <property name="tname" value="Tom"></property>
    <property name="tcourse" ref="course1"></property>
</bean>
<!--  子 Bean -->
<bean id="teacher2" parent="teacher1">
    <property name="tid" value="2"></property>
    <property name="tname" value="Jean"></property>
</bean>
<bean id="course1" class="com.imut.relation.pojo.Course1" >
    <property name="cid" value="2"></property>
```

```
    <property name="cname" value="Java 程序设计"></property>
    <property name="cno" value="001"></property>
</bean>
```

在以上代码中，子 Bean teacher2 继承了父 Bean teacher1。teacher2 从 teacher1 中继承配置，包括 Bean 的属性配置 tcourse，但是覆盖了从 teacher1 继承过来的配置 tid 和 tname。

在 src/main/java 目录的包 com.imut.relation.test 下新建 TeacherTest 测试类，如例 7-48 所示。

例 7-48：TeacherTest.java。

```
public class TeacherTest {
    @Test
    public void test() {
        ApplicationContext context = new
            ClassPathXmlApplicationContext("relationBean.xml");
        Teacher teacher1 = (Teacher) context.getBean("teacher1");
        System.out.println(teacher1);
        Teacher teacher2 = (Teacher) context.getBean("teacher2");
        System.out.println(teacher2);
    }
}
```

执行测试类 TeacherTest，得到的结果如图 7-19 所示。从执行结果可以看到，teacher2 继承了 teacher 的 tcourse 属性，覆盖了 tid 和 tname 属性。

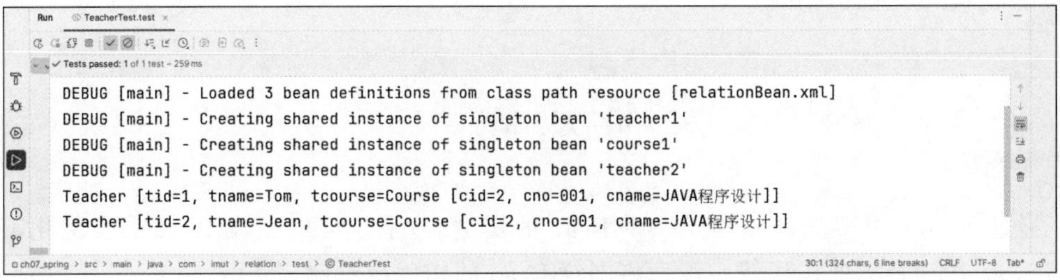

```
Run    TeacherTest.test
✓ Tests passed: 1 of 1 test – 259 ms
DEBUG [main] - Loaded 3 bean definitions from class path resource [relationBean.xml]
DEBUG [main] - Creating shared instance of singleton bean 'teacher1'
DEBUG [main] - Creating shared instance of singleton bean 'course1'
DEBUG [main] - Creating shared instance of singleton bean 'teacher2'
Teacher [tid=1, tname=Tom, tcourse=Course [cid=2, cno=001, cname=JAVA程序设计]]
Teacher [tid=2, tname=Jean, tcourse=Course [cid=2, cno=001, cname=JAVA程序设计]]
ch07_spring > src > main > java > com > imut > relation > test > TeacherTest
```

图 7-19　测试类 TeacherTest 的执行结果

在面向对象的编程原则中，当多个类拥有相同的方法和属性，则可以引入父类消除代码重复。类似地，在 Spring 容器中，如果多个 Bean 存在相同的配置信息，同样也可以定义一个父 Bean，子 Bean 将自动继承父 Bean 的配置信息，从而减少配置的重复工作。

7.5.2　Bean 与 Bean 之间的依赖关系

Spring 允许用户使用 depends-on 属性设定 Bean 前置依赖。这些前置依赖的 Bean 会在当前 Bean 实例化之前创建好。如果前置依赖于多个 Bean，则可以通过逗号、空格的方式配置 Bean 的名称。Bean 的依赖关系不会进行赋值操作。Bean 的依赖关系配置核心代码如例 7-49 所示。

例 7-49：relationBean.xml。

```
<!-- Bean 的依赖关系 -->
```

```
<bean id="teacher3" class="com.imut.relation.pojo.Teacher" depends-on=
"course1">
    <property name="tid" value="3"></property>
    <property name="tname" value="Lily"></property>
    <!--<property name="tcourse" ref="course1"></property>-->
</bean>
```

执行测试类 teacherTest 的结果如图 7-20 所示。从执行结果可以看出，teacher3 依赖 Bean 对象 course1，但是 course1 的值并没有注入 teacher3 的属性 tcourse 中。

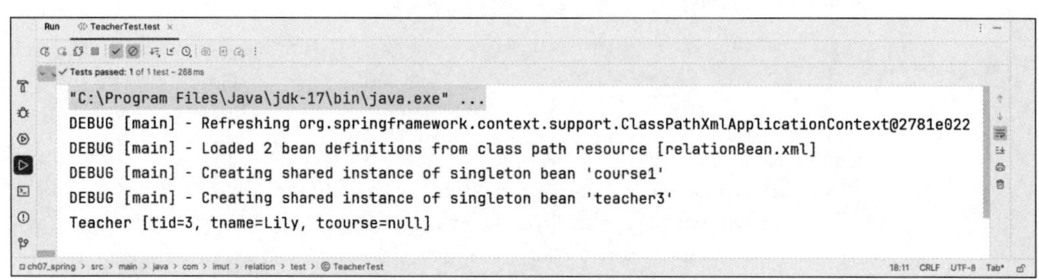

图 7-20　teacherTest 测试类的执行结果

由 depends-on 定义的依赖关系和 ref 的引用关系有所不同。ref 标签用于表示 Bean 对其他 Bean 的引用，而 depends-on 属性只是表明依赖关系（不一定会引用）。当使用 depends-on 属性时，这个依赖关系决定了被依赖的 Bean 必定会在依赖 Bean 之前被实例化，这意味着依赖 Bean 在容器启动时会首先实例化。反过来，容器关闭时，依赖 Bean 会在被依赖的 Bean 之前被销毁。

7.6　Bean 的作用域

通过 Spring 框架的容器创建一个 Bean 的实例时，不仅可以完成 Bean 的实例化，还可以为 Bean 指定特定的作用域。下面将围绕 Bean 的作用域讲解。

7.6.1　作用域的种类

在 Spring 框架中定义 Bean，除了可以创建 Bean 实例，并对 Bean 的属性进行注入，还可以为所定义的 Bean 指定一个作用域。这个作用域的取值决定了 Spring 框架创建该组件实例的策略，进而影响程序的运行效率和数据安全。

在 Spring 中，可以在<bean>元素的 scope 属性里设置 Bean 的作用域。Spring 框架中为 Bean 的实例定义了 7 种作用域，如表 7-6 所示。

表 7-6　Bean 的作用域

作用域名称	说　明
singleton(单例)	默认值。使用 singleton 定义的 Bean 在 Spring 框架的容器中只有一个实例，也就是说，无论有多少个 Bean 引用它，始终将指向同一个对象，这也是 Spring 容器默认的作用域

续表

作用域名称	说　明
prototype(原型)	每次通过 Spring 容器获取的 prototype 定义的 Bean 时,容器都将创建一个新的 Bean 实例
request	在一次 HTTP 请求中,容器会返回该 Bean 的同一个实例。对不同的 HTTP 请求,则会产生一个新的 Bean,而且该 Bean 仅在当前 HTTP Request 内有效
session	在一次 HTTP Session 中,容器会返回该 Bean 的同一个实例。对不同的 HTTP 请求,则会产生一个新的 Bean,而且该 Bean 仅在当前 HTTP Session 内有效
globalSession	在一个全局的 HTTP Session 中,容器会返回该 Bean 的同一个实例。仅在使用 portlet 上下文时有效
application	为每个 ServletContext 对象创建一个实例。仅在 Web 相关的 ApplicationContext 中生效
websocket	为每个 websocket 对象创建一个实例。仅在 Web 相关的 ApplicationContext 中生效

Spring 中 Bean 的作用域有 7 种：singleton(单例)、prototype(原型)、request、session、globalSession、application、websocket。其中,singleton 和 prototype 是最常用的两种。默认情况下,Spring 为每个在 IoC 容器里声明的 Bean 创建唯一一个实例,这是 singleton(单例)作用域。而 prototype(原型)作用域则会在每次 getBean()调用时创建新的 Bean 实例。request、session、globalSession、application 和 websocket 作用域是扩展的,仅在 Web 应用中使用。request 表示一次请求与响应期间,session 表示一次会话期间,globalSession 是全局作用域,与 portlet 应用相关,而 application 作用域与 Servlet 中的 session 作用域效果相同。

简单来说,Spring 中 Bean 的默认作用域是单例,但也可以通过配置实现其他作用域,如原型、请求、会话等。这些作用域的选择取决于 Bean 的使用场景和需求。下面对这两种基本作用域进行详细讲解。

7.6.2 singleon 作用域

Singleton 是 Spring 默认的作用域。在此作用域下,Spring 容器仅存在一个共享的 Bean 实例。只要请求的 Bean 的 id 与该 Bean 的 id 属性匹配,就会返回同一实例。对于无会话状态的 Bean,如 DAO 层和 Service 组件,Singleton 是最理想的选择。它适用于无须考虑线程安全的组件,因为它可以显著减少对象创建的开销,提高运行效率。

如要将作用域定义成 singleton,只需将 scope 的属性值设置为 singleton 即可,其代码如下。

```
<bean id="course" class="com.imut.scope.pojo.Course" scope="singleton"/>
```

在 Java 工程项目 ch07_spring_01 的 src/main/java 目录中创建一个 com.imut.scope.pojo 包。在包中创建 Course 类,为该类添加无参构造方法。然后在 conf 目录中创建一个配置文件 scope.xml,将上述代码写入配置文件中,最后在 com.imut.scope.test 包中创建测试类 ScopeTest,以测试 singleton 作用域。Course 类的核心代码如例 7-50 所示,配置文件 scope.xml 的核心代码如例 7-51 所示,测试类 ScopeTest 的核心代码如例 7-52 所示。

例 7-50：Course.java。

```
package com.imut.scope.pojo;
public class Course {
    private int cid;
    private String cno;
    private String cname;
    public Course() {
        System.out.println("Address no param Constructor ...");
    }
    //此处省略 getter()、setter()和 toString()方法
}
```

以上代码定义了 Course 类，添加了无参构造方法，在方法中输出打印一句话，用于验证是否成功调用了该方法。

例 7-51：scope.xml。

```
<!-- Bean 的作用域:singleton 单例的  -->
    <bean id="course" class="com.imut.scope.pojo.Course" scope="singleton" >
        <property name="cid" value="1"></property>
        <property name="cno" value="350225001"></property>
        <property name="cname" value="软件工程"></property>
    </bean>
```

在配置文件中创建 Bean 对象，id 为 course，作用域为 singleton。用 setter 方法为其属性注入值。在测试类中，分别用 getBean()方法来获取 id 为 course 的 Bean 实例两次，验证两次获得的是同一个 Bean，即 Spring 只为在 IoC 容器里声明的 Bean 创建唯一一个实例，所有后续的 getBean()调用和 Bean 引用都将返回这个唯一的 Bean 实例。

例 7-52：ScopeTest.java。

```
public class ScopeTest {
    @Test
    public void test() {
        ApplicationContext context = new
            ClassPathXmlApplicationContext("scope.xml");
        Course course1 = context.getBean("course", Course.class);
        System.out.println(course1);
        Course course2 = context.getBean("course", Course.class);
        System.out.println(course2);
    }
}
```

测试类 ScopeTest 的执行结果如图 7-21 所示。从控制台的输出结果可以看到，Course 类的构造方法只执行了一次，使用 getBean()方法两次获取的 Bean 实例是同一个。说明 Spring 容器使用了 singleton 作用域，只创建了一个 Course 类的实例。如果不设置 scope="singleton"，其输出结果也是一个实例，因为 Spring 框架的容器默认的作用域就是 singleton。

```
DEBUG [main] - Creating shared instance of singleton bean 'course'
Address no param Constructor ....
Course [cid=1, cno=350225001, cname=软件工程]
Course [cid=1, cno=350225001, cname=软件工程]

Process finished with exit code 0
```

图 7-21　测试类在 singleton 作用域的执行结果

7.6.3　prototype 作用域

对于存在线程安全问题的组件，不能使用 singleton 作用域。还有那些需要保持会话状态的 Bean(如 Struts 2 的 Action 类)应该使用 prototype 作用域。在使用 prototype 作用域时，Spring 容器会为每个对该 Bean 的请求者都创建一个新的实例。使用 prototype 作用域，可以通过 scope 属性将 Bean 设置为 prototype 作用域，其关键代码如下。

```
<bean id="course" class="com.imut.scope.pojo.Course" scope="prototype"/>
```

这样，Spring 框架每次获取该组件时都会创建一个新的实例，可避免因为共用同一个实例而产生的线程安全问题。将例 7-51 中配置文件中的<bean>标签的 scope 属性更改为 prototype，如上述代码所示后，再次运行测试类 ScopeTest，控制台的输出结果如图 7-22 所示。

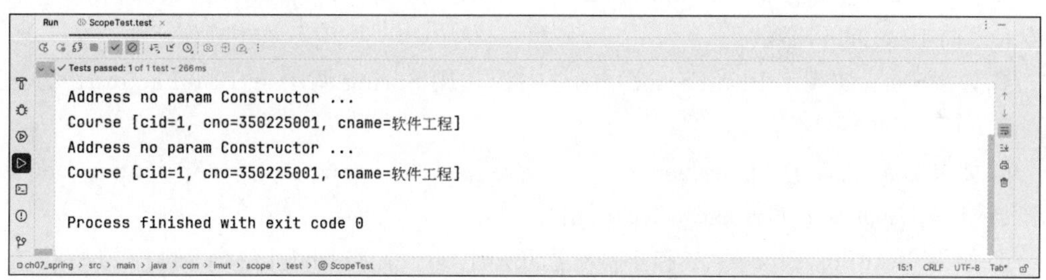

图 7-22　测试类在 prototype 作用域的执行结果

可以看到，虽然 Bean 实例的属性值相同，但是输出了两次构造方法，这说明在 prototype 作用域下创建了两个 Course 实例，两次输出的 Bean 实例并不相同。

7.7　Bean 的生命周期

Spring 容器可管理 singleton 作用域 Bean 的生命周期，包括创建、初始化和销毁。对于 prototype 作用域的 Bean，Spring 仅负责创建，生命周期由客户端代码管理。每次客户端请求 prototype 作用域的 Bean 时，Spring 都会创建新实例，不跟踪其生命周期。了解 Bean 的生命周期意义，可以在特定时刻执行相关操作，如 postinitiation 和 predestruction。Spring 框架提供了方法以编制 Bean 的创建过程，Bean 加载到 Spring 容器时即具有生命周期。

而 Spring 容器在保证一个 Bean 能够使用之前，会做很多工作。在 Spring 容器中，Bean 的生命周期流程如图 7-23 所示。

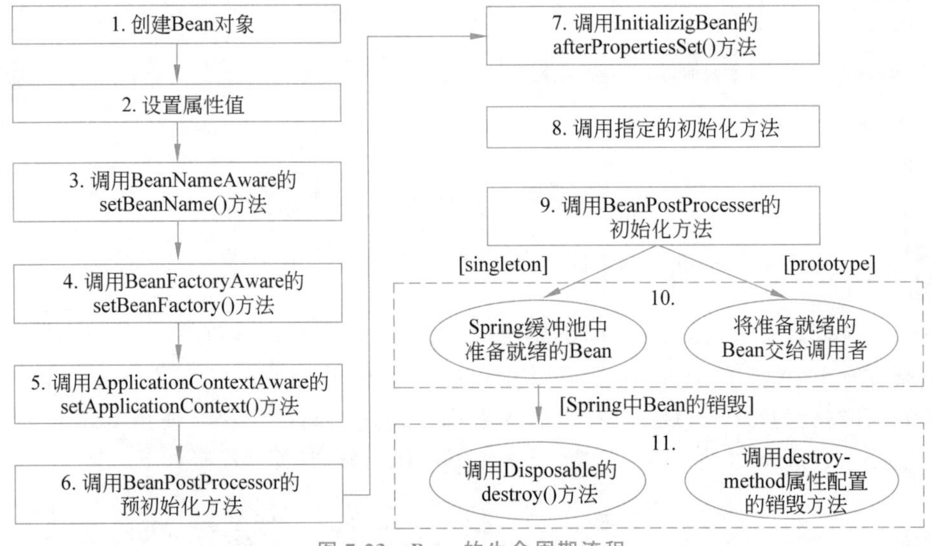

图 7-23　Bean 的生命周期流程

Bean 生命周期的整个执行过程描述如下。

① 根据配置情况调用 Bean 构造方法或工厂方法实例化 Bean。

② 利用依赖注入完成 Bean 中所有属性值的配置注入。

③ 如果 Bean 实现了 BeanNameAware 接口，则 Spring 调用 Bean 的 setBeanName()方法传入当前 Bean 的 id 值。

④ 如果 Bean 实现了 BeanFactoryAware 接口，则 Spring 调用 setBeanFactory()方法传入当前工厂实例的引用。

⑤ 如果 Bean 实现了 ApplicationContextAware 接口，则 Spring 调用 setApplicationContext()方法传入当前 ApplicationContext 实例的引用。

⑥ 如果 BeanPostProcessor 和 Bean 关联，则 Spring 将调用该接口的预初始化方法 postProcessBeforeInitialzation()对 Bean 进行加工操作，这个非常重要，Spring 的 AOP 就是用它实现的。

⑦ 如果 Bean 实现了 InitializingBean 接口，则 Spring 将调用 afterPropertiesSet()方法。

⑧ 如果在配置文件中通过 init-method 属性指定了初始化方法，则调用该初始化方法。

⑨ 如果有 BeanPostProcessor 和 Bean 关联，则 Spring 将调用该接口的初始化方法 postProcessAfterInitialization()。此时，Bean 已经可以被应用系统使用了。

⑩ 如果在<bean>中指定了该 Bean 的作用范围为 scope="singleton"，则将该 Bean 放入 Spring IoC 的缓存池中，将触发 Spring 对该 Bean 的生命周期管理；如果在<bean>中指定了该 Bean 的作用范围为 scope="prototype"，则将该 Bean 交给调用者。由调用者管理该 Bean 的生命周期，而 Spring 不再管理该 Bean。

⑪ 如果 Bean 实现了 DisposableBean 接口，则 Spring 会调用 destroy()方法，将 Spring

中的 Bean 销毁；如果在配置文件中通过 destroy-method 属性指定了 Bean 的销毁方法，则 Spring 将调用该方法进行销毁。

7.7.1 IoC 容器中 Bean 的生命周期方法

Spring 框架为 Bean 提供了细致全面的生命周期过程，通过实现特定的接口或通过<bean>的属性设置，都可以对 Bean 的生命周期过程产生影响。虽然可以自由配置<bean>的属性，但建议不要过多地使用 Bean 实现接口，以减少代码和 Spring 的紧耦合性。

下面通过简单的例子演示 Bean 的生命周期。

(1) 在 Java 工程项目 ch07_spring_01 的 src/main/java 目录中创建一个 com.imut.lifecycle.pojo 包。在该包中创建 Course 类，为该类添加无参构造方法。生命周期的第一个阶段是创建 Bean 对象，所以在构造方法中输出打印一句话，用于提示生命周期的第一阶段完成。

Bean 的生命周期的第二个阶段是给对象的属性赋值，在 Course 类的 setCname()方法中输出打印一句话，用于提示生命周期的第二阶段完成。

Bean 的生命周期的第三阶段是调用 Bean 的初始化方法，在 Course 类中添加 init()方法，并输出打印一句话，用于提示生命周期的第三阶段完成。

Bean 的生命周期的第四个阶段是使用 Bean 实例，例子在测试类中完成第四个阶段。

Course 类中的 destroy()方法是 Bean 生命周期的第五个阶段，即当容器关闭时，调用 Bean 的销毁方法。在 destroy()中输出打印一句话，用于提示生命周期的第五阶段完成。Course 类的定义如例 7-53 所示。

例 7-53：Course.java。

```
package com.imut.lifecycle.pojo;
public class Course {
    private int cid;
    private String cno;
    private String cname;
    /**
     * 生命周期  第一个阶段:调用构造器 创建对象
     */
    public Course() {
        System.out.println("1. 创建 Bean 对象");
    }
    //此处省略 getter()、setter()和 toString()方法
    /**
     * 生命周期  第二个阶段: 给对象的属性赋值
     */
    public void setCname(String cname) {
        System.out.println("2. 给对象的属性注入值或者是引用其他的 Bean");
        this.cname = cname;
    }
    /**
     * 生命周期 第三个阶段: 初始化阶段
     */
```

```
    public void init() {
        System.out.println("3. 初始化");
    }
    /**
     * 生命周期  第五个阶段：销毁
     */
    public void destroy() {
        System.out.println("5. 销毁");
    }
    @Override
    public String toString() {
        return "Course [cid=" + cid + ", cno=" + cno + ", cname=" + cname + "]";
    }
}
```

(2) 在 resources 目录中创建一个配置文件 lifeCycle.xml。在<bean>元素中设置 init-method 属性值为 init，即 Course 类中的 init()方法；设置 destroy-method 属性值为 destroy，即 Course 类中的 destroy()方法。核心代码如例 7-54 所示。

例 7-54：lifeCycle.xml。

```
<!-- Course -->
    <bean id="course" class="com.imut.lifecycle.pojo.Course"
        init-method="init"  destroy-method="destroy">
        <property name="cid" value="1"></property>
        <property name="cno" value="350225001"></property>
        <property name="cname" value="软件工程"></property>
    </bean>
```

(3) 在 com.imut.lifecycle.test 包中创建测试类 lifeCycleTest，以验证 Bean 的生命周期的几个主要阶段。调用了 getBean()方法创建 Bean 实例之后，输出打印一句话，表示使用了 Bean 对象，用于提示生命周期的第四阶段完成。为了能够执行 Course 类的 destroy()方法，程序结束前要关闭 Spring IoC 容器。这时要调用 ClassPathXmlApplicationContext 类的 close()方法。因此在创建容器时，容器类型要定义为 ClassPathXmlApplicationContext。核心代码如例 7-55 所示。

例 7-55：lifeCycleTest.java。

```
public class lifeCycleTest {
    @Test
    public void test() {
        ClassPathXmlApplicationContext context =
                new ClassPathXmlApplicationContext("lifeCycle.xml");
        Course course = context.getBean("course",Course.class);
        System.out.println("4. 使用" +course);
        context.close();
    }
}
```

执行测试类 lifeCycleTest，可以看到在控制台依次输出了 Bean 生命周期的主要五个阶段，如图 7-24 所示。

```
DEBUG [main] - Loaded 1 bean definitions from class path resource [lifeCycle.xml]
DEBUG [main] - Creating shared instance of singleton bean 'course'
1. 创建bean对象
2. 给对象的属性注入值或者是引用其他的bean
3. 初始化
4. 使用Course [cid=1, cno=350225001, cname=软件工程]
DEBUG [main] - Closing org.springframework.context.support.ClassPathXmlApplicationContext@2781e022,
 started on Sat Oct 07 10:25:49 CST 2023
5. 销毁
```

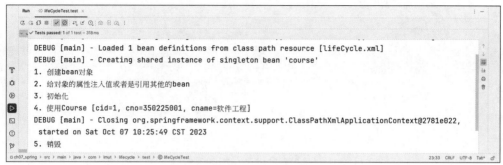

图 7-24 Bean 生命周期

7.7.2 添加 Bean 后置处理器后 Bean 的生命周期

Bean 后置处理器是 Spring 框架的一种机制，用于在 Bean 初始化前后进行额外处理。它处理 IoC 容器中的所有 Bean 实例，而非单一实例。常用于检查 Bean 属性正确性或根据特定标准更改 Bean 属性。为实现 Bean 后置处理器，需实现相应接口。在初始化方法调用前后，Spring 将 Bean 实例传递给上述接口的 postProcessAfterInitialization() 和 postProcessBeforeInitialization()方法，具体如图 7-25 所示。

Object	**postProcessAfterInitialization**(Object bean, String beanName) Apply this BeanPostProcessor to the given new bean instance *after* any afterPropertiesSet or a custom init-method).
Object	**postProcessBeforeInitialization**(Object bean, String beanName) Apply this BeanPostProcessor to the given new bean instance *before* an afterPropertiesSet or a custom init-method).

图 7-25 Bean 后置处理器

添加了 Bean 后置处理器后，Spring IoC 容器对 Bean 的生命周期进行管理的过程发生了变化，调用 Bean 的初始化方法前后会分别调用 postProcessBeforeInitialization()方法和 postProcessAfterInitialization()方法。

下面通过简单的例子演示添加 Bean 后置处理器之后 Bean 的生命周期。

(1) 在 com.imut.lifecycle.pojo 包中创建 MyBeanPostPorcessor 类，实现 BeanPostProcessor 接口，并重写 postProcessBeforeInitialization(Object bean, String beanName)方法和 postProcessAfterInitialization(Object bean, String beanName)方法，分别在两个方法中输出打印一句话，用于验证是否成功调用了该方法。在这两个方法的参数中，Object 类型的 bean 是当前的 bean 对象，String 类型的 beanName 是当前 bean 对象的 id 值。postProcessBeforeInitialization 方法会在初始化方法之前执行，而 postProcessAfterInitialization 会在初始化方法之后执行。MyBeanPostPorcessor 类的核心代码如例 7-56 所示。

例 7-56：MyBeanPostPorcessor.java。

```
public class MyBeanPostPorcessor implements BeanPostProcessor {
    @Override
    public Object postProcessBeforeInitialization(Object bean, String beanName) throws
```

```
            BeansException {
        System.out.println("postProcessBeforeInitialization:" + bean );
        return bean;
    }
    @Override
    public Object postProcessAfterInitialization(Object bean, String beanName)
throws BeansException {
        System.out.println("postProcessAfterInitialization:" + bean );
        return bean;
    }
}
```

(2) 修改在 resources 目录中的配置文件 lifeCycle.xml。为了配置 Bean 的后置处理，在配置文件 lifeCycle.xml 中添加如下代码，Spring 会自动识别该 Bean 是一个后置处理器类。

```
<bean class="com.imut.lifecycle.pojo.MyBeanPostPorcessor"></bean>
```

执行测试类 lifeCycleTest，得到图 7-26 所示的结果。

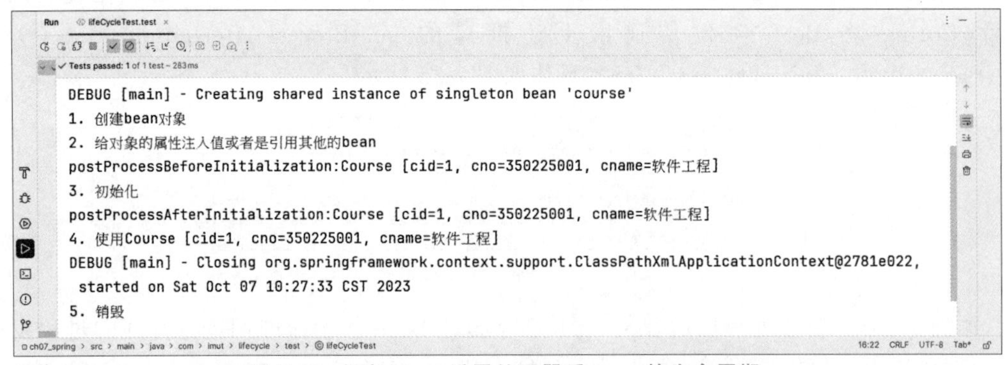

图 7-26　添加 Bean 后置处理器后 Bean 的生命周期

从控制台的输出结果可以看到，和前一次执行结果相比，在执行 initial()方法前后，分别执行了 postProcessBeforeInitialization 方法和 postProcessAfterInitialization 方法。

Bean 的后置处理器在初始化方法的前后可以分别执行一个方法，进行额外的处理。Bean 的后置处理器会对 IoC 容器中所有的 bean 对象的生命周期都起作用。

7.8 综合案例

本章案例主要实现通过注解装配获取学生信息及其所选课程信息的功能。

7.8.1 案例设计

案例使用 Spring 的控制反转，通过注解定义类的实例。本章案例使用的主要文件如表 7-7 所示。

表 7-7 本章案例使用的文件

文　　件	所在包/路径	功　　能
log4j.properties	src/	日志输出配置文件
applicationContext.xml	src/	Spring 的配置文件
Student.java	com.imut.pojo	封装学生信息的类
Course.java	com.imut.pojo	封装课程信息的类
StudentTest.java	com.imut.test	测试类

7.8.2 案例演示

使用 Junit4 测试执行 StudentTest.java 中的方法。运行 test()方法，在控制台输出的结果如图 7-27 所示。

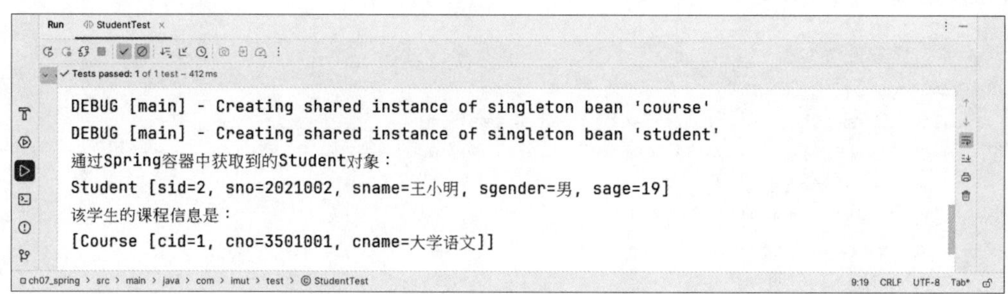

图 7-27 运行结果

7.8.3 代码实现

1. 创建项目，添加依赖

将 Maven 工程 ch06_spring 复制为 ch07_spring，在 pom.xml 文件中增加 Spring 的 AOP 依赖，代码如下。

```
<dependency>
    <groupId>org.springframework</groupId>
    <artifactId>spring-aop</artifactId>
    <version>6.0.11</version>
</dependency>
```

2. 编写配置文件

修改 Spring 配置文件 applicationContext.xml，编写创建 Spring 容器时要扫码的包。配置文件 applicationContext.xml 代码如下。

```
<?xml version="1.0" encoding="UTF-8"?>
<beans xmlns="http://www.springframework.org/schema/beans"
      xmlns:context="http://www.springframework.org/schema/context"
```

```
    xmlns:aop="http://www.springframework.org/schema/aop"
    xmlns:xsi="http://www.w3.org/2001/XMLSchema-instance"
    xsi:schemaLocation="http://www.springframework.org/schema/beans
    http://www.springframework.org/schema/beans/spring-beans.xsd
    http://www.springframework.org/schema/context
    http://www.springframework.org/schema/context/spring-context.xsd
    http://www.springframework.org/schema/aop
    http://www.springframework.org/schema/aop/spring-aop.xsd">
    <context:component-scan base-package="com.imut.pojo"/>
</beans>
```

3. 创建实体类

修改实体类 Student.java 文件，增加注解配置。在 com.imut.pojo 包下新建实体类 Course.java。

修改后的程序 Student.java 代码如下。

```
//使用注解配置实例
@Component("student")
public class Student implements Serializable{
    @Value("2")
    private int sid;
    @Value("2021002")
    private String sno;
    @Value("王小明")
    private String sname;
    @Value("男")
    private String sgender;
    @Value("19")
    private int sage;
    @Autowired
    private List<Course> courses;
    //此处省略 getter()、setter()和 toString()方法
}
```

程序 Course.java 代码如下。

```
@Component
public class Course {
    @Value("1")
    private int cid=1;
    @Value("3501001")
    private String cno="3501001";
    @Value("大学语文")
    private String cname="大学语文";
    @Autowired
    private List<Student> students;
    //此处省略 getter()、setter()和 toString()方法
}
```

4. 编写测试类

修改 com.imut.test 包下的测试类 StudentTest.java，测试通过注解装配获取学生信息及其所选课程信息的功能。

程序 StudentTest.java 代码如下。

```
public class StudentTest {
    @Test
    public void test() {
        ApplicationContext context=new
            ClassPathXmlApplicationContext("applicationContext.xml");
        Student student=(Student) context.getBean("student");
        System.out.println("通过 Spring 容器中获取到的 Student 对象:");
        System.out.println(student);
        System.out.println("该学生的课程信息是:");
        System.out.println(student.getCourses());
    }
}
```

7.9 习题

1. 选择题

(1) 在 BeanFactory 接口的方法中，用于判断是否为单实例的是(　　)。

A. isSingleton(String name)　　B. isPrototype(String name)

C. String[]getAliases(String name)　　D. boolean containsBean(String name)

(2) Spring 框架的<bean>元素中的 autowire 属性取值不包括(　　)。

A. default　　B. byName　　C. byType　　D. byld

(3) Spring 框架容器支持多种形式的 Bean 的装配方式，不包括(　　)。

A. 基于 XML 的装配　　B. 基于 properties 的装配

C. 基于注解(Annotation)的装配　　D. 自动装配

(4) 以下关于 Spring 框架对 Bean 生命周期管理，说法错误的是(　　)。

A. Spring 框架的容器可以管理 singleton 作用域的 Bean 的生命周期。Spring 框架能够精确地知道该 Bean 何时被创建、何时初始化完成，以及何时被销毁

B. 对于 prototype 作用域的 Bean，Spring 只负责创建，当容器创建 Bean 实例后，Bean 的实例就交给客户端代码来管理，Spring 容器将不再跟踪其生命周期

C. 每次客户端请求 singleton 作用域的 Bean 时，Spring 框架的容器都会创建一个新的实例，并且不会管那些被配置成 singleton 作用域的 Bean 的生命周期

D. 了解 Bean 生命周期的意义就在于，可以在某个 Bean 生命周期的某些指定时刻完成一些相关操作

2. 填空题

(1) 依赖注入主要有两种方式，它们分别是________和________。

(2) 在 Spring 配置文件<bean>元素的属性中，________属性指定 Bean 对应类的全路径。

(3) 在 Spring 配置文件<bean>元素的属性中，________属性指定 Bean 对象的作用域。

(4) 在默认情况下，Spring 中 Bean 实例的作用域是________。

(5) 当需要在 Bean 实例初始化后执行指定行为时，可通过实现________完成。

3. 简单题

(1) 简述 Spring 中 Bean 的作用域。

(2) 简述 Spring 中 Bean 的生命周期。

第8章 面向切面编程

面向切面编程(AOP)是 Spring 的核心功能,它补充和完善了面向对象编程(OOP)。AOP 通过横向抽取降低代码冗余和功能耦合。在实际开发中,AOP 常用于数据库事务、日志记录、性能监控和权限检查等功能。

8.1 Spring AOP 的基本概念

8.1.1 AOP 简介

AOP 是 OOP 的补充,通过横向隔离相同业务逻辑,将重复逻辑抽取到独立模块,提高程序复用性和开发效率。它适用于横切逻辑场合,如访问控制、事务管理和性能监测。面向对象编程通过对象协作实现功能,引入抽象、封装、继承等概念,构建纵向的类结构体系。

随着软件规模的不断扩大,系统中出现了一些面向对象编程难以彻底解决的问题。例如,系统的某个类中有若干个方法都包含事务管理的业务逻辑,如图 8-1 所示。

图 8-1 操作用户信息

从图 8-1 中可以看出,用户信息查询、修改和删除的方法中均涉及事务管理逻辑,导致

重复代码和增加维护成本。通过横向抽取机制，利用 AOP 将事务管理逻辑抽取到可重用模块，解决重复逻辑问题，降低耦合，减少重复代码。

8.1.2 理解 AOP

什么是横切逻辑呢？先来看下面的程序代码。

在该段代码中，UserService 的 addNewUSer()方法根据需求增加了日志和事务功能。

```
public class UserServiceImpl implements UserService {
    private static final Logger = Logger.getLogger(UserserviceImpl.class);
    public boolean addNewUser(User user) {
        log.info("添加用户" + user.getUserName());                        //记录日志
        SqlSession sqlSession = null;
        boolean flag = false;
        //异常处理
        try {
            sqlSession = MyBatisUtil.createSqlSession();
            if(sqlSession.getMapper(UserMapper.class).add(user) > 0)
                flag = true;
            sqlSession.commit();                                          //事务控制
        }catch(Exception e) {
            log.error("添加用户" + user.getUsername() + "失败", e); //记录日志
            sqlSession.rollback();                                        //事务控制
            flag = false;
        }finally {
            MyBatisUtil.closeSqlSession(sqlSession);
        }
        return false;
    }
}
```

为了保证业务系统的健壮性和可用性，业务方法中经常需要编写相似的代码，如日志记录、异常处理和事务控制等，导致代码复杂。修改日志格式、安全验证规则或添加新功能时，需要频繁修改大量业务代码。

业务系统中存在散布在各处的必要处理，即“横切逻辑”或“切面”。怎样专注于真正的业务逻辑，不受额外代码干扰？一种方法是抽取重复性代码至专门的类和方法。但这无法实现业务和横切逻辑的彻底解耦合，因为业务代码中仍需保留方法调用代码，增减横切逻辑仍需修改业务方法中的调用代码。

AOP 旨在解决此问题，使系统能在需要时“自动”调用所需功能，无须显式编写调用。

在 AOP 思想中，类与切面的关系如图 8-2 所示。

可以看出，通过 Aspect(切面)分别在 Classl 和 Class2 的方法中加入了事务、日志、权限和异常等功能。

AOP 思想的示意图如图 8-3 所示。在编写业务逻辑时，可以专心于核心业务，而不用过多地关注其他业务逻辑的实现，这不但提高了开发效率，还增强了代码的可维护性。

目前流行的 AOP 框架有两个，分别为 Spring AOP 和 AspectJ。Spring AOP 使用纯 Java 实现，不需要专门的编译过程和类加载器，在运行期间通过代理方式向目标类织入增

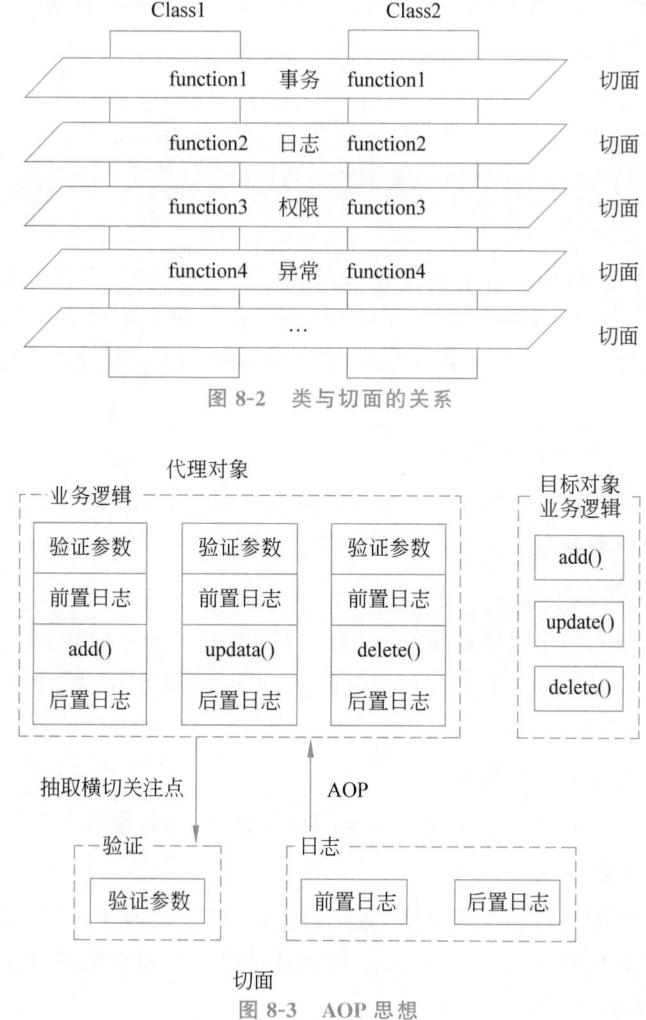

图 8-2 类与切面的关系

图 8-3 AOP 思想

强的代码。AspectJ 是一个基于 Java 语言的 AOP 框架，从 Spring 2.0 版开始，Spring AOP 引入了对 AspectJ 的支持，AspectJ 扩展了 Java 语言，提供了一个专门的编译器，在编译时提供横向代码的织入。

8.1.3 AOP 的术语

学习使用 AOP 之前，先要了解 AOP 的专业术语。这些术语包括 Aspect、Joinpoint、Pointcut、Advice、Target Object、Proxy、AOP proxy 和 Weaving，其解释具体如下。

1. Aspect 类

Aspect 类是 Spring AOP 的核心概念之一，可以看作是一个拦截器。要使其被 Spring 容器识别为切面，需要在配置文件中通过＜bean＞元素指定。

2. Joinpoint（连接点）

连接点是程序执行过程中的某个阶段点，通常指的是对象的一个操作，如类方法的调用前后、方法抛出异常等。在 Spring AOP 中，连接点主要指的是方法的调用。通过切点的正

则表达式，可以判断哪些方法是抛出异常连接点，从而织入对应的通知。

3. Pointcut（切入点）

切入点是切面与程序流程的交叉点，用于定义需要处理的连接点。通常指的是类或者方法名，可以通过书写切入点表达式来选择所需的方法作为切入点。切入点和连接点不是一一对应的关系，一个切入点可以匹配多个连接点。

4. Advice（通知/增强处理）

通知/增强处理是 AOP 框架在特定切入点执行的增强处理，即在定义好的切入点处所要执行的程序代码。通知/增强是一段功能性代码，用于执行拦截到连接点之后的特点操作。Spring AOP 提供了 5 种类型的通知，包括前置通知、后通知、返回后通知、抛出异常后通知和环绕通知。

5. Target Object（目标对象）

目标对象是指所有被通知的对象，也称为被增强对象。如果 AOP 框架采用的是动态的 AOP 实现，那么该对象就是一个被代理对象。

6. Proxy（代理）

代理是在将通知应用于目标对象后动态创建的对象，是将 AOP 增强应用到目标类后生成的结果类。代理类融合了原始类和增强逻辑，可以使用与调用原始类相同的方式来调用代理类。

7. AOP proxy（AOP 代理）

AOP 代理指由 AOP 框架所创建的对象，实现执行增强处理方法等功能。

8. Weaving（织入）

织入是将通知添加到目标类具体连接点的过程，可以在编译时、类加载时和运行时完成。Spring 采用动态代理织入，而 AspectJ 采用编译期织入和类装载器织入。Spring AOP 负责实施切面，完成织入工作。

8.2 Spring AOP 的实现机制

Spring AOP 采用动态代理解决非核心业务需求（验证、日志等）带来的代码混乱和代码分散的问题。动态代理是指客户通过代理类来调用其他对象的方法，并且是在程序运行时根据需要动态创建目标类的代理对象。通过学习已经知道 AOP 中的代理就是由 AOP 框架动态生成的一个对象，该对象可以作为目标对象使用。

AOP 的核心思想就是在不改变原程序的基础上为代码段增加新的功能，对代码段进行增强处理。它的设计思想来源于代理设计模式。代理设计模式的原理是使用一个代理将原始对象包装起来，并取代原始对象。任何对原始对象的调用都要通过代理。代理对象决定是否以及何时将方法调用转到原始对象上。

通常情况下调用对象的方法如图 8-4 所示。在代理模式中，可以为该对象设置一个代理对象，代理对象为 fun()提供一个代理方法，当通过代理对象的 fun()方法调用原对象的

fun()方法时，就可以在代理方法中添加新的功能，即增强处理。增强的功能既可以插到原对象的 fun()方法前面，也可以插到其后面，如图 8-5 所示。

图 8-4 直接调用对象的方法

图 8-5 通过代理对象调用的方法

这种模式是在原有代码乃至原业务流程都不修改的情况下直接在业务流程中切入新代码，增加新功能，这就是所谓的面向切面编程。Spring 的 AOP 代理提供了两种实现方式，即 JDK 动态代理和 CGLIB 动态代理。

在 Spring 中，AOP 代理的创建和依赖关系管理由 Spring 的 IoC 容器负责。因此 Spring AOP 可以直接将 IoC 容器中的其他 Bean 实例作为目标对象。在默认情况下，Spring AOP 使用 JDK 动态代理创建代理对象。而当目标对象是一个类并且这个类没有实现接口时，Spring 会切换为使用 CGLIB 代理。

下面将结合相关案例演示这两种代理方式的使用。

8.2.1 JDK 动态代理

JDK 动态代理基于接口实现，通过 Proxy.newInstance 获取代理对象，需指定目标类实现的接口。Spring 框架默认使用 JDK 动态代理实现 AOP。主要涉及 InvocationHandler 和 Proxy 两个 API，分别定义横切逻辑和生成代理对象。代理类实现指定接口，与目标类方法一致，调用目标类方法实际上是调用 invoke 方法。获取到的代理类对象是接口类型，可以通过接口接收，不能使用具体实现类接收，因为代理对象与目标类不能相互转换。

下面通过一个计算器的例子来演示 Spring 框架中 JDK 动态代理的实现过程，具体步骤如下。

(1) 创建一个名为 ch08_spring_01 的 Java 项目，使用 Maven 管理 Spring 核心容器的 4 个基础包、Spring 依赖包 commons-logging-1.2.jar 以及 spring-aop 依赖包，并在 pom.xml 文件中引入依赖坐标。

(2) 在 src/main/java 目录下创建一个 com.imut.jdk 包，在该包下创建接口 ArithmeticCalculator，并在该接口中编写加、减、乘、除的方法，代码如例 8-1 所示。

例 8-1：ArithmeticCalculator.java。

```
package com.imut.jdk;
public interface ArithmeticCalculator {
    public int add(int a , int b );
    public int sub(int a , int b );
    public int mul(int a , int b );
    public int div(int a , int b );
}
```

(3) 在 com.imut.jdk 包中创建 ArithmeticCalculator 接口的实现类 ArithmeticCalculatorImpl，

分别实现接口中的方法，并在每个方法中添加一条输出语句，验证方法执行成功，代码如例 8-2 所示。

例 8-2：ArithmeticCalculatorImpl.java。

```
package com.imut.jdk;
public class ArithmeticCalculatorImpl implements ArithmeticCalculator{
    @Override
    public int add(int a, int b) {
        int result = a + b ;
        System.out.println ("The method add ends with :" + a + "+" + b + "=" +
result);
        return  result ;
    }
    @Override
    public int sub(int a, int b) {
        int result = a - b ;
        System.out.println("The method sub ends with :" + a + "-" + b + "="  +
result);
        return result ;
    }
    @Override
    public int mul(int a, int b) {
        int result = a * b ;
        System.out.println("The method mul ends with :" + a + "*" + b + "="  +
result);
        return result ;
    }
    @Override
    public int div(int a, int b) {
        int result = a / b ;
        System.out.println("The method div ends with :" + a + "+" + b + "="  +
result);
        return result ;
    }
}
```

需要注意的是，本例中会将实现类 ArithmeticCalculatorImpl 作为目标类，对其中的方法进行增强处理。

（4）在 src/main/java 目录下，在 com.imut.jdk 包下创建切面类 MyAspect，在该类中定义一个模拟验证参数的方法和一个模拟记录日志的方法，这两个方法就表示切面中的通知，代码如例 8-3 所示。

例 8-3：MyAspect.java。

```
package com.imut.jdk;
public class MyAspect {
    public void check_parameters() {
        System.out.println("模拟验证参数...");
    }
```

```
    public void log() {
        System.out.println("模拟记录日志");
    }
}
```

(5) 在 com.imut.jdk 包下创建代理类 ArithmeticCalculatorProxy，该类需要实现 InvocationHandler 接口，并编写代理方法。在代理方法中，需要通过 Proxy 类实现动态代理，代码如例 8-4 所示。

例 8-4：ArithmeticCalculatorProxy.java。

```
package com.imut.jdk;
import java.lang.reflect.InvocationHandler;
import java.lang.reflect.Method;
import java.lang.reflect.Proxy;
public class ArithmeticCalculatorProxy implements InvocationHandler {
    //被代理的对象：目标对象
    private ArithmeticCalculator target ;
    //使用构造器将目标对象传入
    public ArithmeticCalculatorProxy(ArithmeticCalculator target) {
        this.target = target ;
    }
    //获取代理对象
    public ArithmeticCalculator getProxy() {
        //代理对象
        ArithmeticCalculator proxy ;
        /**
         * 1. Loader:类加载器对象,加载类 xxxx.class(字节码文件)===>Class 对象。
         * 2. Interfaces:接口。目的是获取到目标对象的所有方法(只有接口中的方法才能被
代理),也就是 JDK 的动态代理是基于接口来实现的
         * 3. this: 实现 InvocationHandler 接口的对象,用于完成动态代理的整个代理过程
        **/
        ClassLoader loader = target.getClass().getClassLoader();
        Class []  interfaces = target.getClass().getInterfaces();
            proxy  =  (ArithmeticCalculator)  Proxy. newProxyInstance ( loader,
interfaces, this);
        return proxy ;
    }
    @Override
    /**
     * invoke:将来代理对象去调用代理方法时,会回来调用 invoke 方法
     * proxy:代理对象
     * method:正在被调用的方法
     * args:方法的参数
    **/
    public Object invoke(Object proxy, Method method, Object[] args) throws
Throwable {
        //获取方法名
        String methodName = method.getName();
```

```
        //声明切面
        MyAspect myaspect = new MyAspect();
        //前增强
        myaspect.check_parameters();
        //目标对象方法调用
        Object result = method.invoke(target, args);
                                                    //真正的执行目标对象的+ 、- 、* 、/方法
        //后增强
        myaspect.log();
        return result;
    }
}
```

在示例代码中，ArithmeticCalculatorProxy 类实现了 InvocationHandler 接口，以及接口中的 invoke 方法。所有动态代理类所调用的方法都会交由 invoke 方法处理。在创建的代理方法 getProxy()中，使用 Proxy 类的 newProxyInstance()方法来创建代理对象。newProxyInstance()方法包含 3 个参数，其中第 1 个参数是当前类的类加载器，第 2 个参数是被代理对象实现的所有接口，第 3 个参数 this 代表的就是代理类 ArithmeticCalculatorProxy 本身。在 invoke()方法中，目标类方法执行的前后会分别执行切面类中的 check_parameters()方法和 log()方法。

(6) 在 com.imut.test 包中创建测试类 JdkTest。在该类的 test()方法中创建目标对象 target 和代理对象 proxy，然后从代理对象中获得对目标对象增强后的对象，最后调用该对象中的加和除方法，代码如例 8-5 所示。

例 8-5：JdkTest.java。

```
public class JdkTest {
    @Test
    public void test() {
        //创建目标对象
        ArithmeticCalculator target = new ArithmeticCalculatorImpl();
        //获取代理对象
        ArithmeticCalculator proxy = new ArithmeticCalculatorProxy(target).
getProxy();
        System.out.println("proxy: " + proxy.getClass().getName());
        //代理对象调用方法。
        //代理对象调用方法,会回去调用 Invocationhandler 中的 invoke()方法
        Object result = proxy.add(1,1);
        System.out.println("Main Result: " +result );
        result = proxy.div(2,1);
        System.out.println("Main Result: " +result );
    }
}
```

执行测试类 JdkTest，得到的执行结果如图 8-6 所示。

可以看出，使用 roxy. newProxyInstance()方法成功创建的代理对象，成功调用了 ArithmeticCalculator 实例中的加和除的方法，并且在调用前后分别增加了验证参数和记录

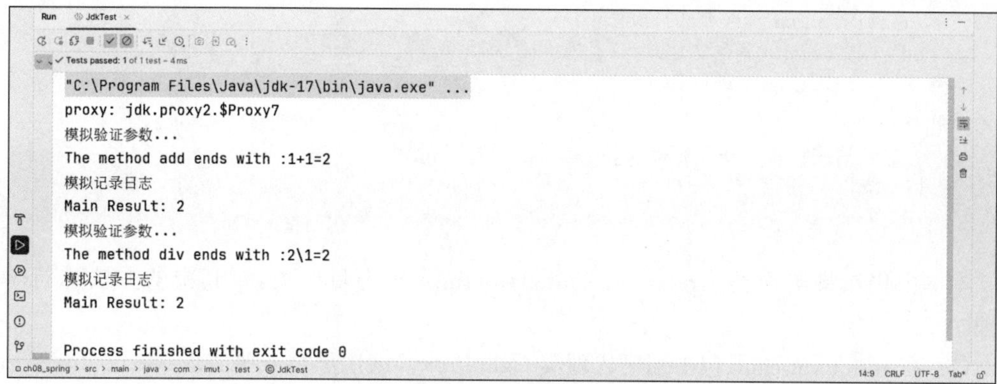

图 8-6 测试类 JdkTest 的执行结果

日志的功能。这种基于接口的代理方式就是 Spring 中的 JDK 动态代理。

8.2.2 CGLIB 动态代理

JDK 动态代理易用，但有局限性，仅限于接口代理。对于类代理，应使用 CGLIB。CGLIB 不需要实现接口，利用字节码技术继承创建代理。CGLIB 是高性能开源库，为指定类生成增强子类。Spring 核心包已集成所需包，无须额外导入。通过 Enhancer 创建 CGLIB 代理实例，利用 MethodInterceptor 织入增强方法。以下通过一个例子演示 CGLIB 动态代理技术的代码实现，具体步骤如下。

(1) 在 ch08_spring_01 项目的 src/main/java 目录下创建一个 com.imut.cglib 包，在该包下重新创建类 ArithmeticCalculator，并在该类中编写加、减、乘、除的方法，代码如例 8-6 所示。

例 8-6：ArithmeticCalculator.java。

```
public class ArithmeticCalculator {
    public int add(int a, int b) {
        int result = a + b ;
        System.out.println ("The method add ends with :" + a + "+" + b + "=" +
result);
        return  result ;
    }
    public int sub(int a, int b) {
        int result = a - b ;
        System.out.println("The method sub ends with :" + a + "-" + b + "="  +
result);
        return result ;
    }
    public int mul(int a, int b) {
        int result = a * b ;
        System.out.println("The method mul ends with :" + a + " * " + b + "="  +
result);
        return result ;
    }
```

```
    public int div(int a, int b) {
        int result = a / b ;
        System.out.println("The method div ends with :" + a + "\\" + b + "=" +
result);
        return result ;
    }
}
```

本例中仍然将实现类 ArithmeticCalculatorImpl 作为目标类，对其中的方法进行增强处理。

（2）在 com.imut.cglib 包中创建代理类 CglibProxy，该代理类需要实现 MethodInterceptor 接口，并实现接口中的 intercept()方法，代码如例 8-7 所示。

例 8-7：CglibProxy.java。

```
package com.imut.cglib;
import java.lang.reflect.Method;
import org.springframework.cglib.proxy.Enhancer;
import org.springframework.cglib.proxy.MethodInterceptor;
import org.springframework.cglib.proxy.MethodProxy;
import com.imut.jdk.MyAspect;
public class CglibProxy implements MethodInterceptor {
    public Object getProxy(Object target) {
        Enhancer enhancer = new Enhancer();
        enhancer.setSuperclass(target.getClass());
        enhancer.setCallback(this);
        return enhancer.create();
    }
    /**
     * proxy:CGLIB 根据指定父类生成的代理对象
     * method:要拦截的方法
     * args:拦截方法的参数数组
     * methodProxy:方法的代理对象,用于执行父类的方法
     */
    @Override
    public Object intercept(Object proxy, Method method, Object[] args,
MethodProxy methodProxy) throws Throwable {
        MyAspect myAspect = new MyAspect();
        myAspect.check_parameters();
        Object obj = methodProxy.invokeSuper(proxy, args);
        myAspect.log();
        return obj;
    }
}
```

在上述代码中，CglibProxy 类实现了接口 MethodInterceptor，并提供了一个生成代理对象的方法 getProxy(Object target)。该方法首先创建一个动态类对象 Enhancer，它是 CGLIB 的核心类。然后调用 Enhancer 类的 setSuperdass()方法来确定目标对象。接下来调用 Enhancer 类的 setCallback()方法添加回调函数，其中的参数 this 代表的就是代理类

CglibProxy 本身。最后通过 return 语句将创建的代理类对象返回。同时,CglibProxy 类还实现了 MethodInterceptor 接口的 intercept()方法。intercept()方法会在程序执行目标方法时被调用,该方法运行时将执行切面类中的增强方法。

(3) 在 com.imut.test 包中创建测试类 cglibTest。在该类的 test()方法中先创建代理对象和目标对象,然后从代理对象中获得增强后的目标对象,最后调用对象的加和除方法,hex 代码如例 8-8 所示。

例 8-8: cglibTest。

```
public class cglibTest {
    @Test
    public void test() {
        CglibProxy cglibProxy = new CglibProxy();
        ArithmeticCalculator ac = new ArithmeticCalculator();
        ArithmeticCalculator acproxy = (ArithmeticCalculator) cglibProxy.
getProxy(ac);
        System.out.println("proxy: " + acproxy.getClass().getName());
        acproxy.add(1, 1);
        acproxy.div(2,1);
    }
}
```

在以上代码中,程序通过 cglibTest 测试类创建了代理对象,程序可以通过代理对象调用目标对象的方法。测试类 cglibTest 的执行结果如图 8-7 所示。

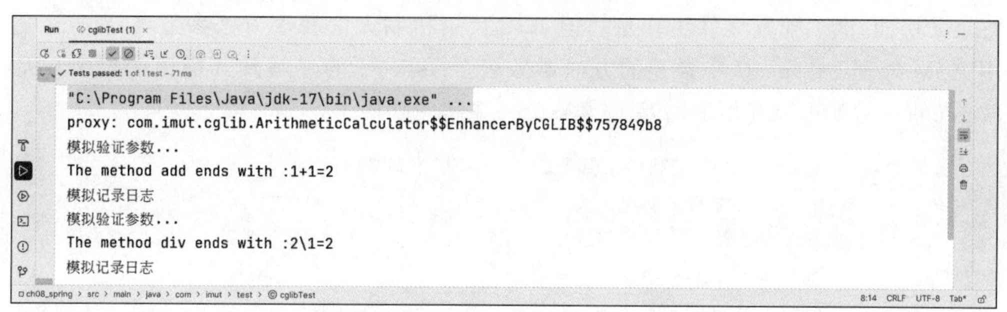

图 8-7　测试类 cglibTest 的执行结果

从图 8-7 中可以看出,与 JDK 动态代理的结果相同,当执行代理对象的 add()和 div()方法时,原接口中的 add()和 div()方法和增强的方法分别都被执行。

8.3 基于注解开发 Spring AOP

8.2 节介绍了 Spring AOP 的实现机制,接下来讲解 Spring AOP 的开发方法。为了便于理解,给出 Spring AOP 的流程图,如图 8-8 所示。

Spring AOP 的两种常用开发方法是基于 XML 和基于注解的 AspectJ。由于@AspectJ 注解已成为主流,因此先讨论其开发方式。深入理解@AspectJ 注解后,其他方式也类似。

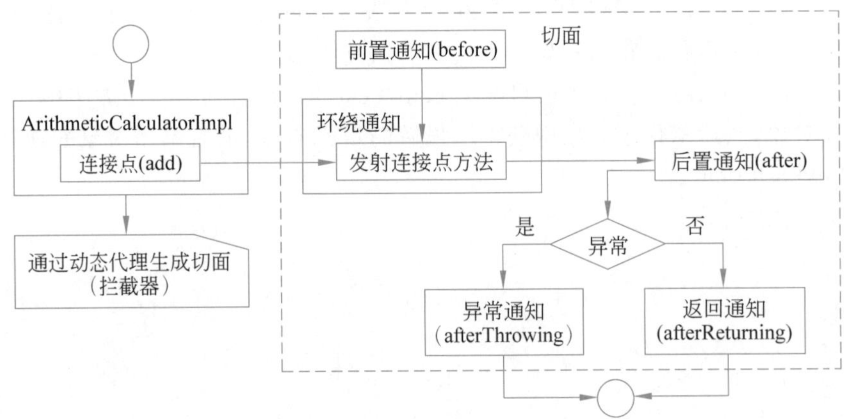

图 8-8 Spring AOP 的流程图

8.3.1 @AspectJ 简介

AspectJ 是 Java 社区中流行且完整的 AOP 框架，Spring 2.0 及以上版本支持 AspectJ，支持基于注解或 XML 配置的 AOP。AspectJ 是面向切面的框架，扩展 Java 语言，定义 AOP 语法，编译期提供代码织入，有专门的编译器生成遵守字节编码规范的 Class 文件。

AspectJ 5 新增了@AspectJ 功能，使用 JDK 5.0 及以上版本的注解技术和正规切入点表达式语言描述切面。使用前需确保 JDK 版本符合要求。Spring 框架集成 AspectJ，通过注解定义切面，减少配置文件工作量。但 Java 反射机制无法获取方法参数名，Spring 还需利用 ASM 处理@AspectJ 中描述的方法参数名。AspectJ 的注解及其描述如表 8-1 所示。了解 Aspectj 后就可以开始编写基于@Aspect 注解的切面了。

表 8-1 AspectJ 的注解及其描述

注解名称	描　述
@Aspect	用于定义一个切面
@Pointcut	用于定义切入点表达式。使用时还需定义一个包含名字和任意参数的方法签名来表示切入点名称。实际上，这个方法签名就是一个返回值为 void，且方法体为空的普通方法
@Before	用于定义前置通知，相当于 BeforeAdvice。使用时，通常需要指定一个 value 属性值，该属性值用于指定一个切入点表达式(既可以是已有的切入点，也可以直接定义切入点表达式)
@AfterReturning	用于定义后置通知，相当于 AfterReturningAdvice。使用时可以指定 pointcut/value 和 returning 的属性，其中 pointcut/value 这两个属性的作用一样，都用于指定切入点表达式。 returning 属性值用于表示 Advice 方法中可定义与此同名的形参，该形参可用于访问目标方法的返回值。目标方法正常执行结束后，返回通知，能获取到目标方法
@Around	用于定义环绕通知，相当于 MethodInterceptor。使用时需要指定一个 value 属性，该属性用于指定该通知被织入的切入点。环绕着整个目标方法执行，类似于动态代理的整个过程

续表

注解名称	描述
@AfterThrowing	用于定义异常通知来处理程序中未处理的异常，相当于 ThrowAdvice。使用时可指定 pointcut/value 和 throwing 的属性，其中 pointcut/value 这两个属性用于指定切入点表达式，而 throwing 属性值用于指定一个形参名来表示 Advice 方法中可定义与此同名的形参，该形参可用于访问目标方法抛出的异常
@After	用于定义最终(final)通知，不管是否异常，该通知都会执行。使用时需要指定一个 value 属性，该属性用于指定通知被织入的切入点。获取不到目标方法的返回值
@DeclareParents	用于定义引介通知，相当于 IntroductionInterceptor(不要求掌握)

8.3.2 使用注解的切面编程

1. 在 Spring 中启用 AspectJ 注解支持

要在 Spring5.0 应用中使用 AspectJ 注解，需包含 AspectJ 类库：aopalliance.jar、aspectj.weaver.jar 和 sf.cglib.jar，并添加 aop schema 到＜beans＞根元素。启用 AspectJ 注解支持，只需在 Bean 配置文件中定义一个空的 XML 元素＜aop:aspectj-autoproxy＞。Spring IoC 容器会自动为匹配的 Bean 创建代理。

2. 用 AspectJ 注解声明切面

在 Spring 中声明 AspectJ 切面，只需将切面声明为 Bean 实例。Spring IoC 容器会自动为匹配的 Bean 创建代理。切面是一个带有@Aspect 注解的 Java 类。AspectJ 支持 5 种类型的通知注解：@Before(前置通知)、@After(后置通知)、@AfterRunning(返回通知)、@AfterThrowing(异常通知)和@Around(环绕通知)。

接下来，本节通过前面小节定义的计算器 ArithmeticCalculator 类为例演示使用注解的切面编程的代码实现，代码如例 8-9 所示。

(1) 在 Java 项目 ch08_spring_01 中增加依赖包，在 pom.xml 文件中追加 AspectJ 注解需要的 aopalliance.jar、aspectj.weaver.jar 和 sf.cglib.jar 依赖，如图 8-9 所示。

(2) 在配置文件 applicatioContext 里配置 aop 命名空间和 context 命名空间。然后使用＜context＞元素设置需要扫描的包，自动扫描组件，使注解生效。

(3) 在配置文件 applicatioContext 里用＜aop:aspectj-autoproxy/＞配置 AspectJ，自动生成代理。

代码如例 8-9 所示。

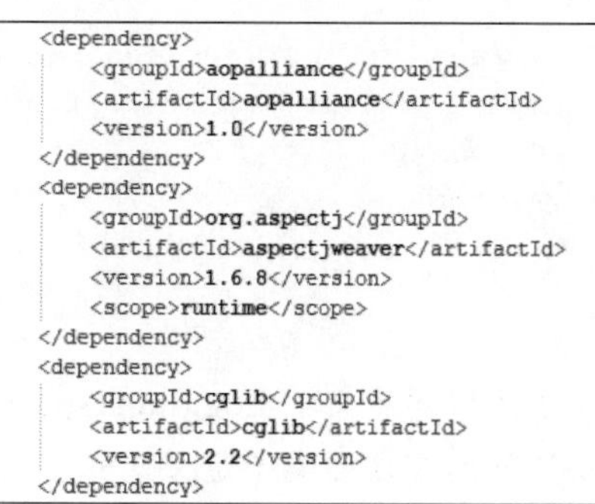

```
<dependency>
    <groupId>aopalliance</groupId>
    <artifactId>aopalliance</artifactId>
    <version>1.0</version>
</dependency>
<dependency>
    <groupId>org.aspectj</groupId>
    <artifactId>aspectjweaver</artifactId>
    <version>1.6.8</version>
    <scope>runtime</scope>
</dependency>
<dependency>
    <groupId>cglib</groupId>
    <artifactId>cglib</artifactId>
    <version>2.2</version>
</dependency>
```

图 8-9 导入 AspectJ 注解所需的包

例 8-9：applicationContext.xml。

```
<?xml version="1.0" encoding="UTF-8"?>
<beans xmlns="http://www.springframework.org/schema/beans"
    xmlns:xsi="http://www.w3.org/2001/XMLSchema-instance"
```

```
    xmlns:context="http://www.springframework.org/schema/context"
    xmlns:aop="http://www.springframework.org/schema/aop"
    xsi:schemaLocation="http://www.springframework.org/schema/beans
        http://www.springframework.org/schema/beans/spring-beans.xsd
        http://www.springframework.org/schema/context
        http://www.springframework.org/schema/context/spring-context-4.0.xsd
        http://www.springframework.org/schema/aop
        http://www.springframework.org/schema/aop/spring-aop-4.0.xsd">
    <context:component-scan base-package = "com.imut.aop.annotationAspectj" >
</ context : component-scan >
    <!-- 基于注解配置 AspectJ: 自动生成代理. -->
    <aop:aspectj-autoproxy/>
</beans>
```

在上述代码中，配置文件中先要导入 aop 命名空间，并且引入 context 约束信息。然后使用<context>元素设置需要扫描的包，使注解生效。由于此案例中的目标类位于 com.imut.aop.annotationAspectj 包中，所以这里设置 base-package 的值为 com.imut.aop.annotationAspectj。

要在 Spring IoC 容器中启用 AspectJ 注解支持，只需在 Bean 配置文件中定义一个空的 XML 元素<aop:aspectj-autoproxy>。当 Spring IoC 容器侦测到 Bean 配置文件中的<aop:aspectj-autoproxy>元素时，会自动为与 AspectJ 切面匹配的 Bean 创建代理。

(4) 在 src/main/java 目录下创建包 com.imut.aop.annotationAspectj，将例 8-1 和例 8-2 的 ArithmeticCalculator 接口和 ArithmeticCalculatorImpl 类导入到该包中。通过@Component 将 ArithmeticCalculatorImpl 类标识为一个受容器管理的组件。

(5) 在 com.imut.aop.annotationAspectj 包中新建 LoggingAspect 类。首先通过@Component 将该类标识为一个受容器管理的组件，然后通过@Aspect 标识该类为一个切面。代码如例 8-10 所示。

例 8-10：LoggingAspect.java。

```
package com.imut.aop.annotationAspectj;
import org.springframework.stereotype.Component;
@Component                                   //标识为一个组件
@Aspect                                      //标识为一个切面
public class LoggingAspect {
}
```

上述代码定义了 LoggingAspect 类，作为日志切面。切面编程中的主要对象是通知，下面讲述如何在切面类中定义通知。

8.3.3 通知/增强 Advice

在 AspectJ 注解中，切面只是一个带有@Aspect 注解的 Java 类。通知是标注有某种注解的简单的 Java 方法。AspectJ 支持 5 种类型的通知注解，@Before 是前置通知，在方法执行之前执行。@After 是后置通知，在方法执行之后执行。@AfterRunning 是返回通知，在方法返回结果之后执行。@AfterThrowing 是异常通知，在方法抛出异常之后执行。

@Around 是环绕通知,围绕着方法执行。

1. 前置通知

前置通知是在方法执行之前执行的通知。前置通知使用@Before 注解,并且可以将切入点表达式的值作为注解值。

为日志切面 LoggingAspectj 添加前置通知方法,核心代码如例 8-11 所示。

例 8-11:LoggingAspectj 的前置通知。

```
@Component                                      //标识为一个组件
@Aspect                                         //标识为一个切面
public class LoggingAspect {
      @ Before ( " execution ( public  int  com. imut. aop. annotationAspectj.
ArithmeticCalculatorImpl.add(int,int))")
    public void beforeMethod(JoinPoint joinPoint) {
        String methodName = joinPoint.getSignature().getName();
        Object [] args= joinPoint.getArgs();
        System.out.println("LoggingAspect:前置通知===>The method "+methodName
+" begin with " +Arrays.asList(args));
    }
```

在 src/main/java 目录下的包创建测试类 aopTest,用于测试为计算器 arithmeticCalculator 创建的切面的前置通知。调用计算器 arithmeticCalculator 的加法时,会执行 LoggingAspect 切面的前置通知。

@Before 标识 beforeMethod()方法是个前置通知。@Before 通过方法签名定义切入点参数,表示将前置通知织入到 ArithmeticCalculator 接口的 add()方法中。符号'*'代表匹配任意修饰符及任意返回值,参数列表中的符号'..'匹配任意数量的参数。切入点参数 execution(public * *(..)表示匹配所有目标类的 public 方法,第一个'*'代表返回类型,第二个'*'代表方法签名,而'..'代表任意参数。如果希望给 ArithmeticCalculator 接口的所有方法都织入 LoggingAspect 切面的前置通知,可以将切入点参数修改为 execution(public int com.imut.aop.annotationAspectj. ArithmeticCalculatorImpl. *(..))。execution(public int com. imut. aop. annotationAspectj. *. *(..))则表示为包 com. imut. aop. annotationAspectj 下所有类的方法织入该前置通知。

为了验证切面的通知,定义测试类 aopTest,代码如例 8-12 所示。

例 8-12:测试类 aopTest。

```
@Test
public void test() {
    ApplicationContext ctx =
        new ClassPathXmlApplicationContext("applicationContext.xml");
        ArithmeticCalculator   ac =
          ctx.getBean("arithmeticCalculatorImpl",ArithmeticCalculator.class);
        System.out.println(ac.getClass().getName());  //代理对象
        int result = ac.add(1, 1);
        System.out.println("Main Result: " + result );
    }
}
```

测试类 aopTest 的执行结果如图 8-10 所示。从控制台输出的结果可以看出，执行加法 add()之前执行了前置通知。

```
jdk.proxy2.$Proxy20
LoggingAspect:前置通知===>The method add begin with [1, 1]
Main Result: 2

Process finished with exit code 0
```

图 8-10　测试类 aopTest 前置通知的执行结果

2. 后置通知

后置通知是在目标方法执行后执行的。使用@After 将一个方法标注为后置通知。后置通知的特点是不管方法有没有抛出异常都会执行，但是后置通知获取不到方法的返回值。一个切面可以包括一个或多个后置通知。

为日志切面 LoggingAspectj 添加后置通知，核心代码如例 8-13 所示。

例 8-13：LoggingAspectj 的后置通知。

```
@After("execution(* com.imut.aop.annotationAspectj.ArithmeticCalculatorImpl.
*(..))")
public void afterMethod(JoinPoint joinPoint) {
    String methodName = joinPoint.getSignature().getName();
    Object [] args= joinPoint.getArgs();
    System.out.println("LoggingAspect:后置通知===>The method "+methodName+
" ends .");
}
```

在测试类 aopTest 中调用 ArithmeticCalculator 接口的 div(4,0)，程序会抛出异常，但是后置通知仍然会执行。测试类 aopTest 的核心代码如例 8-14 所示，执行结果如图 8-11 所示。

例 8-14：测试类 aopTest。

```
@Test
public void test() {
    ApplicationContext ctx =
            new ClassPathXmlApplicationContext("applicationContext.xml");
    ArithmeticCalculator  ac =
            ctx.getBean("arithmeticCalculatorImpl",ArithmeticCalculator.class);
    System.out.println(ac.getClass().getName());     //代理对象
    int result = ac.add(1, 1);
    System.out.println("Main Result: " + result );
    result = ac.div(4, 1);
    System.out.println("Main Result: " + result );
}
```

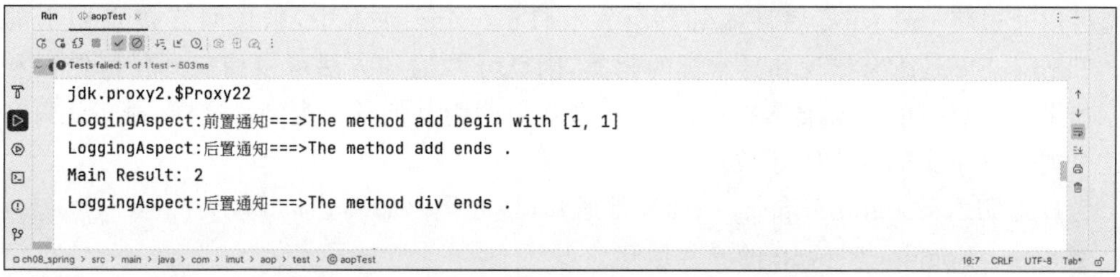

```
jdk.proxy2.$Proxy22
LoggingAspect:前置通知===>The method add begin with [1, 1]
LoggingAspect:后置通知===>The method add ends .
Main Result: 2
LoggingAspect:后置通知===>The method div ends .
```

图 8-11　测试类 aopTest 后置通知的执行结果

3. 返回通知

返回通知是在目标方法成功执行并返回后执行的。使用@AfterReturning 来将一个方法标注为返回通知。返回通知的特点是可以获取到方法的返回值。如果只想在连接点返回的时候记录日志,应使用返回通知代替后置通知。

在返回通知中,只要将 returning 属性添加到@AfterReturning 注解中,就可以访问连接点(目标方法)的返回值。该属性的值即为用来传入返回值的参数名称。必须在返回通知方法的签名中添加一个同名参数。运行时,Spring AOP 会通过这个参数传递返回值。

原始的切点表达式需要出现在 pointcut 属性中。

为日志切面 LoggingAspectj 添加返回通知,核心代码如例 8-15 所示。

例 8-15：LoggingAspectj 的返回通知。

```
/**
 * returning: 用于指定接收目标方法的返回值,必须与返回通知方法的形参名一致
 */
@AfterReturning(value="execution(* com.imut.aop.annotationAspectj.*.*(..))",
returning = "result")
public void afterReturningMethod(JoinPoint joinPoint,Object result) {
    String methodName = joinPoint.getSignature().getName();
    Object[] args= joinPoint.getArgs();
    System.out.println("LogginAspect:返回通知===>The method "+methodName+
" ends with : "+result);
}
```

执行例 8-12 的测试类 aopTest,得到的结果如图 8-12 所示。

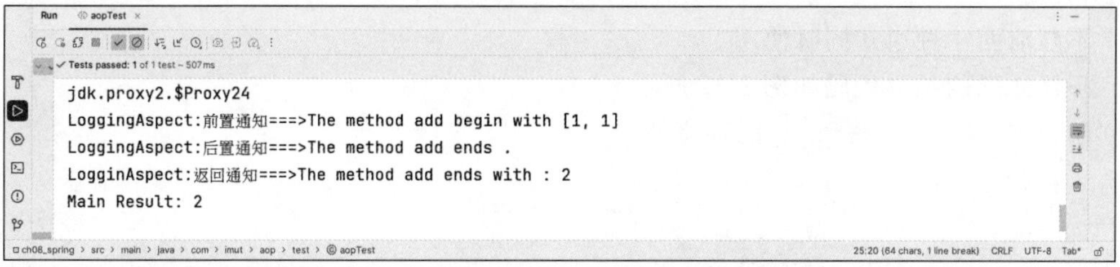

```
jdk.proxy2.$Proxy24
LoggingAspect:前置通知===>The method add begin with [1, 1]
LoggingAspect:后置通知===>The method add ends .
LogginAspect:返回通知===>The method add ends with : 2
Main Result: 2
```

图 8-12　测试类 aopTest 返回通知的执行结果

4. 异常通知

异常通知是在目标方法抛出异常后执行。使用@AfterThrowing 将一个方法标注为异常通知。

可以通过将 throwing 属性添加到@AfterThrowing 注解中，来访问连接点抛出的异常。Throwable 是所有错误和异常类的超类，因此异常通知方法可以捕获到任何错误和异常。如果只对特定的异常类感兴趣，可以将参数声明为其他异常类型。这样，通知就只在抛出这个类及其子类的异常时才被执行。

为日志切面 LoggingAspectj 添加异常通知，核心代码如例 8-16 所示。

例 8-16：LoggingAspectj 的异常通知。

```
@AfterThrowing(value="execution(* com.imut.aop.annotationAspectj.*.*(..))",
throwing = "ex")
public void afterThrowingMethod(JoinPoint joinPoint,ArithmeticException ex) {
    String methodName = joinPoint.getSignature().getName();
    Object [] args= joinPoint.getArgs();
    System.out.println("LoggingAspect:异常通知===>The method " + methodName +
" occurs exception: " +ex);
}
```

执行例 8-14 的测试类 aopTest，得到的结果如图 8-13 所示。可以看到，由于调用的 div(4,0) 方法抛出了异常，所以执行了异常通知，同时后置通知不管方法有没有抛出异常都会执行。

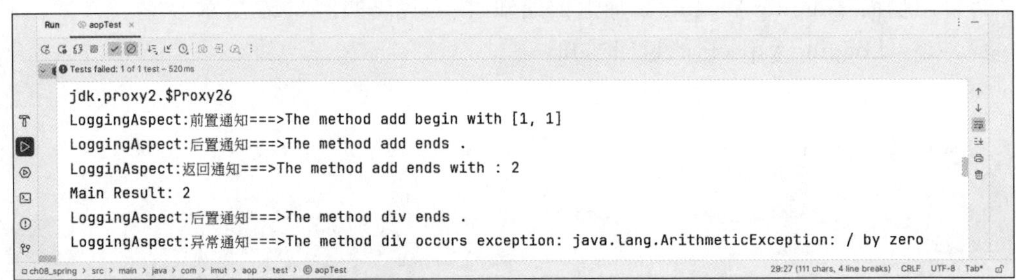

图 8-13 测试类 aopTest 异常通知的执行结果

5. 环绕通知

环绕通知通过@Around 注解标记，围绕目标方法执行，具有强大的功能，能全面控制连接点，甚至决定是否执行。它类似于 JDK 动态代理，结合了前置、后置、返回和异常通知的功能。环绕通知的方法参数为 ProceedingJoinPoint，需调用其 proceed()方法执行被代理方法，否则目标方法不会执行。方法需返回 proceed()的结果，否则可能出现空指针异常。通常不与前面 4 种通知同时使用。

环绕通知的核心代码如例 8-17 所示。

例 8-17：LoggingAspectj 的环绕通知。

```
@Around("execution(* com.imut.aop.annotationAspectj.*.*(..))")
public Object aroundMethod(ProceedingJoinPoint pjp)  {
    String methodName = pjp.getSignature().getName();
    Object [] args= pjp.getArgs();
    try {
        System.out.println("*LoggingAspect:前置===>The method "+methodName+
" begin with "+Arrays.asList(args));
```

```
            Object result= pjp.proceed();
            System.out.println("*LogginAspect:返回===>The method "+methodName+
" ends with : "+result);
            return result;
        } catch (Throwable e) {
            System.out.println("*LoggingAspect:异常===>The method " + methodName +
" occurs exception: "+e.getMessage());
            e.printStackTrace();
        }finally {
            System.out.println("*LoggingAspect:后置===>The method "+methodName+
" ends .");
        }
        return null;
    }
```

8.3.4 连接点对象

连接点在 AOP 中的作用就是用来判断是否要拦截对应的方法。比如上面的 5 个注解可以理解成为连接点。连接点对象中包含了与当前目标方法相关的一些信息,可以在通知方法的形参中声明该对象,并用来获取相关的目标方法信息。

比如@After("execution(* com.imut.aop.annotationAspectj.ArithmeticCalculatorImpl.*(..))"),execution 代表执行方法的时候触发,第一个 * 代表方法可以是任意返回类型,"com.imut.aop.annotationAspectj.ArithmeticCalculatorImpl"表示对应的方法提供者的全限定名,第二个 * 表示被拦截的方法为该类中的全部方法,(..)表示是任意类型的参数。

JoinPoint 对象封装了切面方法的信息,在切面方法中添加 JoinPoint 参数,就可以获取封装了该方法信息的 JoinPoint 对象。然后就能访问方法名称和参数值等链接细节,如例 8-11、例 8-13 等。通过调用连接点对象 JoinPoint 的 getSignature()方法获取被拦截方法的名称,调用连接点对象 JoinPoint 的 getArgs()获取方法的参数。ProceedingJoinPoint 对象是 JoinPoint 的子接口,该对象只用在@Around 的切面方法中。JoinPoint 对象的常用方法如表 8-2 所示。

表 8-2 JoinPoint 对象的常用方法

方 法 名	功 能
Signature getSignature()	获取封装了署名信息的对象,在该对象中可以获取到目标方法名所属类的 Class 等信息
Object[ ] getArgs()	获取传入目标方法的参数对象
Object getTarget()	获取被代理的对象
Object getThis()	获取代理对象

8.3.5 重用切入点表达式

在 AOP 中,连接点始终代表方法的执行。切入点是与连接点匹配的,切入点表达语言是以编程方式描述切入点的方式。切入点是定义了在"什么地方"进行切入,哪些连接点会得到通知。切入点一定是连接点。

切点是通过@PointCut注解和切点表达式定义的。@PointCut注解可以在一个切面内定义可重用的切点。上面定义的这个切面LoggingAspect是比较复杂的，每一个连接点都把自己的配置重复写了一遍，如图8-14所示。

```
@Before("execution(public int  com.imut.aop.annotationAspectj.ArithmeticCalculatorImpl.add(int,int))")
@After("execution(* com.imut.aop.annotationAspectj.ArithmeticCalculatorImpl.*(..))")
@AfterReturning(value="execution(* com.imut.aop.annotationAspectj.*.*(..))",returning="result")
@AfterThrowing(value="execution(* com.imut.aop.annotationAspectj.*.*(..))",throwing="ex")
```

图 8-14　切入点表达式的重复配置

其实这个完全没有必要，可以进行简化。具体步骤如下。

(1) 在切面中使用@PointCut()来标注一个方法，定义一个切入点表达式。

(2) 在本切面具体的通知注解中，例如@Before()中，直接指定定义切入点表达式的方法名(需要带上圆括号)即可。

在LoggingAspect切面类中定义切入点及在@Before()定义切点表达式，代码如例8-18所示。

例 8-18：重用切入点表达式。

```
@Pointcut("execution(* com.imut.aop.annotationAspectj.*.*(..))")
public void declarePointCut() {}
@Before("declarePointCut()")
public void beforeMethod(JoinPoint joinPoint) {
    String methodName = joinPoint.getSignature().getName();
    Object [] args= joinPoint.getArgs();
    System.out.println("LoggingAspect:前置通知===>The method "+methodName+"
begin with "+Arrays.asList(args));
}
```

重新执行例8-14的测试类aopTest，执行结果与前次相同。同样也可以在@After()、@AfterReturning()、@AfterThrowing()和@Around()使用@PointCut定义的切点表达式，而不用重复写配置。

编写AspectJ切面时，可以直接在通知注解中书写切入点表达式。但同一个切点表达式可能会在多个通知中重复出现。

在AspectJ切面中，可以通过@PointCut注解将一个切入点声明成简单的方法。切入点的方法体通常是空的，因为将切入点定义与应用程序逻辑混在一起是不合理的。

切入点方法的访问控制符同时也控制着这个切入点的可见性。如果切入点要在多个切面中共享，最好将它们集中在一个公共的类中。在这种情况下，切入点方法必须被声明为public。引入这个切入点时，必须将类名也包括在内。如果切面所在的包与切入点方法所在的包不同，还必须包括包名。

其他通知可以通过方法名称引入该切入点。

8.3.6　多个切面的优先级

在实际开发中，多个切面可能需要在同一个连接点上定义多个通知，这就要考虑多个切面的优先级。在同一个连接点上应用多个切面时，除非明确指定，否则它们的优先级是不确定的。切面的优先级可以通过实现 Ordered 接口或利用@Order 注解指定。实现 Ordered 接口，getOrder()方法的返回值越小，优先级越高。若使用@Order 注解，序号出现在注解中，序号越小，优先级越高。本节介绍使用@Order 注解指定优先级。

通过一个例子简述注解的形式实现 Aspect 的优先级，具体步骤如下。

(1) 在项目 ch08_spring_02 的包 com.imut.aop.annotationAspectj 中创建一个新的切面类 ValidateAspect。ValidateAspect 切面类用来为计算器 ArithmeticCalculator 验证参数。

(2) 为 ValidateAspect 切面类添加前置通知，@Before 的切入点表达式使用 LoggingAspect 中定义的 declarePointCut()。代码如例 8-19 所示。

例 8-19：ValidateAspect 切面类。

```
package com.imut.aop.annotationAspectj;
import org.aspectj.lang.annotation.Aspect;
import org.aspectj.lang.annotation.Before;
import org.springframework.core.annotation.Order;
import org.springframework.stereotype.Component;
@Component
@Aspect
public class ValidateAspect {
    @Before("LoggingAspect.declarePointCut()")
    public void beforeMethod() {
        System.out.println("[[ValidateAspect]]===>The method xxx begin . ");
    }
}
```

(3) 由于有两个切面中定义的多个通知织入同一个切入点，所以需要为切面指定优先级。使用@Order 注解指定优先级。核心代码如例 8-20 所示。

例 8-20：使用@Order 注解指定优先级。

```
//为 ValidateAspect 切面指定优先级
@Component                                    //标识为一个组件
@Aspect                                       //标识为一个切面
@Order(2)                                     //指定优先级为 2
public class ValidateAspect {
…
}
//
@Component                                    //标识为一个组件
@Aspect                                       //标识为一个切面
@Order(3)                                     //指定优先级为 3
public class LoggingAspect {
…
}
```

再次执行例 8-14 的测试类 aopTest，执行结果如图 8-15 所示。

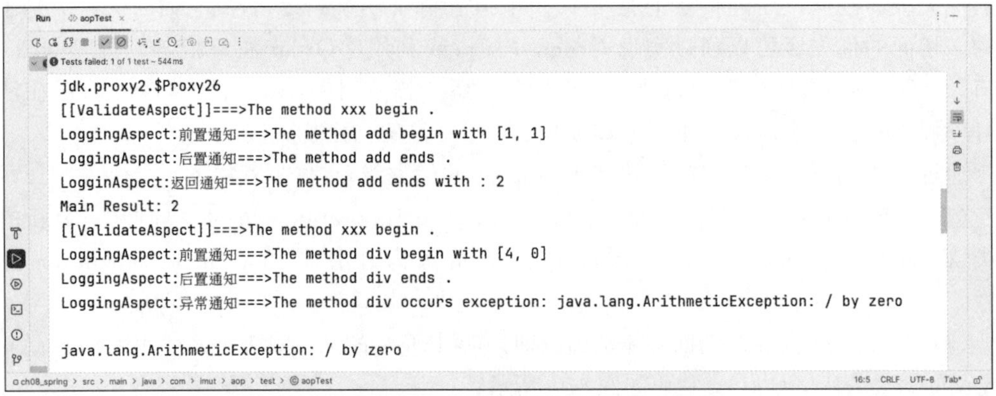
Run
aopTest
Tests failed: 1 of 1 test – 544 ms
jdk.proxy2.$Proxy26
[[ValidateAspect]]===>The method xxx begin .
LoggingAspect:前置通知===>The method add begin with [1, 1]
LoggingAspect:后置通知===>The method add ends .
LogginAspect:返回通知===>The method add ends with : 2
Main Result: 2
[[ValidateAspect]]===>The method xxx begin .
LoggingAspect:前置通知===>The method div begin with [4, 0]
LoggingAspect:后置通知===>The method div ends .
LoggingAspect:异常通知===>The method div occurs exception: java.lang.ArithmeticException: / by zero
java.lang.ArithmeticException: / by zero
ch08_spring > src > main > java > com > imut > aop > test > aopTest
16:5 CRLF UTF-8 Tab*

图 8-15　为多个切面指定优先级的执行结果

从执行结果可以看到，ValidateAspect 切面的优先级被指定为 2，LoggingAspect 切面的优先级被指定为 3。序号越小，优先级越高，即 ValidateAspect 切面的优先级高于 LoggingAspect 切面，ValidateAspect 切面的通知先于 LoggingAspect 切面的通知执行。

8.4　基于 XML 配置开发 Spring AOP

除了使用 AspectJ 注解声明切面，Spring 也支持在 Bean 配置文件中声明切面。这种声明是通过 aop schema 中的 XML 元素完成的。

通常情况下，基于注解的声明要优先于基于 XML 的声明。通过 AspectJ 注解，切面可以与 AspectJ 兼容，而基于 XML 的配置方式则是 Spring 专有的。由于 AspectJ 得到越来越多的 AOP 框架支持，以注解风格编写的切面将会有更多重用的机会。

8.3 节详细讨论了基于注解 AOP 的开发，本节介绍使用 XML 方式开发 AOP。其实它们的原理是相同的，所以这里主要介绍一些用法。这里需要在 XML 中引入 AOP 的命名空间，所以先来了解一下 AOP 可配置的元素，如表 8-3 所示。

表 8-3　基于 XML 配置 AOP 的元素

AOP 配置元素	用　　途	备　　注
aop:advisor	定义 AOP 的通知器	一种较老的方式，目前很少使用，所以本书不再论述
aop:aspect	定义一个切面	
aop:before	定义前置通知	
aop:after	定义后置通知	
aop:around	定义环绕方式	
aop:after-returning	定义返回通知	
aop:after-throwing	定义异常通知	
aop:config	顶层的 AOP 配置元素	AOP 的配置是以它为开始的

续表

AOP 配置元素	用　　途	备　　注
aop:declare-parents	给通知引入新的额外接口,增强功能	
aop:pointcut	定义切点	

学习了@AspectJ 注解驱动的切面开发之后,对于基于 XML 配置的切面开发,也是依据图 8-8 的织入流程来操作的。首先定义要拦截的类和方法。尽管 Spring 不强制定义接口使用 AOP,有接口使用 JDK 动态代理,没有接口则使用 CGLIB 动态代理。但建议使用接口,这有利于实现和定义相分离,使得系统更为灵活。

(1) 在 ch08_spring_01 项目工程下新建包 com.imut.aop.xmlAspectj,将例 8-1 和例 8-2 定义的计算器 arithmeticCalculator 接口及其实现类复制到该包下,并且删除代码中的@component 注解。

(2) 修改例 8-19 定义的 ValidateAspect 切面类,将其放到 com.imut.aop.xmlAspectj 包下。修改后的 ValidateAspect 切面类代码如例 8-21 所示。

例 8-21:基于 XML 配置的 ValidateAspect 切面类。

```
package com.imut.aop.xmlAspectj;
public class ValidateAspect {
    public void beforeMethod() {
        System.out.println("[[ValidateAspect]]===>The method xxx begin . ");
    }
}
```

对比例 8-19,上述代码只是将例 8-19 中的所有注解都删掉了,后面将通过 XML 来配置切面。对于 LoggingAspect 切面类做同样的操作,这里不再赘述。

(3) 在 src/main/java 目录下创建配置文件 applicationContext-xml.xml,配置 aop 命名空间。在配置文件中分别配置目标对象、切面类,核心代码如例 8-22 所示。

例 8-22:配置目标对象与切面类。

```
<!-- 目标对象 -->
< bean  id =" arithmeticCalculatorImpl "  class = " com. imut. aop. xmlAspectj.
ArithmeticCalculatorImpl"></bean>
<!-- 切面类 -->
<bean id="loggingAspect" class="com.imut.aop.xmlAspectj.LoggingAspect">
</bean>
<bean id="validateAspect" class="com.imut.aop.xmlAspectj.ValidateAspect">
</bean>
```

(4) 在配置文件中使用＜aop:config＞标签完成 AOP 的配置。首先使用＜aop:poincut＞标签配置切入点表达式,然后使用＜aop:aspect＞标签配置 LoggingAspect 切面,并为其分别配置前置通知、后置通知、返回通知和异常通知。通知标签中的 pointcut-ref 属性值为＜aop:poincut＞定义得到的切入点表达式。由于环绕通知一般不与前面 4 种通知同时定义,这里不再给出其配置,方法与前述通知类似。最后配置 ValidateAspect 切面及通知。核心代码如例 8-23 所示。

例 8-23：配置 AOP。

```
<!-- 配置 AOP -->
<aop:config>
    <!-- 1. 配置切入点表达式  -->
    <aop:pointcut expression="execution(* com.imut.aop.xmlAspectj.*.*(..))"
id="myPointCut"/>
    <!-- 2. 配置切面 -->
    <aop:aspect ref="loggingAspect" order="2" >
        <!-- 3. 通知 -->
        <aop:before method="beforeMethod"  pointcut-ref="myPointCut" />
        <aop:after method="afterMethod" pointcut-ref="myPointCut"/>
        <aop:after-returning method="afterReturningMethod" pointcut-ref="myPointCut"
                returning="result"/>
        <aop:after-throwing method="afterThrowingMethod" pointcut-ref="myPointCut"
                throwing="ex"/>
        <!-- <aop:around method=""/> -->
    </aop:aspect>
    <aop:aspect ref="validateAspect" order="1">
        <aop:before method="beforeMethod" pointcut-ref="myPointCut"/>
    </aop:aspect>
</aop:config>
```

基于 XML 配置的 AOP，多个切面的优先级可以由<aop:aspect>标签的 order 属性设置，同样值越小，优先级越高。

将例 8-12 定义的测试类 aopTest 中的配置文件改为 applicationContext-xml.xml，并引用 com.imut.aop.xmlAspectj 包中的计算器 arithmeticCalculatorImpl。执行测试类，得到图 8-16 的执行结果。

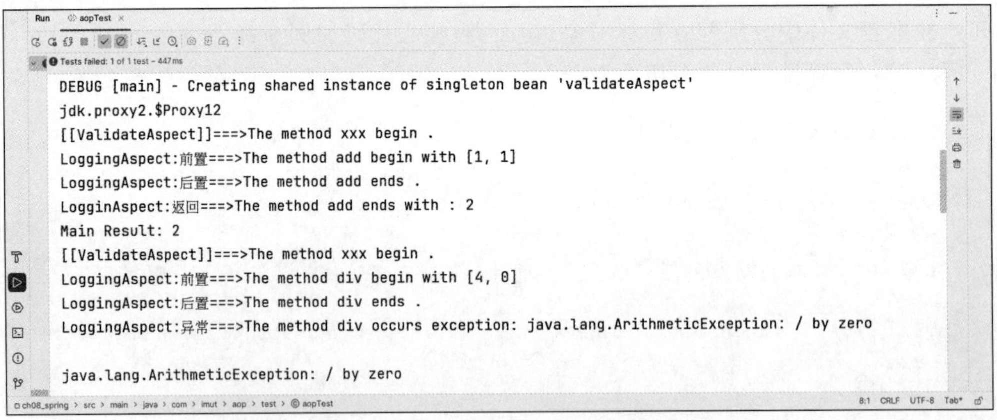
Run
aopTest
Tests failed: 1 of 1 test – 447 ms
DEBUG [main] - Creating shared instance of singleton bean 'validateAspect'
jdk.proxy2.$Proxy12
[[ValidateAspect]]===>The method xxx begin .
LoggingAspect:前置===>The method add begin with [1, 1]
LoggingAspect:后置===>The method add ends .
LogginAspect:返回===>The method add ends with : 2
Main Result: 2
[[ValidateAspect]]===>The method xxx begin .
LoggingAspect:前置===>The method div begin with [4, 0]
LoggingAspect:后置===>The method div ends .
LoggingAspect:异常===>The method div occurs exception: java.lang.ArithmeticException: / by zero
java.lang.ArithmeticException: / by zero
ch08_spring > src > main > java > com > imut > aop > test > aopTest
8:1 CRLF UTF-8 Tab*

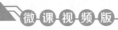

图 8-16　基于 XML 配置 AOP 的测试类执行结果

从图 8-15 和图 8-16 可以看出，基于注解的方式与基于 XML 的方式的执行结果相同。相对来说，使用注解的方式更加简单、方便，所以在实际开发中推荐使用注解的方式进行 AOP 开发。

Spring 框架为同一问题提供多种选择,可能使初学者困惑。应根据项目情况选择。若使用 JDK 5.0 以上版本,可考虑使用@AspectJ 注解减少配置工作;若不愿使用注解或 JDK 版本低,可选择<aop:aspect>配合普通 JavaBean。

8.5 综合案例

本章案例主要使用 Spring AOP 实现模拟日志记录的功能。

8.5.1 案例设计

本章案例划分为持久层和业务逻辑层。持久层包含实体类,业务逻辑层包含 Service 接口和对应的实现类。业务逻辑层的接口统一使用“Service”结尾,实现类的命名规则是在接口名后添加“Impl”。记录日志的功能被定义为一个切面类,并将业务逻辑层中添加学生的方法作为切入点。本章案例使用的主要文件如表 8-4 所示。

表 8-4 本章案例使用的主要文件

文 件	所在包/路径	功 能
log4j.properties	src/	日志输出配置文件
applicationContext.xml	src/	Spring 的配置文件
Student.java	com.imut.pojo	封装学生信息的类
StudentService.java	com.imut.service	定义操作学生类业务层的接口
StudentServiceImpl.java	com.imut.service.impl	操作学生类业务层接口的实现类
Logger.java	com.imut.aspect	切面类
StudentTest.java	com.imut.test	测试类

8.5.2 案例演示

运行 StudentTest.java 中的 test()方法,在控制台输出的结果如图 8-17 所示。

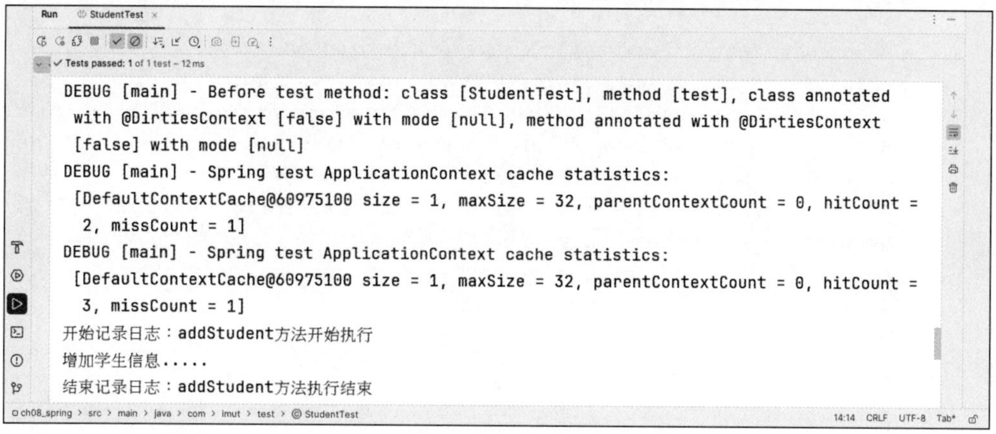

```
Run   StudentTest ×
✓ Tests passed: 1 of 1 test – 12 ms
DEBUG [main] - Before test method: class [StudentTest], method [test], class annotated
 with @DirtiesContext [false] with mode [null], method annotated with @DirtiesContext
 [false] with mode [null]
DEBUG [main] - Spring test ApplicationContext cache statistics:
 [DefaultContextCache@60975100 size = 1, maxSize = 32, parentContextCount = 0, hitCount =
  2, missCount = 1]
DEBUG [main] - Spring test ApplicationContext cache statistics:
 [DefaultContextCache@60975100 size = 1, maxSize = 32, parentContextCount = 0, hitCount =
  3, missCount = 1]
开始记录日志：addStudent方法开始执行
增加学生信息.....
结束记录日志：addStudent方法执行结束
ch08_spring > src > main > java > com > imut > test > StudentTest        14:14  CRLF  UTF-8  Tab*
```

图 8-17 运行结果

8.5.3 代码实现

1. 创建项目，添加依赖

将 Maven 工程 ch07_spring 复制为 ch08_spring，在 pom.xml 中增加 Spring 的 aspects、test、aopalliance 以及 aspectjweaver 依赖，增加的代码如下所示。

```
<dependency>
    <groupId>org.springframework</groupId>
    <artifactId>spring-aspects</artifactId>
    <version>6.0.11</version>
</dependency>
<dependency>
    <groupId>org.aspectj</groupId>
    <artifactId>aspectjweaver</artifactId>
    <version>1.9.7</version>
</dependency>
<dependency>
    <groupId>aopalliance</groupId>
    <artifactId>aopalliance</artifactId>
    <version>1.0</version>
</dependency>
<dependency>
    <groupId>org.springframework</groupId>
    <artifactId>spring-test</artifactId>
    <version>6.0.11</version>
</dependency>
```

2. 编写配置文件

修改 Spring 配置文件 applicationContext.xml，增加 AOP 的约束，增加对@Aspectj 注解支持的配置。

配置文件 applicationContext.xml 代码如下。

```
<?xml version="1.0" encoding="UTF-8"?>
<beans xmlns="http://www.springframework.org/schema/beans"
      xmlns:context="http://www.springframework.org/schema/context"
      xmlns:aop="http://www.springframework.org/schema/aop"
      xmlns:xsi="http://www.w3.org/2001/XMLSchema-instance"
      xsi:schemaLocation="http://www.springframework.org/schema/beans
      http://www.springframework.org/schema/beans/spring-beans.xsd
      http://www.springframework.org/schema/context
      http://www.springframework.org/schema/context/spring-context.xsd
      http://www.springframework.org/schema/aop
      http://www.springframework.org/schema/aop/spring-aop.xsd">
      <context:component-scan base-package="com.imut"/>
      <aop:aspectj-autoproxy/>
</beans>
```

3. 创建实体类

实体类 Student.java 文件与第 7 章综合案例中的 Student.java 文件内容一致，不再赘述。

4. 编写 Service 层接口及其实现类

在/src/main/java 下新建 com.imut.service 包，在该包下新建接口文件 StudentService.java。在/src/main/java 下新建 com.imut.service.impl 包，在该包下新建接口实现类 StudentServiceImpl.java。

程序 StudentService.java 的代码如下。

```
public interface StudentService {
    public void addStudent();
}
```

程序 StudentServiceImpl.java 的代码如下。

```
@Component
public class StudentServiceImpl implements StudentService {
    @Override
    public void addStudent() {
        System.out.println("增加学生信息...");
    }
}
```

5. 编写切面类

在/src/main/java 下新建 com.imut.aspect 包，在该包下新建切面类文件 Logger.java。程序 Logger.java 的代码如下。

```
@Component                                  //将该类交给 Spring 容器管理
@Aspect                                     //表示当前类是一个切面类
public class Logger {
    @Pointcut("execution(* com.imut.service.*.*(..))")
    public void pt() { }
    @Around("pt()")
    public Object log(ProceedingJoinPoint pjp) throws Throwable {
        System.out.println("开始记录日志:"+pjp.getSignature().getName()+"方法
开始执行");
        Object obj = pjp.proceed();
        System.out.println("结束记录日志:"+pjp.getSignature().getName()+"方法
执行结束");
        return obj;
    }
}
```

6. 编写测试类

修改 com.imut.test 包下的测试类 StudentTest.java，测试模拟日志记录的功能。

程序 StudentTest.java 的代码如下。

```
@RunWith(SpringJUnit4ClassRunner.class)
@ContextConfiguration(locations="/applicationContext.xml")
public class StudentTest {
    @Autowired
    private StudentService studentService;
    @Test
    public void test() {
        studentService.addStudent();
    }
}
```

8.6 习题

1. 选择题

(1) 以下关于 Spring AOP 的介绍，错误的是(　　)。

A. AOP 的全称是 Aspect-Oriented Programming，即面向切面编程(也称面向方面编程)

B. AOP 采取横向抽取机制，将分散在各个方法中的重复代码提取出来，这种横向抽取机制的方式 OOP 思想是无法办到的

C. AOP 是一种新的编程思想，采取横向抽取机制，它是 OOP 的升级替代品

D. 目前流行的 AOP 框架有两个，分别为 Spring AOP 和 AspectJ

(2) 关于 AOP 的基本术语，下列选项错误的是(　　)。

A. 通知(Advice)是在目标类连接点上执行的一段代码

B. 目标对象(Target)是指通知所作用的目标业务类

C. 代理(Proxy)是指通知所作用的目标业务类

D. 织入(Weaving)是将通知添加到目标类具体链接点上的过程

(3) 关于 Spring AOP 配置文件(XML 文件)中的元素，下列选项错误的是(　　)。

A. <aop:config>元素是 AOP 配置的根元素

B. <aop:aspect>元素用于指定切面

C. <aop:pointcut>元素用于指定切点

D. <aop:after-returning>用于指定异常通知

(4) 在下列注解中，可以指定切面优先级的是(　　)。

A. @Aspect　　B. @Pointcut　　C. @Around　　D. @Order

2. 填空题

(1) AOP 指的是________。

(2) Spring AOP 支持的通知类型共有________种。

(3) Spring AOP 的实现机制有两种，它们分别是________和________。

(4) 在默认情况下,Spring AOP 的实现方式是________。

(5) 当目标对象是一个没有实现任何接口的类时,Spring AOP 通过________实现。

3. 思考题

(1) 简述 AOP 的概念。

(2) 列举你所知道的 AOP 专业术语并解释。

(3) 列举你所知道的 Spring 框架通知类型并解释。

第9章 Spring框架的数据库编程

Spring 框架简化了 JavaEE API 的使用，包括数据库编程。数据库用于持久化业务数据，操作数据库是应用程序的常见任务。本章将详细讲解 Spring JDBC 相关知识。

9.1 Spring JDBC 基础

9.1.1 Spring JDBC 简介

Spring JDBC 的核心是 JdbcTemplate 类，它提供简化的 JDBC 模板类、统一的 DAO 类，封装 SQLException 为 DataAccessException 异常，以及集成其他数据库访问框架，简化了 JDBC 操作，处理资源建立和释放，避免常见错误。JdbcTemplate 类提供模板方法，控制整个 JDBC 操作过程，并允许覆盖特定任务，以最低的工作量实现数据库存取。Spring 的 JDBC 模块由 core、object、DataSource 和 support 4 个包组成，它们相互协作，共同支撑 Spring 的 JDBC 功能。

9.1.2 为什么要使用 Spring 的 JdbcTemplate

使用 JdbcTemplate 相较于直接使用 JDBC，可省去加载驱动、释放资源、异常处理等烦琐步骤，使代码更直观。只需提供 SQL 语句并提取结果，即可完成数据库操作。此外，JdbcTemplate 还可将数据对象映射为实体类，无须手动从 ResultSet 中提取和赋值，提高开发效率。

9.1.3 Spring JdbcTemplate 的解析

JdbcTemplate 类是 Spring JDBC 的核心，是其他高层抽象类的基础。它继承自 JdbcAccessor，并实现 JdbcOperations 接口，具有简单的继承关系，如图 9-1 所示。

JdbcAccessor 类为子类提供了访问数据库所需的公共属性，包括 DataSource 和 SQLExceptionTranslator。DataSource 用于获取数据库连接，并提供缓冲池和分布式事务支持。SQLExceptionTranslator 接口负责转译 SQLException，使 JdbcTemplate 可以委托其实现类完成转译工作。

JdbcOperations 接口定义了 JdbcTemplate 类中的操作集合，包括添加、修改、查询和删除等。

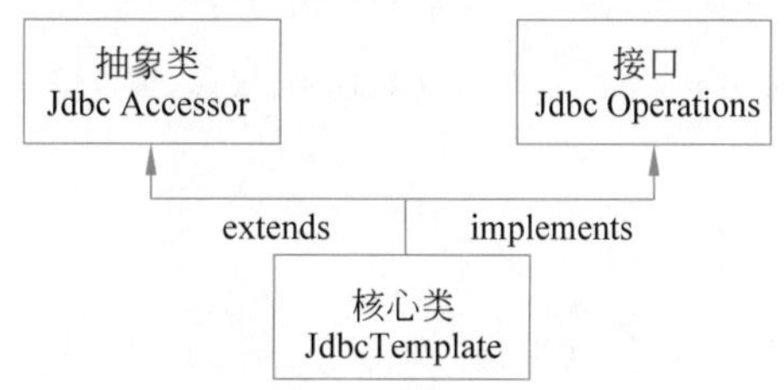

图 9-1　JdbcTemplate 类的继承关系

9.1.4 Spring JdbcTemplate 类

JdbcTemplate 类是一个模板类，Spring JDBC 中更高层次的抽象类均在 JdbcTemplate 类基础上构建。JdbcTemplate 类包含了所有操作数据库的基本方法，包括添加、删除、查询、更新等。

使用 JdbcTemplate 类对象操作数据库之前，需要为其提供数据源。为了便于操作数据库，JdbcTemplate 类还提供了一系列方法，其中常用的如表 9-1 所示。在实际开发中，开发者可根据具体需要选择使用。

表 9-1　JdbcTemplate 类的常用方法

方　　法	描　　述
void execute(String sql)	发出单个 SQL 执行，通常是 DDL 语句
<T>T execute(String sql, PreparedStatementCallback<T> psc)	执行 JDBC 数据访问操作，实现 JDBC PreparedStatement 上的回调操作
<T>T execute(String callString, CallableStatementCallback<T> csc)	执行 JDBC 数据访问操作，实现为处理 JDBC CallableStatement 的回调操作
List query(String sql, Object[]args, RowMapper<T> rowMapper)	查询给定 SQL 以从 SQL 创建预准备语句以及绑定到查询的参数列表，通过 RowMapper 将每一行映射到 Java 对象
List query (String sql, RowMapper<T> rowMapper)	执行给定静态 SQL 的查询，通过 RowMapper 将每一行映射到 Java 对象
List query(String sql, RowMapper<T> rowMapper, Object args)	查询给定 SQL 以从 SQL 创建预准备语句以及绑定到查询的参数列表，通过 RowMapper 将每一行映射到 Java 对象
TqueryForObject (String sql, Class <T> requiredType)	执行 SQL 语句，返回结果对象
List queryForList(String sql)	执行 SQL 语句，返回包含有执行结果的 List 集合
List queryForList (String sql,Class<T> elementType)	执行 SQL 语句，返回包含有执行结果的 List 集合

续表

方　　法	描　　述
List queryForList (String sql, Class<T> elementType, Object args)	查询给定 SQL 以从 SQL 创建预准备语句以及绑定到查询的参数列表,期望结果列表
ListqueryForList(String sql, Object args)	查询给定 SQL 以从 SQL 创建预准备语句以及绑定到查询的参数列表,期望结果列表
int update(String sql)	发出单个 DML 操作
int update(String sql, Object args)	发出单个 DML 操作,绑定给定的参数

9.1.5 Spring JDBC 的配置

Spring JDBC 模块主要由 4 个包组成,分别是 core(核心包)、dataSource(数据源包)、object(对象包)和 support(支持包),这 4 个包的具体说明如表 9-2 所示。

表 9-2　Spring JDBC 模块的主要包及说明

包	说　　明
core	包含 JDBC 的核心功能,即 JdbcTemplate 类、SimpleJdbclnsert 类、SimpleJdbcCall 类、NamedParameterJdbcTemplate 类
dataSource	访问数据源的实用工具类,它有多种数据源的实现,可以在 JavaEE 容器外部测试 JDBC 代码
object	以面向对象的方式访问数据库,它允许执行查询并将返回结果作为业务对象,可以在数据表的列和业务对象的属性之间映射查询结果
support	包含 core 包和 object 包的支持类,如提供异常转换功能的 SQLException 类

Spring 框架对数据库的操作都封装在这 4 个包中了,要使用 SpringJDBC 时,就需要对这 4 个包进行配置。在 Spring 框架中,JDBC 的配置是在配置文件 applicationContext.xml 中完成的,其配置模板如例 9-1 所示。

例 9-1: applicationContext.xml。

```
<?xml version="1.0" encoding="UTF-8"?>
<beans xmlns="http://www.springframework.org/schema/beans"
    xmlns:xsi="http://www.w3.org/2001/XMLSchema-instance"
    xsi:schemaLocation="http://www.springframework.org/schema/beans
    http://www.springframework.org/schema/beans/spring-beans-3.2.xsd">
    <!-- 1 配置数据源 -->
    <bean id="dataSource" class=
     "org.springframework.jdbc.datasource.DriverManagerDataSource">
        <property name="driverClassName" value="com.mysql.jdbc.Driver" />
        <property name="url" value="jdbc:mysql://localhost:3306/db_ssm" />
        <property name="username" value="root" />
        <property name="password" value="123456" />
    </bean>
    <!-- 2 配置 JDBC 模板 -->
    <bean id="jdbcTemplate" class="org.springframework.jdbc.core.JdbcTemplate">
```

```
        <property name="dataSource" ref="dataSource" />
    </bean>
    <bean id="employeeDao" class="com.imut.dao.employeeDaoImpl">
        <property name="jdbcTemplate" ref="jdbcTemplate" />
    </bean>
</beans>
```

上述代码定义了 3 个 Bean，分别是 DataSource、JdbcTemplate 和需要注入类的 Bean。其中 DataSource 对应的 org.springframework.jdbc.datasource.DriverManagerDataSource 类，用于对数据源进行配置。DataSource 的配置就是连接数据库时需要的 4 个属性，如表 9-3 所示。

表 9-3　dataSource 的 4 个属性

属 性 名	含 义
driverClassName	所使用的驱动名称，对应驱动 JAR 包中的 Driver 类
url	数据源所在地址
username	访问数据库的用户名
password	访问数据库的密码

这 4 个属性需要根据数据库类型和机器配置进行设置。如果使用不同数据库，需根据实际更改属性信息。JdbcTemplate 配置为数据源，创建时需注入 DataSource。其他使用 JdbcTemplate 的 Bean 也需注入 JdbcTemplate。通常注入到 DAO 类中，用于执行数据库操作。

9.2 JdbcTemplate 操作数据库

JdbcTemplate 类中提供了大量更新和查询数据库的方法，本节将详细讲解一些常用方法。

9.2.1 JdbcTemplate 类实现 DDL 操作

Execute(String sql)方法能够完成执行 SQL 语句的功能。通常情况下，可以通过 Execute(String sql)方法完成创建数据库表的 DDL 操作，接下来通过一个实例演示 DDL 操作。

(1) 在 MySQL 数据库中创建名为 db_ssm 的数据库。登录数据库的用户名为 root，密码是 123456。在 db_ssm 库中创建表 tbl_employee。数据表 tbl_employee 的字段信息如图 9-2 所示。

名	类型	长度	小数点	不是 null	键	注释
id	int	11		☑	🔑1	
last_name	varchar	50		☐		
email	varchar	50		☐		
gender	char	1		☐		

图 9-2　tbl_employee 表的结构

(2) 在 IDEA 中创建一个名为 ch09_spring_01 的 Java 项目，使用 Maven 管理项目中所需的依赖。在 pom.xml 文件中，MySQL 数据库的驱动 JAR 包、Spring JDBC 的 JAR 包等的配置如下所示。

```
<dependency>
        <groupId>mysql</groupId>
        <artifactId>mysql-connector-java</artifactId>
        <version>8.0.11</version>
</dependency>
<dependency>
    <groupId>org.springframework</groupId>
    <artifactId>spring-jdbc</artifactId>
    <version>5.2.25.RELEASE</version>
</dependency>
<dependency>
    <groupId>org.springframework</groupId>
    <artifactId>spring-tx</artifactId>
    <version>5.2.25.RELEASE</version>
</dependency>
```

(3) 在 resources 目录下创建名为 applicationContext.xml 的配置文件，在该文件中配置 id 为 dataSource 的数据源 Bean 和 id 为 jdbcTemplate 的 JDBC 模板 Bean，并将数据源注入 JDBC 模板中，如例 9-1 所示。

(4) 在 src/main/java 目录下创建一个名为 com.imut.test 的包，并在该包中创建测试类 JdbcTemplateTest 。先在该类中通过 Spring 容器获取配置文件中定义的 JdbcTemplate 实例，然后使用该实例的 execute(String sql)方法执行创建数据表的 SQL 语句，代码如例 9-2 所示。

例 9-2：JdbcTemplateTest.java。

```
package com.imut.test;
import org.junit.Before;
import org.junit.Test;
import org.springframework.context.ApplicationContext;
import org.springframework.context.support.ClassPathXmlApplicationContext;
import org.springframework.jdbc.core.JdbcTemplate;
public class JdbcTemplateTest {
    private ApplicationContext ctx = null ;
    private JdbcTemplate jdbcTemplate = null ;
    @Before  //在每个@Test 的方法之前执行
    public void init() {
        ctx = new ClassPathXmlApplicationContext("applicationContext.xml");
        jdbcTemplate = ctx.getBean("jdbcTemplate",JdbcTemplate.class);
    }
    @Test
    public void testExcute() {
        jdbcTemplate.execute("create table tbl_employee(" + "id int(11) primary
key auto_increment," + "last_name varchar(50)," +  "email varchar(50)," + "
gender char(1));");
```

```
        System.out.println("账户表 employee 创建成功!");
    }
}
```

9.2.2 JdbcTemplate 类实现 DML 操作

在 JdbcTemplate 类提供的方法中，update()方法、batchUpdate()方法用于实现 DML 操作。DML 操作包括对数据表中数据的添加、删除和更新。

1. 用 SQL 语句和参数更新数据库

使用 update()方法执行 SQL 语句，往数据库表中插入一条记录。在测试类 JdbcTemplateTest 中添加代码，如例 9-3 所示。

例 9-3：testUpdate。

```
//update()：完成增删改操作
@Test
public void testUpdate(){
    String sql = " insert into tbl _ employee (last _ name, email, gender)
values(?,?,?)";
    jdbcTemplate.update(sql, "Jerry","jerry@sina.com","0");
}
```

执行测试类 JdbcTemplateTest 之后，在 MySQL 的可视化工具中查询 db_ssm 数据库的 tbl_employee 表，可以得到图 9-3 所示的结果。

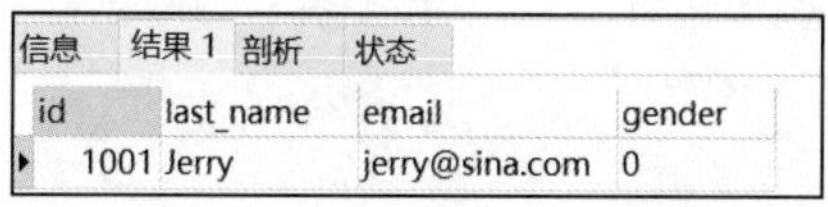

图 9-3　更新数据库的结果

2. 批量更新数据库

使用 batchUpdate()方法执行 SQL 语句，实现批量更新数据库。这里通过一个实例演示批量插入数据。在测试类 JdbcTemplateTest 中添加代码，如例 9-4 所示。

例 9-4：testBatchUpdate。

```
@Test
public void testBatchUpdate() {
    String sql = "insert into tbl_employee(last_name,email,gender)
values(?,?,?)";
    List<Object[]> batchArgs = new ArrayList<Object[]> ();
    batchArgs.add(new Object[] {"Lily","Lily@60.com","1"});
    batchArgs.add(new Object[] {"Tom","Tom@163.com","0"});
    batchArgs.add(new Object[] {"Ali","Ali@126.com","1"});
    jdbcTemplate.batchUpdate(sql, batchArgs);
}
```

从上述代码可以看出，在一次数据库会话中，使用 batchUpdate()方法可以实现定义一个 insert 语句，插入多条记录。执行测试类 JdbcTemplateTest，在 MySQL 的可视化工具中查询 db_ssm 数据库的 tbl_employee 表，可以得到图 9-4 所示的结果。

id	last_name	email	gender
1001	Tom	Tom@sina.com	1
1002	Lily	Lily@60.com	1
1003	Tom	Tom@163.com	0
1004	Ali	Ali@126.com	1

图 9-4　批量更新数据库的结果

9.2.3　JdbcTemplate 类实现 DQL 操作

JdbcTemplate 类提供 query()、queryForObject()、queryForList()方法用于 DQL 操作。查询单条记录时，常用 query()或 queryForObject()，多条记录则用 query()或 queryForList()。处理结果集时，若需映射为自定义类，需使用 RowMapper<T>接口。该接口定义了 mapRow()方法，用于字段值和类属性的映射。Spring 提供了 BeanPropertyRowMapper<T>实现类，简化了开发过程。

1. 查询单条记录

使用 queryForObject()方法查询单行记录，在测试类 JdbcTemplateTest 中添加代码，如例 9-5 所示。

例 9-5：查询单条记录。

```
@Test
public void testQueryForObject() {
    String sql = "select id,last_name ,email, gender from tbl_employee where id = ? ";
    RowMapper<Employee> rowMapper = new
                BeanPropertyRowMapper<>(Employee.class);
    Employee employee = jdbcTemplate.queryForObject(sql, rowMapper, 1001);
    System.out.println(employee);
}
```

执行测试类 JdbcTemplateTest，在 IDEA 的控制台得到图 9-5 所示的结果。

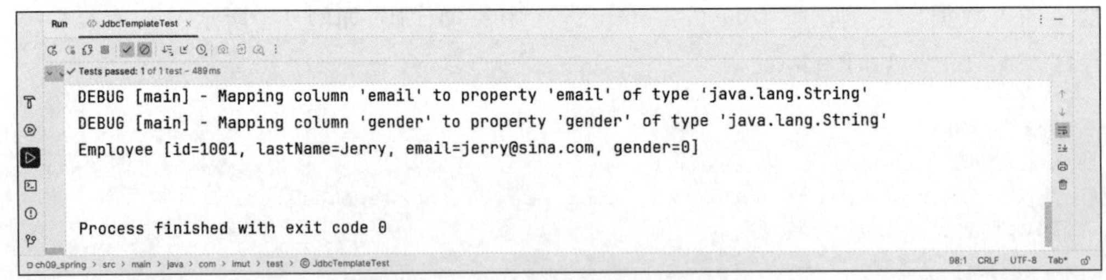

图 9-5　查询单条记录的执行结果

从以上运行结果可以看出，程序输出 id 为 1001 的员工对象信息。由此可见，BeanPropertyRowMapper <Employee>类对象实现了表记录和 Employee 类对象的映射。

2. 查询多条记录

以上介绍了 JdbcTemplate 类对单条记录的查询方法，接下来学习对多条记录的查询方法。使用 query()方法查询多条记录，在测试类 JdbcTemplateTest 中添加代码，如例 9-6 所示。

例 9-6：查询多条记录。

```
@Test
public void testQuery() {
    String sql = "select id ,last_name ,email, gender from tbl_employee";
    RowMapper<Employee> rowMapper =
                    new BeanPropertyRowMapper<> (Employee.class);
    List<Employee> emps = jdbcTemplate.query(sql, rowMapper);
    System.out.println(emps );
}
```

执行 JdbcTemplateTest 类，执行结果如图 9-6 所示。

```
DEBUG [main] - Mapping column 'gender' to property 'gender' of type 'java.lang.String'
[Employee [id=1001, lastName=Jerry, email=jerry@sina.com, gender=0]
, Employee [id=1002, lastName=Lily, email=Lily@60.com, gender=1]
, Employee [id=1003, lastName=Tom, email=Tom@163.com, gender=0]
, Employee [id=1004, lastName=Ali, email=Ali@126.com, gender=1]
]
```

图 9-6　查询多条记录的执行结果

从以上运行结果可以看出，程序输出数据表 employee 中的所有员工信息。由此可见，JdbcTemplate 类实现了多条记录的查询。

3. 查询单值

使用 queryForObject()返回查询数据库表的结果值，在测试类 JdbcTemplateTest 中添加代码，如例 9-7 所示。

例 9-7：查询单值。

```
@Test
public void testQueryForObjectReturnSingleValue() {
    String sql = "select count(id) from tbl_employee";
    Integer result = jdbcTemplate.queryForObject(sql, Integer.class);
    System.out.println(result );
}
```

在以上代码中，程序查询数据表 employee 中所有记录的数量。由于查询结果是 int 类型，因此可使用 Integer 包装类映射查询结果。

执行 JdbcTemplateTest 类，执行结果如图 9-7 所示。

从以上运行结果可以看出，程序输出数据表 employee 中所有记录的数量。

4. 使用具名参数

在经典 JDBC 中，SQL 参数用占位符“?”表示，受位置限制，参数顺序变化需调整绑定。Spring JDBC 则提供了具名参数(named parameter)方法，以冒号开头的名称指定，不受位

```
DEBUG [main] - Fetching JDBC Connection from DataSource
DEBUG [main] - Creating new JDBC DriverManager Connection to
 [jdbc:mysql://localhost:3306/db_ssm?useSSL=false&serverTimezone=Asia/Shanghai]
4
```

图 9-7　查询单值的执行结果

置限制，易于维护，提升可读性。Spring 运行时将具名参数替换为占位符，此方法主要由 NamedParameterJdbcTemplate 类支持。使用具名参数时，可通过 Map 或 SqlParameterSource 提供参数值，批量更新时，可提供 Map 或 SqlParameterSource 的数组作为参数值。

为了使用具名参数 NamedParameterJdbcTemplate 类，需要在 applicationContext 配置文件中配置其 Bean 组件，并配置 dataSource 数据源，核心代码如例 9-8 所示。

例 9-8：具名参数的配置。

```
<!-- NamedParameterJdbcTemplate -->
<bean id="namedParameterJdbcTemplate"
    class="org.springframework.jdbc.core.namedparam.
NamedParameterJdbcTemplate">
    <constructor-arg ref="dataSource"></constructor-arg>
</bean>
```

修改测试类 JdbcTemplateTest，在程序中使用具名参数查询数据库，核心代码如例 9-9 所示。

例 9-9：使用具名参数查询。

```
public class JdbcTemplateTest {
    private ApplicationContext ctx = null ;
    private JdbcTemplate jdbcTemplate = null ;
    private NamedParameterJdbcTemplate namedParameterJdbcTemplate = null ;
    @Before  //会在每个@Test 的方法之前执行
    public void init() {
        ctx = new ClassPathXmlApplicationContext("applicationContext.xml");
        jdbcTemplate = ctx.getBean("jdbcTemplate",JdbcTemplate.class);
        namedParameterJdbcTemplate = ctx.getBean("namedParameterJdbcTemplate",
NamedParameterJdbcTemplate.class);
    }
    //测试具名参数的 JDBC 模板
    @Test
    public void testNamedParameterJdbcTemplate() {
        String sql = "insert into tbl_employee (last_name,email,gender)
                values(:ln,:email,:gender)";
        Map<String,Object> paramMap = new HashMap<String,Object>();
        paramMap.put("ln", "Angel");
        paramMap.put("email", "Angel@sina.com");
        paramMap.put("gender", "0");
```

```
        namedParameterJdbcTemplate.update(sql, paramMap);
    }
}
```

以上代码在测试类 JdbcTemplateTest 中新增了 NamedParameterJdbcTemplate 类的成员属性,并在 init()方法中初始化。NamedParameterJdbcTemplate 类的实例化在例 9-9 的配置文件中进行。在 SQL 语句中,具名参数以冒号开头,由 Map 给出。具名参数解决在占位符很多的情况下关注占位符位置的问题。使用具名参数,程序员只要根据具名参数指定具体的值即可。

执行测试类 JdbcTemplateTest,在 MySQL 的可视化工具中查询 db_ssm 数据库的 tbl_employee 表,可以得到图 9-8 所示的结果。

在 Spring JDBC 框架中,还可以使用 BeanPropertySqlParameterSource 完成 Java 对象到 SQL 语句具名参数的映射。使用该类时,要求 SQL 语句的具名参数名与 Java 对象的属性名保持一致。使用 queryForObject()返回查询数据库表的结果值,在测试类 JdbcTemplateTest 中添加代码,如例 9-10 所示。

例 9-10:使用 BeanPropertySqlParameterSource 的具名参数。

```
@Test
public void testNamedParameterJdbcTemplate2() {
    String sql = "insert into tbl_employee(last_name ,email,gender )
                values(:lastName,:email,:gender)";
    //从 Service 层传入到持久层的对象
    Employee employee = new Employee();
    employee.setLastName("Rose");
    employee.setEmail("Rose@sina.com");
    employee.setGender("0");
    SqlParameterSource paramSource = new
                BeanPropertySqlParameterSource(employee);
    namedParameterJdbcTemplate.update(sql, paramSource);
}
```

执行测试类 JdbcTemplateTest,在 MySQL 的可视化工具中查询 db_ssm 数据库的 tbl_employee 表,可以得到图 9-9 所示的结果。

信息 结果 1 剖析 状态

id	last_name	email	gender
1001	Tom	Tom@sina.com	1
1002	Lily	Lily@60.com	1
1003	Tom	Tom@163.com	0
1004	Ali	Ali@126.com	1
1005	Angel	Angel@sina.com	0

图 9-8 具名参数的执行结果

信息 结果 1 剖析 状态

id	last_name	email	gender
1001	Tom	Tom@sina.com	1
1002	Lily	Lily@60.com	1
1003	Tom	Tom@163.com	0
1004	Ali	Ali@126.com	1
1005	Angel	Angel@sina.com	0
1006	Rose	Rose@sina.com	0

图 9-9 使用 BeanPropertySqlParameterSourcede 具名参数

9.3 使用 Spring JDBC 完成 DAO 封装

在实际开发中，程序与数据库的交互通常由 DAO 完成。为了能够在 DAO 中使用 JdbcTemplate 类对象，需要将 JdbcTemplate 类对象注入 DAO 类中，进而通过 JdbcTemplate 类对象完成对数据库的操作。同时，也可以使用上面提到的具名参数来完成对数据库的操作。

下面通过一个实例演示使用 Spring JDBC 完成 DAO 封装，具体步骤如下。

(1) 在 Java 项目 ch09_spring_01 的 com.imut.pojo 包中创建 employeeDao 类，核心代码如例 9-11 所示。

例 9-11：EmployeeDao.java。

```
public class EmployeeDao {
    private NamedParameterJdbcTemplate namedParameterJdbcTemplate;
    public void insertEmployee(Employee employee) {
        String sql = "insert into tbl_employee(last_name,email,gender)
                values(:lastName,:email,:gender)";
        SqlParameterSource paramSource = new
                BeanPropertySqlParameterSource(employee);
        namedParameterJdbcTemplate.update(sql, paramSource);
    }
    public void setNamedParameterJdbcTemplate(NamedParameterJdbcTemplate
                namedParameterJdbcTemplate) {
        this.namedParameterJdbcTemplate = namedParameterJdbcTemplate;
    }
}
```

以上代码声明了 EmployeeDao 类，封装了对数据库的进行操作的方法。这里以插入数据记录为例，在 EmployeeDao 类中添加 NamedParameterJdbcTemplate 属性，在 insertEmployee 方法中通过具名参数的方法将 employee 对象的信息插入数据库表中。

(2) 在配置文件中定义 Employee 类和 EmployeeDao 类的 Bean 组件，核心代码如例 9-12 所示。

例 9-12：配置 Bean 组件。

```
<bean id="employee" class="com.imut.pojo.Employee">
    <property name="lastName" value="Meimei"/>
    <property name="email" value="Meimei@126.com"/>
    <property name="gender" value="0"/>
</bean>
<bean id="employeeDao" class="com.imut.pojo.EmployeeDao">
    <property name="namedParameterJdbcTemplate"
          ref="namedParameterJdbcTemplate"/>
</bean>
```

修改测试类 JdbcTemplateTest，添加 Employee 类的对象和 EmployeeDao 类的对象作为成员属性，并在 init()方法中获取它们的 Bean 对象。最后，定义 testEmployeeDao()方法，调用 employeeDao 的 insertEmployee()方法，验证该方法的有效性。核心代码如例 9-13 所示。

例 9-13：测试类 JdbcTemplateTest.java。

```
private Employee employee = null;
private EmployeeDao employeeDao = null;
@Before       //会在每个@Test 的方法之前执行
public void init() {
    ctx = new ClassPathXmlApplicationContext("applicationContext.xml");
    jdbcTemplate = ctx.getBean("jdbcTemplate",JdbcTemplate.class);
    namedParameterJdbcTemplate  =
            ctx.getBean("namedParameterJdbcTemplate",
                        NamedParameterJdbcTemplate.class);
    employee = ctx.getBean("employee", Employee.class);
    employeeDao = ctx.getBean("employeeDao", EmployeeDao.class);
}
@Test
public void testEmployeeDao() {
    employeeDao.insertEmployee(employee);
}
```

执行测试类 JdbcTemplateTest，在 MySQL 的可视化工具中查询 db_ssm 数据库的 tbl_employee 表，可以得到图 9-10 所示的结果。

信息 结果 1 剖析 状态

id	last_name	email	gender
1001	Tom	Tom@sina.com	1
1002	Lily	Lily@60.com	1
1003	Tom	Tom@163.com	0
1004	Ali	Ali@126.com	1
1005	Angel	Angel@sina.com	0
1006	Rose	Rose@sina.com	0
1007	Meimei	Meimei@126.cor	0

图 9-10 封装 DAO 类的执行结果

Spring 的 DaoSupport 类可以实现 DAO 封装，但不推荐使用。Spring 提供了丰富的数据操作功能，如 JdbcTemplate 用于简化 JDBC 编程，TransactionTemplate 提供事务支持。在持久层方面，MyBatis 则采用社区开发的 Spring 集成包。

9.4 综合案例

本章案例主要实现“学生管理”模块中根据 sid 查询学生信息、查询所有学生信息、增加学生信息、更新学生信息以及删除学生信息的功能，使用 Spring 的 JDBC 完成数据库的资源管理。

9.4.1 案例设计

本章案例分为持久层、业务逻辑层和数据访问层 3 部分。持久层主要包括实体类。数据访问层则由接口和接口实现类构成，通常接口的名称以“Dao”结尾，而实现类的名称在接口名称后添加“Impl”。业务逻辑层包括 Service 接口和相应的实现类，通常 Service 接口的名称以“Service”结尾，而实现类的名称在接口名后面加上“Impl”。表 9-4 列出了本章案例中主要使用的文件和类。

表 9-4　本章案例使用的文件

文　　件	所在包/路径	功　　能
log4j.properties	src/	日志输出配置文件
applicationContext.xml	src/	Spring 的配置文件
db.properties	src/	数据库连接的属性文件
Student.java	com.imut.pojo	封装学生信息的类
StudentDao.java	com.imut.dao	定义了操作学生表的接口
StudentDaoImpl.java	com.imut.dao.impl	实现了操作学生表接口的方法
StudentService.java	com.imut.service	定义操作学生类业务层的接口
StudentServiceImpl.java	com.imut.service.impl	操作学生类业务层接口的实现类
StudentTest.java	com.imut.test	测试类

9.4.2　案例演示

运行 StudentTest.java 中的 getStudentByIdTest()方法，根据 sid 查询学生信息，在控制台输出的结果如图 9-11 所示。执行 listStudentsTest()方法查询所有学生信息，控制台输出结果如图 9-12 所示。执行 addStudentTest()方法，增加一条姓名为“韩梅”的记录，控制台输出的结果如图 9-13 所示。执行 updateStudentTest()方法，将 sid 为 6 的学生姓名由“韩梅”修改为“韩梅梅”，控制台输出的结果如图 9-14 所示。执行 deleteStudentTest()方法，删除 sid 为 6 的学生信息，控制台输出的结果如图 9-15 所示。

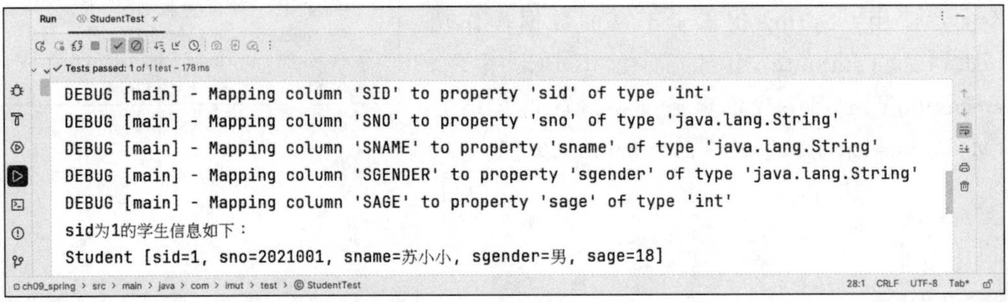

```
Run   StudentTest ×
✓ Tests passed: 1 of 1 test – 178 ms
DEBUG [main] - Mapping column 'SID' to property 'sid' of type 'int'
DEBUG [main] - Mapping column 'SNO' to property 'sno' of type 'java.lang.String'
DEBUG [main] - Mapping column 'SNAME' to property 'sname' of type 'java.lang.String'
DEBUG [main] - Mapping column 'SGENDER' to property 'sgender' of type 'java.lang.String'
DEBUG [main] - Mapping column 'SAGE' to property 'sage' of type 'int'
sid为1的学生信息如下：
Student [sid=1, sno=2021001, sname=苏小小, sgender=男, sage=18]
ch09_spring > src > main > java > com > imut > test > StudentTest    28:1  CRLF  UTF-8  Tab*
```

图 9-11　查询 sid 为 1 的学生信息的运行结果

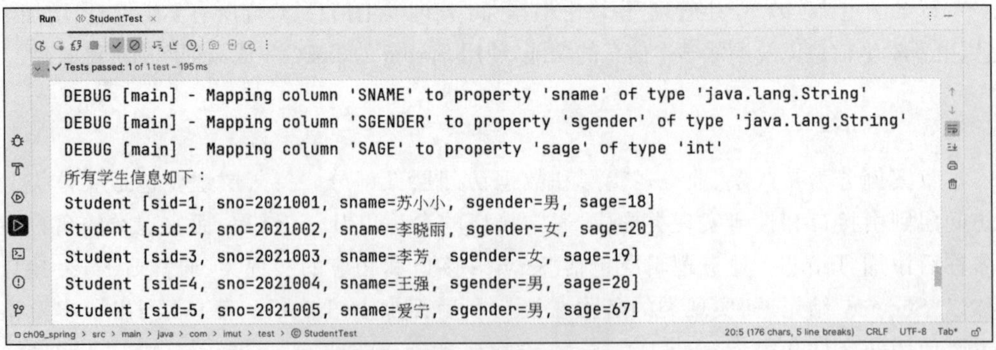

```
Run   StudentTest ×
✓ Tests passed: 1 of 1 test – 195 ms
DEBUG [main] - Mapping column 'SNAME' to property 'sname' of type 'java.lang.String'
DEBUG [main] - Mapping column 'SGENDER' to property 'sgender' of type 'java.lang.String'
DEBUG [main] - Mapping column 'SAGE' to property 'sage' of type 'int'
所有学生信息如下：
Student [sid=1, sno=2021001, sname=苏小小, sgender=男, sage=18]
Student [sid=2, sno=2021002, sname=李晓丽, sgender=女, sage=20]
Student [sid=3, sno=2021003, sname=李芳, sgender=女, sage=19]
Student [sid=4, sno=2021004, sname=王强, sgender=男, sage=20]
Student [sid=5, sno=2021005, sname=爱宁, sgender=男, sage=67]
ch09_spring > src > main > java > com > imut > test > StudentTest    20:5 (176 chars, 5 line breaks)  CRLF  UTF-8  Tab*
```

图 9-12　查询所有学生信息的运行结果

```
Run    StudentTest ×
Tests passed: 1 of 1 test – 167 ms
DEBUG [main] - Executing prepared SQL update
DEBUG [main] - Executing prepared SQL statement [insert into student (sno,sname,sgender,
 sage) values(?,?,?,?)]
DEBUG [main] - Fetching JDBC Connection from DataSource
DEBUG [main] - Creating new JDBC DriverManager Connection to
 [jdbc:mysql://localhost:3306/db_student?characterEncoding=utf8&serverTimezone=GMT&useSSL
 =false]
添加数据成功！
DEBUG [main] - Spring test ApplicationContext cache statistics:
ch09_spring > src > main > java > com > imut > test > StudentTest        46:3  CRLF  UTF-8  Tab*
```

图 9-13　增加学生信息的运行结果

```
Run    StudentTest ×
Tests passed: 1 of 1 test – 232 ms
DEBUG [main] - Fetching JDBC Connection from DataSource
DEBUG [main] - Creating new JDBC DriverManager Connection to
 [jdbc:mysql://localhost:3306/db_student?characterEncoding=utf8&serverTimezone=GMT&useSSL
 =false]
DEBUG [main] - Mapping column 'SID' to property 'sid' of type 'int'
DEBUG [main] - Mapping column 'SNO' to property 'sno' of type 'java.lang.String'
DEBUG [main] - Mapping column 'SNAME' to property 'sname' of type 'java.lang.String'
DEBUG [main] - Mapping column 'SGENDER' to property 'sgender' of type 'java.lang.String'
DEBUG [main] - Mapping column 'SAGE' to property 'sage' of type 'int'
DEBUG [main] - Executing prepared SQL update
DEBUG [main] - Executing prepared SQL statement [update student set sno=?,sname=?,
 sgender=?,sage=? where sid=?]
DEBUG [main] - Fetching JDBC Connection from DataSource
ch09_spring > src > main > java > com > imut > test > StudentTest        54:3  CRLF  UTF-8  Tab*
```

图 9-14　更新学生信息的运行结果

```
Run    StudentTest ×
Tests passed: 1 of 1 test – 154 ms
DEBUG [main] - Executing prepared SQL update
DEBUG [main] - Executing prepared SQL statement [delete from student where sid=?]
DEBUG [main] - Fetching JDBC Connection from DataSource
DEBUG [main] - Creating new JDBC DriverManager Connection to
 [jdbc:mysql://localhost:3306/db_student?characterEncoding=utf8&serverTimezone=GMT&useSSL
 =false]
删除数据成功！
ch09_spring > src > main > java > com > imut > test > StudentTest        60:1  CRLF  UTF-8  Tab*
```

图 9-15　删除学生信息的运行结果

执行增加、更新、删除操作后数据库表的结果如图 9-16 所示。

SIE	SNO	SNAM	SGI	SAG
1	2021001	苏小小	男	18
2	2021002	李晓丽	女	20
3	2021003	李芳	女	19
4	20210104	王强	男	20
5	20210105	爱宁	男	67
6	202101008	韩梅	女	18

SIE	SNO	SNAM	SGI	SAG
1	2021001	苏小小	男	18
2	2021002	李晓丽	女	20
3	2021003	李芳	女	19
4	20210104	王强	男	20
5	20210105	爱宁	男	67
6	202101008	韩梅梅	女	18

SIE	SNO	SNAM	SGI	SAG
1	2021001	苏小小	男	18
2	2021002	李晓丽	女	20
3	2021003	李芳	女	19
4	20210104	王强	男	20
5	20210105	爱宁	男	67

图 9-16　执行添加、更新、删除操作后数据库中表的结果

9.4.3 代码实现

1. 创建项目，添加依赖

将 Maven 工程 ch08_spring 复制为 ch09_spring，在 pom.xml 文件中增加 Spring 的 jdbc 包、Spring 的 tx 包以及 MySQL 数据库的驱动的包的依赖。增加的代码如下所示。

```
<dependency>
    <groupId>org.springframework</groupId>
    <artifactId>spring-jdbc</artifactId>
    <version>6.0.11</version>
</dependency>
<dependency>
    <groupId>org.springframework</groupId>
    <artifactId>spring-tx</artifactId>
    <version>6.0.11</version>
</dependency>
<dependency>
    <groupId>com.mysql</groupId>
    <artifactId>mysql-connector-j</artifactId>
    <version>8.0.33</version>
</dependency>
```

2. 编写配置文件

编写 db.properties 配置文件。修改 Spring 配置文件 applicationContext.xml，增加对数据源的配置。

配置文件 db.properties 的代码如下。

```
jdbc.driver=com.mysql.jdbc.Driver
jdbc.url=jdbc:mysql://localhost:3306/db_student?characterEncoding=
utf8&serverTimezone=GMT&useSSL=false
jdbc.username=root
jdbc.password=123456
```

配置文件 applicationContext.xml 的代码如下。

```
<?xml version="1.0" encoding="UTF-8"?>
<beans xmlns="http://www.springframework.org/schema/beans"
      xmlns:context="http://www.springframework.org/schema/context"
      xmlns:aop="http://www.springframework.org/schema/aop"
      xmlns:xsi="http://www.w3.org/2001/XMLSchema-instance"
      xsi:schemaLocation="http://www.springframework.org/schema/beans
      http://www.springframework.org/schema/beans/spring-beans.xsd
      http://www.springframework.org/schema/context
      http://www.springframework.org/schema/context/spring-context.xsd
      http://www.springframework.org/schema/aop
      http://www.springframework.org/schema/aop/spring-aop.xsd">
   <context:component-scan base-package="com.imut"/>
```

```
    <aop:aspectj-autoproxy/>
    <context:property-placeholder location="classpath:db.properties"/>
     < bean id =" dataSource " class =" org. springframework. jdbc. datasource.
DriverManagerDataSource">
        <property name="driverClassName" value="${jdbc.driver}"></property>
        <property name="url" value="${jdbc.url}"></property>
        <property name="username" value="${jdbc.username}"></property>
        <property name="password" value="${jdbc.password}"></property>
    </bean>
    <bean id="jdbcTemplate" class="org.springframework.jdbc.core.JdbcTemplate">
        <property name="dataSource" ref="dataSource"></property>
    </bean>
</beans>
```

3. 创建实体类

实体类 Student.java 文件与第 7 章综合案例中的 Student.java 文件内容一致，这里不再赘述。

4. 新建 DAO 层接口及其实现类

在/src/main/java 下新建 com.imut.dao 包，在该包下新建接口文件 StudentDao.java，在/src/main/java 下新建 com.imut.dao.impl 包，在该包下新建接口文件的实现类 StudentDaoImpl.java。

程序 StudentDao.java 的代码如下。

```
public interface StudentDao {
    public Student getStudentById(int sid);
    public List<Student> listStudents();
    public void addStudent(Student student);
    public void updateStudent(Student student);
    public void deleteStudent(int sid);
}
```

程序 StudentDaoImpl.java 的代码如下。

```
@Component
public class StudentDaoImpl implements StudentDao {
    @Autowired
    private JdbcTemplate jdbcTemplate;
    @Override
    public Student getStudentById(int sid) {
        String sql="select * from student where sid=?";
        Student student=null;
        student=jdbcTemplate.queryForObject(sql, new BeanPropertyRowMapper
<Student>(Student.class), sid);
        return student;
    }
    @Override
```

```
    public List<Student> listStudents() {
        String sql="select * from student";
        List<Student> list=jdbcTemplate.query(sql, new BeanPropertyRowMapper
<Student>(Student.class));
        return list;
    }
    @Override
    public void addStudent(Student student) {
        String sql="insert into student (sno,sname,sgender,sage) values(?,?,?,?)";
        Object[] params=new Object[]{student.getSno(),student.getSname(),
                student.getSgender(),student.getSage()};
        jdbcTemplate.update(sql, params);
    }
    @Override
    public void updateStudent(Student student) {
        String sql="update student set sno=?,sname=?,sgender=?,sage=? where
sid=?";
        Object[] params=new Object[]{student.getSno(),student.getSname(),
                student.getSgender(),student.getSage(),student.getSid()};
        jdbcTemplate.update(sql, params);
    }
    @Override
    public void deleteStudent(int sid) {
        String sql="delete from student where sid=?";
        jdbcTemplate.update(sql, sid);
    }
}
```

5. 修改 Service 层接口及其实现类

修改 com.imut.service 包下的接口文件 StudentService.java，修改 com.imut.service.impl 包下的接口实现类 StudentServiceImpl.java 文件。

修改后的程序 StudentService.java 代码如下。

```
public interface StudentService {
    public Student getStudentById(int sid);
    public List<Student> listStudents();
    public void addStudent(Student student);
    public void updateStudent(Student student);
    public void deleteStudent(int sid);
}
```

修改后的程序 StudentServiceImpl.java 代码如下。

```
@Component
public class StudentServiceImpl implements StudentService {
    @Autowired
    private StudentDao studentDao;
    public void setStudentDao(StudentDao studentDao) {
```

```
        this.studentDao = studentDao;
    }
    @Override
    public Student getStudentById(int sid) {
        return studentDao.getStudentById(sid);
    }
    @Override
    public List<Student> listStudents() {
        return studentDao.listStudents();
    }
    @Override
    public void addStudent(Student student) {
        studentDao.addStudent(student);
    }
    @Override
    public void updateStudent(Student student) {
        studentDao.updateStudent(student);
    }
    @Override
    public void deleteStudent(int sid) {
        studentDao.deleteStudent(sid);
    }
}
```

6. 编写测试类

修改 com.imut.test 包下的测试类 StudentTest.java，测试根据 sid 查询学生信息、查询所有学生信息、增加学生信息、更新学生信息以及删除学生信息的功能。

程序 StudentTest.java 的代码如下。

```
@RunWith(SpringJUnit4ClassRunner.class)
@ContextConfiguration(locations="/applicationContext.xml")
public class StudentTest {
    @Autowired
    StudentService studentService;
    @Test
    public void getStudentByIdTest() {
        Student student=studentService.getStudentById(1);
        System.out.println("sid为1的学生信息如下:");
        System.out.println(student);
    }
    @Test
    public void listStudentsTest() {
        List<Student> list=studentService.listStudents();
        System.out.println("所有学生信息如下:");
        for(Student student:list) {
            System.out.println(student);
        }
    }
```

```
    @Test
    public void addStudentTest() {
        Student student=new Student();
        student.setSno("202101008");
        student.setSname("韩梅");
        student.setSgender("女");
        student.setSage(18);
        studentService.addStudent(student);
    }
    @Test
    public void updateStudentTest() {
        Student student=studentService.getStudentById(6);
        student.setSname("韩梅梅");;
        studentService.updateStudent(student);
    }
    @Test
    public void deleteStudentTest() {
        studentService.deleteStudent(6);
    }
}
```

9.5 习题

1. 选择题

(1) 关于 Spring JDBC,下列选项错误的是(　　)。

A. Spring JDBC 是对传统 JDBC 的改善和增强

B. Spring JDBC 和传统 JDBC 完全没有关联

C. Spring JDBC 的 core 包负责提供核心功能

D. Spring JDBC 的 support 包负责提供支持类

(2) 下面关于 update()方法,描述错误的是(　　)。

A. update()方法可以完成插入、更新、删除和查询数据的操作

B. 在 JdbcTemplate 类中,提供了一系列的 update()方法

C. update()方法执行后,会返回受影响的行数

D. update()方法返回的参数是 int 类型

(3) Spring JDBC 模块主要由 4 个包组成,其中不包括(　　)。

A. core(核心包)　　B. dataSource(数据源包)

C. driverClass(数据库驱动包)　　D. support(支持包)

2. 填空题

(1) Spring JDBC 的核心类指的是________。

(2) 当执行 DQL 操作时,通常调用 JdbcTemplate 类的________方法和________方法。

(3) 当执行 DML 操作时,通常调用 JdbcTemplate 类的________方法。

(4) 当执行 DDL 操作时，通常调用 JdbcTemplate 类的________方法。

(5) 使用 Spring JDBC 封装 DAO 时，可以通过继承________类的方式实现。

3. 简答题

简述 Spring JDBC 是如何进行配置的。

4. 编程题

请通过 Spring JDBC 编写程序，并完成以下步骤。

(1) 在数据库 chapter09 中创建一张名称为 teacher 的数据表，具体表结构如表 9-5 所示。

表 9-5 teacher 数据表结构

字段名称	数据类型	备注说明
tid	INT	教师编号，主键，自增长
tname	VARCHAR(20)	教师姓名
age	VARCHAR(20)	教师年龄
course	VARCHAR(20)	课程

(2) 向数据表 teacher 中插入一条数据，具体字段值如表 9-6 所示。

表 9-6 插入数据

tname	age	course
ZhangSan	32	Java

(3) 查询数据表 teacher 中的记录条数。

第10章 Spring框架的数据库事务管理

在项目开发中，数据库事务与程序的并发性能与程序正确、安全读取数据息息相关。随着项目规模的扩大和业务逻辑的增多，事务管理成为开发者编码过程中的一项耗时且烦琐的工作。为了简化事务管理的过程，提升开发效率，Spring 提供了通用事务管理的解决方案，这也是 Spring 的优势所在。本章将对 Spring 管理数据库事务涉及的相关知识进行详细讲解。

10.1 事务简介

事务管理是企业应用程序开发中确保数据完整性和一致性的关键技术。事务是一系列对系统数据进行访问和更新的操作逻辑单元，要么全部完成，要么全部不起作用。事务特别指数据库事务，具有隔离机制，防止多个应用程序同时访问数据库时相互干扰。事务为数据库操作序列提供从失败中恢复的方法，保持数据的一致性。

10.1.1 数据库事务 ACID 特性

数据库事务的 ACID 特性包括原子性、一致性、隔离性和持久性。

(1) 原子性：事务中的所有操作要么全部完成，要么全部不完成，不能停滞在中间。如果事务执行过程中发生错误，会回滚到事务开始前的状态。

(2) 一致性：事务执行的结果必须是使数据库从一个一致性状态转变到另一个一致性状态。如果数据库在运行过程中发生故障，未完成的事务对数据库所做的修改可能导致数据库处于不一致的状态。

(3) 隔离性：在并发环境中，不同事务之间是相互隔离的，一个事务的执行不能被其他事务干扰。标准 SQL 规范中定义了 4 个事务隔离级别，包括未授权读取、授权读取、可重复读取和序列化。不同级别对事务的处理不同，其中序列化是最严格的事务隔离级别。

(4) 持久性：一旦事务提交，其对数据库中相应数据的状态更改应该是永久的。即使发生系统崩溃或机器宕机等故障，只要数据库能够重新启动，就能够将其恢复到事务成功结束时的状态。

10.1.2 事务管理的不足

在JavaEE程序开发中，事务管理是一个影响范围较广的领域。在程序与数据库交互时，保证事务的正确执行尤为重要。在实际开发中，事务管理存在诸多弊端。下面是一段常见的事务管理代码。

```
public void purchase(String isbn, String usename) {
    Connection conn = null;
    try {
        conn = dataSource.getConnection();
        conn.setAutoCommit(false);
        //...
        conn.commit();
    }catch(SQLException e) {
        e.printStackTrace();
        if(conn != null) {
            try {
                conn.rollback();
            }catch(SQLException e1) {
                e1.printStackTrace();
            }
        }
        throw new RuntimeException(e);
    }finally {
        if(conn != null) {
            try {
                conn.close();
            }catch(SQLException e) {
                e.printStackTrace();
            }
        }
    }
}
```

在实际开发中，程序与数据库交互时，必须为不同的方法重复写类似的样板代码，比较烦琐。同时由于这段代码是特定于JDBC的，一旦选择了其他数据库存取技术，还需要做出相应的修改。作为企业级应用程序框架，Spring在不同的事务管理API之上定义了一个抽象层，而应用程序开发者不必了解底层的事务管理API，就可以使用Spring的事务管理机制。

10.2 Spring 事务管理概述

10.2.1 Spring 对事务管理的支持

Spring 框架为事务管理提供了解决方案,通过事务抽象和统一编程模型实现了跨环境的事务管理。这使得 Spring 能在多种情况下无缝配置和使用事务,包括 Spring JDBC 和 ORM 框架。Spring 支持编程式和声明式事务管理。编程式事务管理需要手动控制事务的提交和回滚,而声明式事务管理则通过配置实现事务管理,使开发者更专注于业务逻辑。声明式事务管理使用 Spring AOP 框架,将事务管理代码与业务方法分离,以声明的方式实现事务管理。这种方式简化了事务管理,提高了开发效率。

10.2.2 事务管理的核心接口

Spring 主要通过 3 个接口实现事务抽象,它们分别是 PlatformTransactionManager、TransactionDefinition 和 TransactionStatus,都位于 org.springframework.transaction 包中。其中,PlatformTransactionManager 是根据属性管理事务,TransactionDefinition 用于定义事务的属性,TransactionStatus 用于界定事务的状态。

1. PlatformTransactionManager

PlatformTransactionManager 接口是 Spring 事务管理的核心接口,它真正执行了事务管理的职能,提供了一系列方法用于管理事务,并针对不同的持久化技术封装了对应的实现类。

PlatformTransactionManager 接口主要有 3 个事务操作的方法,具体如表 10-1 所示。

表 10-1 PlatformTransactionManager 接口的方法

方　法	描　述
TransactionStatus getTransaction (TransactionDefinition definition)	根据事务定义获取一个已存在的事务或创建一个新的事务,并返回这个事务的状态
void commit(TransactionStatus status)	根据事务的状态提交事务
void rollback(TransactionStatus status)	根据事务的状态回滚事务

在上面 3 个方法中,getTransaction(TransactionDefinition definition)方法会根据 TransactionDefinition 参数返回一个 TransactionStatus 对象,TransactionStatus 对象就表示一个事务,它被关联在当前执行的线程上。

Spring 事务管理是由具体的持久化技术完成的,而 PlatformTransactionManager 接口只提供统一的抽象方法,它并不了解底层是如何管理事务的,具体如何管理事务是由它的实现类完成的。为了应对不同持久化技术的差异性,PlatformTransactionManager 接口为它们提供了许多不同的实现类,常见的实现类如下。

(1) org.springframework.jdbc.datasource.DataSourceTransactionManager:用于配置 JDBC 数据源的事务管理器。

（2）org.springframework.orm.hibemate4.HibernateTransactionManager：用于配置Hibernate的事务管理器。

（3）org.springframework.transaction.jta.JtaTransactionManager：用于配置全局事务管理器。

当底层采用不同的持久层技术时，系统只需使用不同的PlatformTransactionManager实现类即可。

2. TransactionDefinition

TransactionDefinition接口主要用于定义事务的属性，它们包括事务的隔离级别、事务的传播行为、事务的超时时间、是否为只读事务等。

（1）事务的隔离级别。

TransactionDefinition定义了5种隔离级别，具体如表10-2所示。

表10-2 TransactionDefinition定义的隔离级别

隔离级别	值	描述
ISOLATION_ DEFAULT	DEFAULT	采用当前数据库默认的隔离级别
ISOLATION_READ_UNCOMMITTED	READ_UNCOMMITTED	允许读取尚未提交的数据变更，可能会导致脏读、幻读或不可重复读
ISOLATION_READ_COMMITTED	READ_COMMITTED	允许读取已经提交的数据变更，可以避免脏读，无法避免幻读或不可重复读
ISOLATION_REPEATABLE_READ	REPEATABLE_READ	允许可重复读，可以避免脏读、不可重复读，资源消耗上升
ISOLATION_SERIALIZABLE	SERIALIZABLE	事务串行执行，资源消耗最大

表10-2列举了TransactionDefinition定义的5种隔离级别，除了ISOLATION_DEFAULT是TransactionDefinition特有的之外，其余4个分别与java.sql.Connection接口定义的隔离级别相对应。

（2）事务的传播行为。

事务的传播行为是指事务处理过程所跨越的对象将以什么样的方式参与事务，TransactionDefinition定义了7种事务传播行为，具体如表10-3所示。

表10-3 TransactionDefinition定义的事务传播行为

传播行为	值	描述
PROPAGATION_REQUIRED	REQUIRED	默认的事务传播行为。如果当前存在一个事务，则加入该事务；如果当前没有事务，则创建一个新的事务
PROPAGATION_SUPPORTS	SUPPORTS	如果当前存在一个事务，则加入该事务；如果当前没有事务，则直接执行
PROPAGATION_MANDATORY	MANDATORY	当前必须存在一个事务，如果没有，就抛出异常

续表

传播行为	值	描　　述
PROPAGATION_REQUIRES_NEW	REQUIRES_NEW	创建一个新的事务,如果当前已存在一个事务,将已存在的事务挂起
PROPAGATION_NOT_SUPPORTED	NOT_SUPPORTED	不支持事务,在没有事务的情况下执行,如果当前已存在一个事务,则将已存在的事务挂起
PROPAGATION_NEVER	NEVER	永远不支持当前事务,如果当前已存在一个事务,则抛出异常
PROPAGATION_ NESTED	NESTED	如果当前存在事务,则在当前事务的一个子事务中执行

在事务管理过程中,传播行为可以控制是否需要创建事务及如何创建事务。在通常情况下,数据的查询不会影响原数据的改变,所以不需要进行事务管理。但对于数据插入、更新和删除的操作,必须进行事务管理。如果没有指定事务的传播行为,Spring 框架默认传播行为是 REQUIRED。

(3) 事务的超时时间和是否只读。

事务的超时时间是指事务执行的时间界限,超过这个时间界限,事务将会回滚。TransactionDefinition 接口提供了 TIMEOUT_DEFAULT 的常量定义,用来指定事务的超时时间。

当事务的属性为只读时,该事务不修改任何数据,只读事务有助于提升性能。如果在只读事务中修改数据,可能会引发异常。

(4) TransactionDefinition 接口的方法。

TransactionDefinition 接口提供了一系列方法来获取事务的属性,具体如表 10-4 所示。

表 10-4　TransactionDefinition 接口的方法

方　　法	描　　述
IntgetPropagationBehavior()	返回事务的传播行为
int getlsolationLevel()	返回事务的隔离层次
int getTimeout()	返回事务的超时属性
boolean isReadOnly()	判断事务是否为只读
String getName()	返回定义的事务名称

表 10-4 列举了 TransactionDefinition 接口提供的方法,程序可通过 TransactionDefinition 接口的这些方法获取当前事务的属性。

3. TransactionStatus

TransactionStatus 接口主要用于界定事务的状态。在通常情况下,编程式事务中使用该接口较多。TransactionStatus 接口提供的一系列返回事务状态信息的方法,具体如表 10-5 所示。

表 10-5　TransactionStatus 接口的方法

方　　法	描　　述
boolean isNewTransaction()	判断当前事务是否为新事务
boolean hasSavepoint()	判断当前事务是否创建了一个保存点
boolean isRollbackOnly()	判断当前事务是否被标记为 rollback-only
void setRollbackOnly()	将当前事务标记为 rollback-only
boolean isCompleted()	判断当前事务是否已经完成(提交或回滚)
void flush()	刷新底层的修改到数据库

表 10-5 列举了 TransactionStatus 接口提供的方法,事务管理器可以通过该接口提供的方法获取事务运行的状态信息。除此之外,事务管理器还可以通过 setRollbackOnly()方法回滚事务。

10.3　声明式事务管理

Spring 框架的声明式事务管理建立在 Spring AOP 之上,可以通过两种方式来实现,一种是基于 Annotation 的方式,另一种是基于 XML 的方式。下面将详细讲解这两种声明式事务管理方式。

10.3.1　基于注解配置声明式事务

Spring 框架的声明式事务管理可通过注解实现,只需以下两步。

(1) 在 Spring 容器中注册事务注解驱动,代码如下。

```
<tx:annotation-driven transaction-manager="transactionManager"/>
```

(2) 在需要事务管理的 Spring Bean 类或其方法上添加@Transactional 注解。添加在类上,就对整个类的所有方法生效;添加在方法上,就只对该方法生效。在业务实现类上添加该注解可为所有业务方法统一添加事务处理。需要不同事务规则的方法,可单独添加该注解进行设置。对类定义处和方法上的注解,方法上的会覆盖类定义处的。在实际应用中,@Transactional 注解常用于业务实现类上。

@Transactional 注解可配置的参数信息如表 10-6 所示。

表 10-6　@Transactional 注解的属性

属　　性	类　　型	描　　述
value	String 型	用于指定使用的事务管理器
propagation	枚举类型:Propagation	用于指定事务的传播行为,默认为 Propagation.REQUIRED
isolation	枚举类型:Isolation	用于指定事务的隔离级别,默认为 Isolation.DEFAULT(底层事务的隔离级别)

续表

属　　性	类　　型	描　　述
timeout	int 型(以秒为单位)	用于指定事务的超时时间,默认为 TransactionDefinition.TIMEOUT_DEFAULT(底层事务系统的默认时间)
readonly	布尔型	用于指定事务是否为只读,默认为 false
rollbackFor	一组 Class 类的实例,必须是 Throwable 的子类	用于指定导致事务回滚的异常类数组
rollbackForClassName		用于指定导致事务回滚的异常类名称数组
noRollbackFor		用于指定不会导致事务回滚的异常类数组
noRollbackForClassName		用于指定不会导致事务回滚的异常类名称数组

使用@Transactional 注解时,还需在 Spring 的配置文件中引入 tx 命名空间,并通过<tx:annotation-driven>元素配置事务注解驱动。<tx:annotation-driven>元素中有一个常用属性 transaction-manager,该属性用于指定事务管理器。

接下来以一个实例演示使用注解配置声明式事务,具体步骤如下。

(1) 程序和数据库准备。

首先在上一章创建的 MySQL 数据库 db_ssm 中创建 3 张表 account、book、book_stock。其中 account 表用来存储账户信息,book 表用来存放书的信息,book_stock 表用来存放书的库存信息。向表中添加测试数据之后,得到图 10-1 所示的结果。

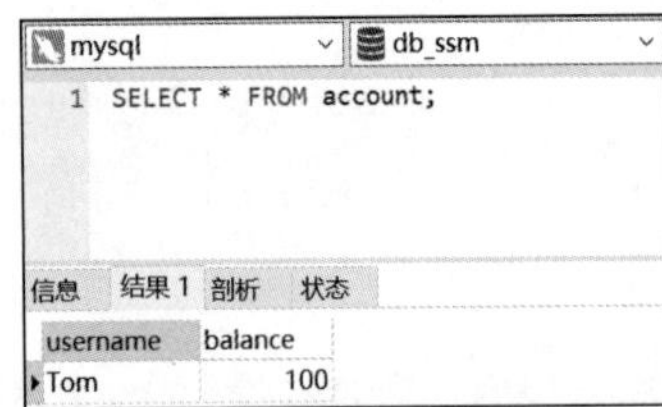

mysql　db_ssm

1 SELECT * FROM book;

isbn	book_name	price
1001	《Java编程思想》	100
1002	《活着》	60

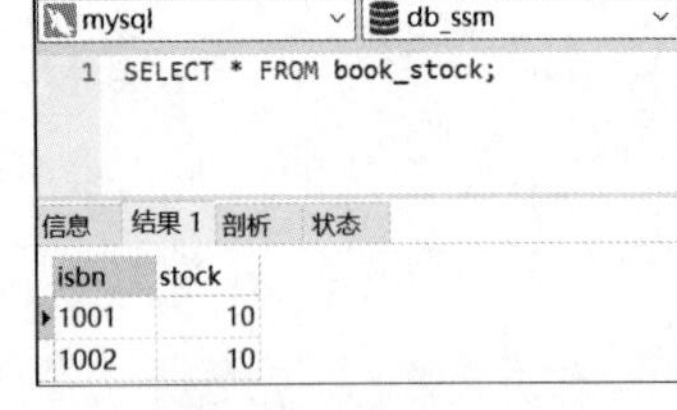

图 10-1　数据库 db_ssm 中的表

然后创建 DAO 层和业务层。在 IDEA 中创建一个名为 ch10_spring_01 的 Java 项目,使用 Maven 管理 Spring 框架所需的基础 JAR 包、Spring AOP 的 JAR 包、MySQL 数据库的驱动 JAR 包、Spring JDBC 的 JAR 包,并在 pom.xml 中引入 JAR 包的依赖。

在 src/main/java 目录下新建 com.imut.txannotation.dao 包,并创建 BookShopDao 接口和 BookShopDaoImpl 类,核心代码如例 10-1、例 10-2 所示。

例 10-1: BookShopDao.java。

```
public interface BookShopDao {
    public Integer findBookPriceByIsbn(String isbn );
    public void updateBookStock(String isbn );
    public void updateUserAccount(String username , Integer price );
}
```

例 10-2: BookShopDaoImpl.java。

```
@Repository
public class BookShopDaoImpl  implements BookShopDao{
    @Autowired
    private JdbcTemplate jdbcTemplate ;
    @Override
    public Integer findBookPriceByIsbn(String isbn) {
        String sql = "select price from book where isbn = ? ";
        return jdbcTemplate.queryForObject(sql, Integer.class, isbn);
    }
    @Override
    public void updateBookStock(String isbn) {
        String sql = "select stock from book_stock where isbn = ?";
        Integer stock = jdbcTemplate.queryForObject(sql, Integer.class, isbn);
        if(stock <=0) {
            throw new BookStockException("库存不足!");
        }
        sql = "update book_stock set stock = stock - 1 where isbn =?";
        jdbcTemplate.update(sql, isbn);
    }
    @Override
    public void updateUserAccount(String username, Integer price) {
        String sql = "select balance from account where username = ? ";
        Integer balance = jdbcTemplate.queryForObject(sql, Integer.class, username);
        if(balance < price ) {
            throw new UserAccountException("余额不足!");
        }
        sql = "update account set balance = balance - ? where  username = ?";
        jdbcTemplate.update(sql, price,username);
    }
}
```

在 src/main/java 目录下新建 com.imut.txannotation.service 包，并创建 BookShopService 接口和 BookShopServiceImpl 类，核心代码如例 10-3、例 10-4 所示。

例 10-3：BookShopService.java。

```
public interface BookShopService {
    public void buyBook(String username , String isbn );
}
```

例 10-4：BookShopServiceImpl.java。

```
@Service
public class BookShopServiceImpl implements BookShopService{
    @Autowired
    private BookShopDao bookShopDao ;
    @Transactional(propagation=Propagation.REQUIRED,
                isolation=Isolation.READ_COMMITTED)
    public void buyBook(String username, String isbn) {
        Integer price = bookShopDao.findBookPriceByIsbn(isbn);
```

```
        bookShopDao.updateBookStock(isbn);
        bookShopDao.updateUserAccount(username, price);
    }
}
```

上述方法已经添加@Transactional 注解，并且使用注解的参数配置了事务详情，各个参数之间要用英文逗号“,”进行分隔。这里设置 propagation 属性为 REQUIRED，isolation 属性为 READ_COMMITTED。

在实际开发中，事务的配置信息通常是在 Spring 的配置文件中完成的，而在业务层类上只需使用@Transactional 注解即可，并不需要配置@Transactional 注解的属性。

为了便于捕获数据库访问异常，定义了图书库存不足的异常 BookStockException 和账户余额不足的异常 UserAccountException。这里不作详细介绍。

(2) 配置声明式事务管理。

在 src/main/java 目录下创建一个 Spring 配置文件 applicationContext-annotation.xml，在该文件中声明事务管理器等配置信息，核心代码如例 10-5 所示。

例 10-5：applicationContext-annotation.xml。

```
<?xml version="1.0" encoding="UTF-8"?>
<beans xmlns="http://www.springframework.org/schema/beans"
    xmlns:xsi="http://www.w3.org/2001/XMLSchema-instance"
    xmlns:context="http://www.springframework.org/schema/context"
    xmlns:tx="http://www.springframework.org/schema/tx"
    xsi:schemaLocation="http://www.springframework.org/schema/beans
        http://www.springframework.org/schema/beans/spring-beans.xsd
        http://www.springframework.org/schema/context
        http://www.springframework.org/schema/context/spring-context-4.0.xsd
        http://www.springframework.org/schema/tx
        http://www.springframework.org/schema/tx/spring-tx-4.0.xsd">
    <context:component-scan base-package = "com.imut.txannotation">
    </context:component-scan>
    <bean id="dataSource"
        class="org.springframework.jdbc.datasource.DriverManagerDataSource">
        <property name="driverClassName" value="com.mysql.jdbc.Driver" />
        <property name="url" value="jdbc:mysql://localhost:3306/db_ssm" />
        <property name="username" value="root" />
        <property name="password" value="123456" />
    </bean>
    <!-- jdbcTemplate -->
    <bean id="jdbcTemplate" class="org.springframework.jdbc.core.JdbcTemplate">
        <property name="dataSource" ref="dataSource"></property>
    </bean>
    <!-- NamedParameterJdbcTemplate -->
    <bean id="namedParameterJdbcTemplate"
    class="org.springframework.jdbc.core.namedparam.
NamedParameterJdbcTemplate">
        <constructor-arg ref="dataSource"></constructor-arg>
    </bean>
```

```
    <bean id="transactionManager"
        class="org.springframework.jdbc.datasource.DataSourceTransactionManager">
        <property name="dataSource" ref="dataSource"></property>
    </bean>
    <tx:annotation-driven/>
</beans>
```

在以上代码中，首先在配置文件头部引入 tx 命名空间。然后用<context:component-scan>标签为注解配置自动扫描，配置数据源和 JdbcTemplate。还为数据库访问配置了具名参数 NamedParameterJdbcTemplate 组件。再配置 JDBC 数据源的事务管理器的 Bean 组件 DataSourceTransactionManager。最后用<tx:annotation-driven/>标签注册事务注解驱动。

(3) 在 src/main/java 目录下创建包 com.imut.test，并在该包中新建测试类 testTransaction，核心代码如例 10-6 所示。

例 10-6：testTransaction.java。

```
public class TestTransaction {
    private ApplicationContext ctx = null ;
    private BookShopDao bookShopDao = null ;
    private BookShopService bookShopService = null ;
    @Before
    public void init() {
            ctx = new ClassPathXmlApplicationContext ( " applicationContext _
annotation.xml");
        bookShopDao = ctx.getBean("bookShopDaoImpl",BookShopDao.class);
        bookShopService = ctx.getBean("bookShopServiceImpl",BookShopService.
class);
    }
    @Test
    public void testBookShopService() {
        bookShopService.buyBook("Tom", "1001");
    }
}
```

以上代码测试了购买图书(包括查询图书价格、更新图书库存和账户余额)的功能，代码执行后，数据表的数据情况如图 10-2 所示。

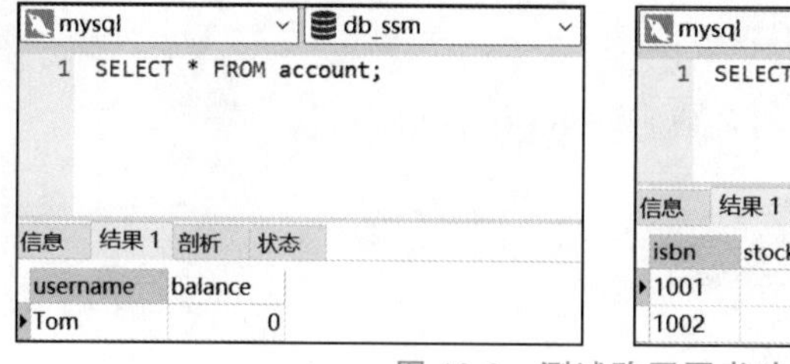

username	balance
Tom	0

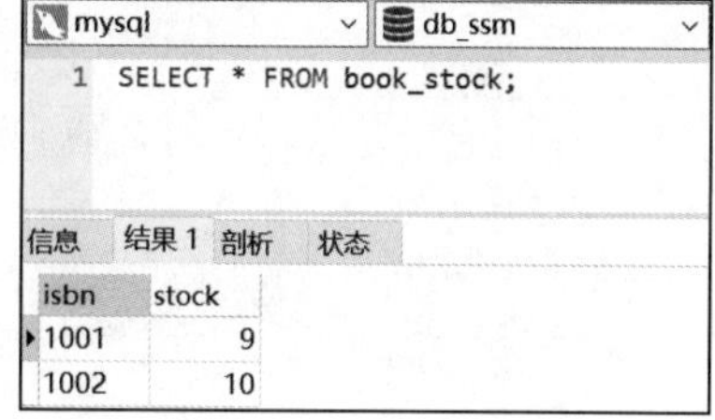

isbn	stock
1001	9
1002	10

图 10-2　测试购买图书功能

从以上结果可以看出，数据表中的相关数据已经被更新。由此可见，事务得到正确执行。下面分别对注解@Transactional的属性进行讨论。

(1) 传播行为属性。

当事务方法被另一个事务方法调用时,必须指定事务应该如何传播。例如,方法可能继续在现有事务中运行,也可能开启一个新事务,并在自己的事务中运行。

事务的传播行为可以由传播属性指定。Spring定义了7种传播行为,如表10-3所示。其中常用的是REQUIRED和REQUIRES_NEW。

为了测试传播行为,在src/main/java目录下的com.imut.txannotation.service包中定义新接口Cashier和CashierImpl实现类,表示客户的结账操作。代码如例10-7、例10-8所示。

例10-7:Cashier.java。

```
public interface Cashier {
    public void checkOut(String username , List<String >isbns);
}
```

例10-8:CashierImpl.java。

```
@Service
public class CashierImpl implements Cashier {
    @Autowired
    private BookShopService bookShopService ;
    @Transactional
    public void checkOut(String username, List<String> isbns) {
        for (String isbn : isbns) {
            bookShopService.buyBook(username, isbn);
        }
    }
}
```

在以上代码中,checkOut方法也添加了@Transactional注解,并使用默认的属性值。

修改测试类testTransaction,添加testCashierImpl()方法,目的是当用户Tom结账时,余额只能支付第一本书,不够支付第二本书,以测试账户余额不足,程序出现异常时事务的执行情况。代码如例10-9所示。

例10-9:testCashierImpl()方法。

```
@Test
public void testCashierImpl() {
    List<String> isbns = new ArrayList<String> ();
    isbns.add("1001");
    isbns.add("1002");
    cashier.checkOut("Tom", isbns);
}
```

① REQUIRED传播行为。

事务传播属性在@Transactional注解的propagation属性中定义。当propagation属

性为 REQUIRED 时,执行测试类 testTransaction 的 testCashierImpl()方法。

之后,查询数据库中的 account 表和 book_stock 表,结果如图 10-3 所示。

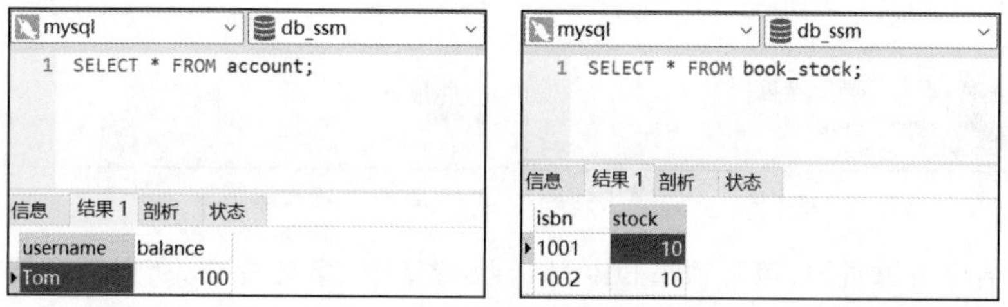

图 10-3 REQUIRED 传播行为的测试结果

从以上运行结果可以看出,数据表中的相关数据都没有更新。当 bookService 的 buyBook()方法被另一个事务方法 checkOut()调用时,默认会在现有的事务内运行,如图 10-4 所示。这个默认的传播行为就是 REQUIRED。因此在 checkOut()方法的开始和终止边界内只有一个事务。这个事务只在 checkOut()方法结束的时候被提交,这个过程中如果有一个操作出现异常,则整个事务都不能提交,结果用户购书失败。所以测试程序执行后,数据表的数据没有更新。

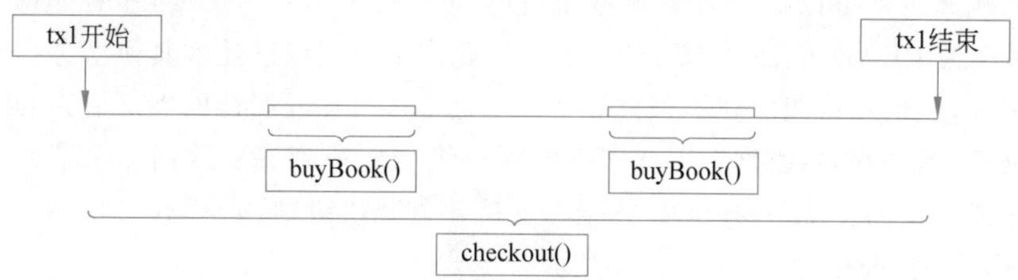

图 10-4 REQUIRED 传播行为的示意图

② REQUIRES_NEW 传播行为。

另一种常见的传播行为是 REQUIRES_NEW。它表示该方法必须启动一个新事务,并在自己的事务内运行。如果有事务在运行,就应该先挂起它。如果把例 10-4 的@Transactional 注解的 propagation 属性改为 REQUIRES_NEW,那么 testCashierImpl()方法的传播行为如图 10-5 所示。每次执行 buyBook()方法时,都会启动一个新的事务。buyBook()方法执行结束后,新的事务提交。

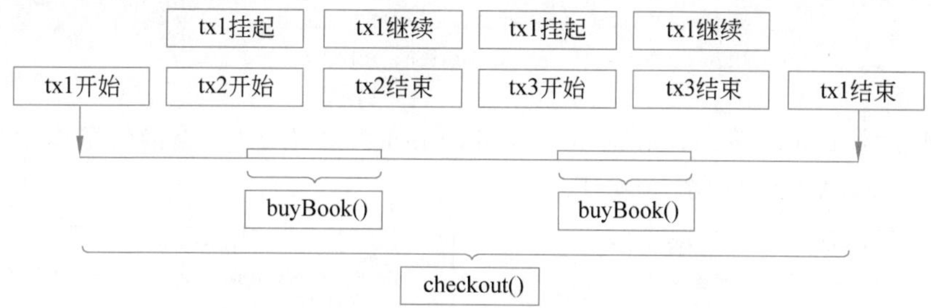

图 10-5 REQUIRES_NEW 传播行为的示意

执行测试类 testTransaction 的 testCashierImpl()方法之后,查询数据库中的 account 表和 book_stock 表,结果如图 10-6 所示。

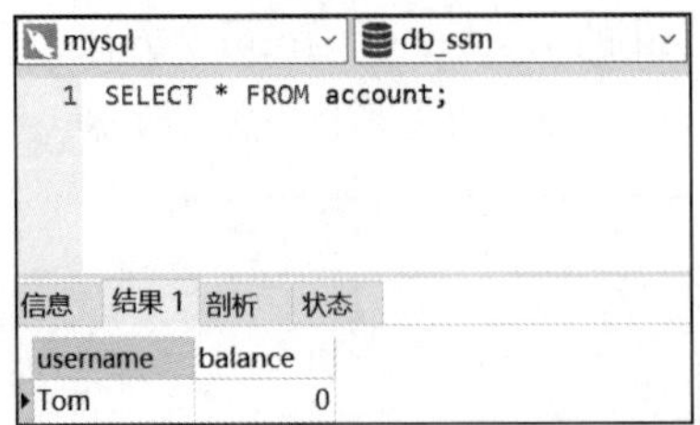

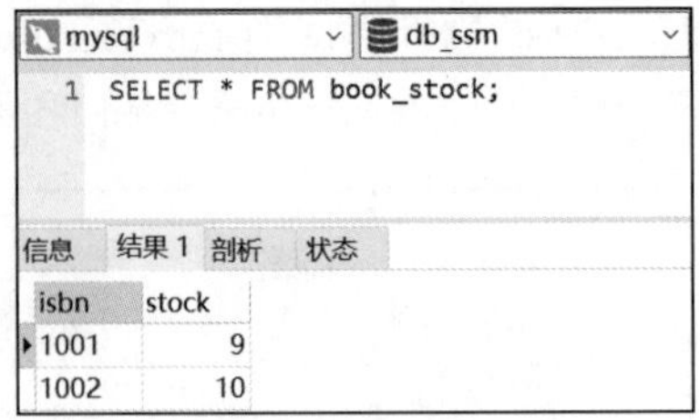

图 10-6　REQUIRES_NEW 传播行为的测试结果

由执行结果可以看出，第一次 buyBook()执行结束，事务提交，所以数据更新了。当第二次执行 buyBook()时，账户余额不足，事务回滚。结果用户买了一本书。

(2) 隔离级别属性。

当同一个应用程序或不同应用程序中的多个事务在同一个数据集上并发执行时，可能会出现许多意外的问题。并发事务导致的问题有 3 种类型，即脏读、不可重复读和幻读。

从理论上来说，事务应该彼此完全隔离，以避免并发事务所导致的问题。然而，那样会对性能产生极大的影响，因为事务必须按顺序运行。在实际开发中，为了提升性能，事务会以较低的隔离级别运行。

事务的隔离级别可以通过隔离事务属性指定。如例 10-4 中，@Transactional 注解的 isolation 属性为 READ_COMMITTED，表示只允许事务读取已经被其他事务提交的变更，可以避免脏读，但仍然可能出现不可重复读和幻读。isolation 属性的默认值为 DEFAULT，即底层数据库默认的隔离级别。因此事务的隔离级别要得到底层数据库引擎的支持，而不是应用程序或框架的支持。MySQL 数据库支持 4 种隔离级别，如表 10-2 所示。

(3) 回滚事务属性。

事务的回滚规则可以通过@Transactional 注解的 rollbackFor 和 noRollbackFor 属性来定义。这两个属性被声明为 Class[]类型，因此可以为这两个属性指定多个异常类。rollbackFor 属性指定遇到时必须进行回滚的异常，而 noRollbackFor 属性指定一组异常类，遇到时必须不回滚。

修改例 10-4 中 buyBook()方法的@Transactional 的属性，添加 noRollbackFor 属性，指定异常 UserAccountException 不回滚，如下所示。

```
@Transactional(propagation=Propagation.REQUIRES_NEW,
            isolation=Isolation.READ_COMMITTED,
            noRollbackFor= {UserAccountException.class})
```

执行测试类 testTransaction 的 testCashierImpl()方法之后，查询数据库中的 account 表和 book_stock 表，结果如图 10-7 所示。

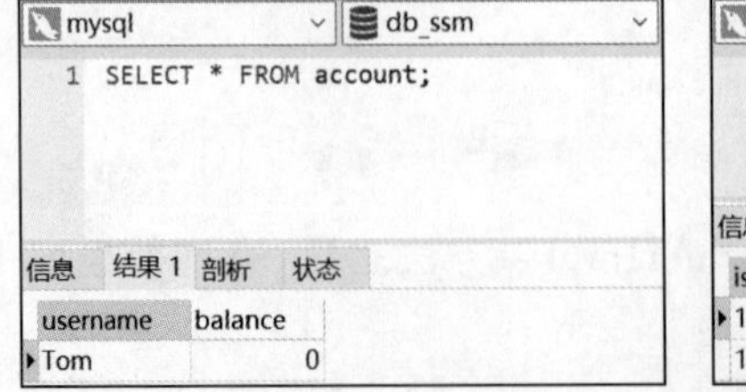

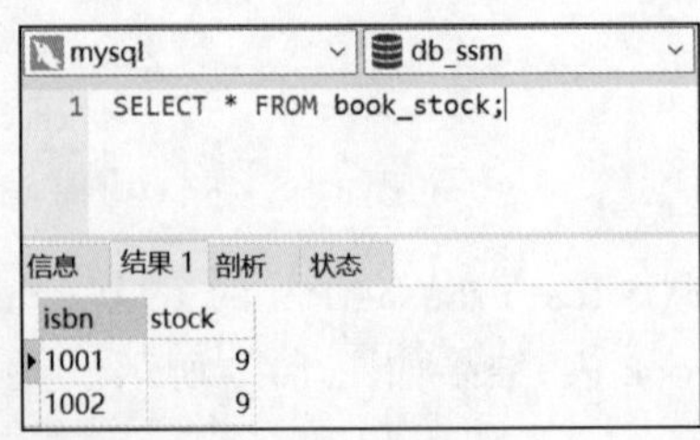

图 10-7　noRollbackFor 属性回滚事务的执行结果

设置了异常 UserAccountException 的 noRollbackFor 属性之后，执行测试类购买两本书。虽然账户余额不足，不能购买，但是图书库存表中的数据更新没有回滚。

(4) 超时和只读属性。

由于事务可以在行和表上获得锁，因此长事务会占用资源，并对整体性能产生影响。可以通过设置超时和只读属性来改善性能。

超时事务属性是指事务在强制回滚之前可以保持多久，这样可以防止长期运行的事务占用资源。

只读事务属性表示这个事务只读取数据但不更新数据，这样可以帮助数据库引擎优化事务。如果一个事物只读取数据，但不做修改，数据库引擎就可以对这个事务进行优化。

在例 10-4 中，可以为 buyBook()方法设置@Transactional 的超时和只读属性，代码如下所示。

```
@Transactional(propagation=Propagation.REQUIRES_NEW,
               isolation=Isolation.READ_COMMITTED,
               readOnly=false, timeout=3)
```

由于 buyBook()方法要更新数据表，所以只读属性设为 false。超时属性设为 3s，如果事务占用数据表超过 3s，将被强制回滚。

10.3.2 基于 XML 配置声明式事务

XML 方式的声明式事务管理通过配置文件定义事务规则实现。Spring 2.0 后，使用 tx 命名空间配置事务，提供<tx:advice>元素配置事务通知。配置后，Spring 通过 AOP 为目标对象生成代理，简化事务管理设置。<tx:advice>包含 id 和 transaction-manager 属性，用于标识和配置 TransactionManager。还包含<tx:attributes>子元素，其内可配置多个<tx:method>子元素，用于配置事务属性，是<tx:advice>的重点。<tx:method>元素的属性具体如表 10-7 所示。

表 10-7 <tx:method>元素的属性

属性名称	描　　述
name	该属性为必选属性，它指定了与事务属性相关的方法名，其属性值支持使用通配符，如“get * ”“select * ”等
propagation	用于指定事务的传播行为，其属性值就是表 10-3 中的值，它的默认值为 REQUIRED
isolation	该属性用于指定事务的隔离级别，其属性值是表 10-2 中的值。默认值为 DEFAULT
read-only	该属性用于指定事务是否只读，默认值为 false
timeout	该属性用于指定事务超时的时间，默认值为－1，即永不超时
rollback-for	该属性用于指定触发事务回滚的异常类，在指定多个异常类时，异常类之间以英文逗号分隔
no-rollback-for	该属性用于指定不触发事务回滚的异常类，在指定多个异常类时，异常类之间以英文逗号分隔

了解了如何在 XML 文件中配置事务，下面通过一个案例来演示使用 XML 方式来实现

Spring 的声明式事务管理。与注解方式不同的是，XML 方式是由 XML 配置文件配置的。所以这里将例 10.3 节中添加在方法上的注解@Transactional 去掉，改由配置文件配置。本案例以 10.3 节的项目代码和数据表为基础，具体步骤如下所示。

(1) 在 ch10_spring_01 项目的 src/main/java 目录下创建包 com.imut.txxml.dao，添加接口 BookShopDao 和实现类 BookShopDaoImpl，创建包 com.imut.txxml.service，添加接口 BookShopService 和实现类 BookShopSerivceImpl。

(2) 在 resources 目录下新建配置文件 applicatioContext_xml，核心代码如例 10-10 所示。

例 10-10：applicatioContext_xml.xml。

```
<context:component-scan base-package="com.imut.txxml"></context:component-scan>
<bean id="dataSource"
    class="org.springframework.jdbc.datasource.DriverManagerDataSource">
    <property name="driverClassName" value="com.mysql.jdbc.Driver" />
    <property name="url" value="jdbc:mysql://localhost:3306/db_ssm" />
    <property name="username" value="root" />
    <property name="password" value="123456" />
</bean>
<!-- jdbcTemplate -->
<bean id="jdbcTemplate" class="org.springframework.jdbc.core.JdbcTemplate">
    <property name="dataSource" ref="dataSource"></property>
</bean>
<!-- NamedParameterJdbcTemplate -->
<bean id="namedParameterJdbcTemplate"
class="org.springframework.jdbc.core.namedparam.NamedParameterJdbcTemplate">
    <constructor-arg ref="dataSource"></constructor-arg>
</bean>
<bean id="transactionManager"
    class="org.springframework.jdbc.datasource.DataSourceTransactionManager">
    <property name="dataSource" ref="dataSource"></property>
</bean>
<tx:advice id="txAdvice" transaction-manager="transactionManager">
    <tx:attributes>
    <!-- 配置哪些方法使用哪些事务属性 -->
    <tx:method name="buyBook" propagation="REQUIRED"
            isolation="READ_COMMITTED" read-only="false" timeout="3"/>
    <tx:method name="checkOut" propagation="REQUIRED"
            isolation="DEFAULT"/>
    </tx:attributes>
</tx:advice>
<!-- 配置事务属性与切入点的结合 -->
<aop:config>
    <aop:pointcut expression="execution(* com.imut.txxml.service.*.*(..))"
            id="txPointCut"/>
    <aop:advisor advice-ref="txAdvice" pointcut-ref="txPointCut"/>
</aop:config>
```

以上代码首先注册了数据源、JdbcTemplate 类、具名参数以及事务管理器。然后在<tx:advice>元素中配置了 id 属性为 txAdvice 的事务通知，并引入了事务管理器

transactionManager。此处需要说明的是，<tx:advice>元素的 transaction-manager 属性值与前面注册的事务管理的 id 值是相同的。<tx:attributes>元素的多个<tx:method>子元素配置了执行事务的方法，这些方法的名称分别是 buyBook 和 checkOut。还可以利用通配符一次性配置一组方法，比如可将查询数据库的方法名定义为以 select 开头，则<tx:method>标签的 name 属性可以设为“select *”。

buyBook()方法和 checkOut()方法适用事务的传播级别为 REQUIRED。因此，如果当前存在事务，这些方法直接在事务中执行；如果当前不存在事务，则创建一个新的事务。使用<tx:advice>元素配置事务通知后，通过<aop:config>元素完成 AOP 声明。

(3) 在 src/main/java 目录下创建包 com.imut.test，在该包中新建测试类 testTransaction，并调用包 com.imut.txxml.dao 和 com.imut.txxml.service 中的方法。核心代码如例 10-6 所示。

执行测试类 testTransaction 之后，得到的结果与 10.3 节注解方式下的结果相同，如图 10-3 所示。由此可见，程序实现了基于 XML 的声明式事务管理。

相较于基于 XML 的声明式事务管理，基于注解的声明式事务管理更简洁一些，实际开发中一般使用基于注解的声明式事务管理。

10.4 综合案例

本章案例主要模拟实现使用事务插入不符合规则的记录实现回滚的功能。

10.4.1 案例设计

案例划分为持久层、业务逻辑层和数据访问层。持久层由实体类组成。数据访问层由接口和接口实现类组成，接口的名称统一以 Dao 结尾，实现类名称在接口名称后加 Impl。业务逻辑层由 Service 接口和实现类组成，业务逻辑层的接口统一使用 Service 结尾，其实现类名称统一在接口名后加 Impl。为了演示事务功能，案例将在测试类的增加方法中使用事务。本章案例使用的主要文件如表 10-8 所示。

表 10-8 本章案例使用的主要文件

文　件	所在包/路径	功　能
log4j.properties	src/	日志输出配置文件
applicationContext.xml	src/	Spring 的配置文件
db.properties	src/	数据库连接的属性文件
Student.java	com.imut.pojo	封装学生信息的类
StudentDao.java	com.imut.dao	定义了操作学生表的接口
StudentDaoImpl.java	com.imut.dao.impl	实现了操作学生表接口的方法
StudentService.java	com.imut.service	定义操作学生类业务层的接口
StudentServiceImpl.java	com.imut.service.impl	操作学生类业务层接口的实现类
StudentTest.java	com.imut.test	测试类

10.4.2 案例演示

运行 StudentTest.java 中的 addStudentTest()方法，测试不使用事务，插入不符合规则的记录，在控制台输出的结果如图 10-8 所示，可以看到插入失败，再查看数据库表，发现第一条记录插入成功，第二条记录插入失败，如图 10-9 所示。运行 addStudentTxTest()方法，测试使用事务，插入不符合规则的记录，在控制台输出的结果如图 10-10 所示，可以看到插入失败，再查看数据库表，发现两条记录都没有插入，如图 10-11 所示。

```
Run   StudentTest ×
✓ Tests passed: 1 of 1 test – 153 ms
DEBUG [main] - Executing prepared SQL update
DEBUG [main] - Executing prepared SQL statement [insert into student (sno,sname,sgender,
 sage) values(?,?,?,?)]
DEBUG [main] - Fetching JDBC Connection from DataSource
DEBUG [main] - Creating new JDBC DriverManager Connection to
 [jdbc:mysql://localhost:3306/db_student?characterEncoding=utf8&serverTimezone=GMT&useSSL
 =false]
DEBUG [main] - JDBC getParameterType call failed - using fallback method instead:
 java.sql.SQLException: Parameter metadata not available for the given statement
DEBUG [main] - Executing prepared SQL update
DEBUG [main] - Executing prepared SQL statement [insert into student (sno,sname,sgender,
 sage) values(?,?,?,?)]
DEBUG [main] - Fetching JDBC Connection from DataSource
ch10_spring > src > main > java > com > imut > test > StudentTest      31:1  CRLF  UTF-8  Tab*
```

图 10-8　不使用事务插入不符合规则记录的运行结果

SID	SNO	SNAME	SGENDER	SAGE
1	2021001	苏小小	男	18
2	2021002	李晓丽	女	20
3	2021003	李芳	女	19
4	20210104	王强	男	20
5	20210105	爱宁	男	67
8	202101008	韩梅梅	女	18
23	20212020	张三	(Null)	0

图 10-9　不使用事务插入不符合规则记录的数据库表结果

```
Run   StudentTest ×
✓ Tests passed: 1 of 1 test – 163 ms
DEBUG [main] - Spring test ApplicationContext cache statistics:
 [DefaultContextCache@176b3f44 size = 1, maxSize = 32, parentContextCount = 0, hitCount =
  4, missCount = 1]
DEBUG [main] - Executing prepared SQL update
DEBUG [main] - Executing prepared SQL statement [insert into student (sno,sname,sgender,
 sage) values(?,?,?,?)]
DEBUG [main] - JDBC getParameterType call failed - using fallback method instead:
 java.sql.SQLException: Parameter metadata not available for the given statement
DEBUG [main] - Executing prepared SQL update
DEBUG [main] - Executing prepared SQL statement [insert into student (sno,sname,sgender,
 sage) values(?,?,?,?)]
DEBUG [main] - JDBC getParameterType call failed - using fallback method instead:
 java.sql.SQLException: Parameter metadata not available for the given statement
ch10_spring > src > main > java > com > imut > test > StudentTest      44:6 (451 chars, 12 line breaks)  CRLF  UTF-8  Tab*
```

图 10-10　使用事务插入不符合规则记录的运行结果

SID	SNO	SNAME	SGENDER	SAGE
1	2021001	苏小小	男	18
2	2021002	李晓丽	女	20
3	2021003	李芳	女	19
4	20210104	王强	男	20
5	20210105	爱宁	男	67
8	202101008	韩梅梅	女	18

图 10-11　使用事务插入不符合规则记录的数据库表结果

10.4.3　代码实现

1. 创建项目

将 Maven 工程 ch09_spring 复制为 ch10_spring。

2. 编写配置文件

修改 Spring 配置文件 applicationContext.xml,增加对事务管理的配置。

配置文件 applicationContext.xml 的代码如下。

```
<?xml version="1.0" encoding="UTF-8"?>
<beans xmlns="http://www.springframework.org/schema/beans"
      xmlns:context="http://www.springframework.org/schema/context"
      xmlns:aop="http://www.springframework.org/schema/aop"
      xmlns:tx="http://www.springframework.org/schema/tx"
      xmlns:xsi="http://www.w3.org/2001/XMLSchema-instance"
      xsi:schemaLocation="http://www.springframework.org/schema/beans
      http://www.springframework.org/schema/beans/spring-beans.xsd
      http://www.springframework.org/schema/context
      http://www.springframework.org/schema/context/spring-context.xsd
      http://www.springframework.org/schema/aop
      http://www.springframework.org/schema/aop/spring-aop.xsd
      http://www.springframework.org/schema/tx
      http://www.springframework.org/schema/tx/spring-tx.xsd">
    <context:component-scan base-package="com.imut"/>
    <aop:aspectj-autoproxy/>
    <context:property-placeholder location="classpath:db.properties"/>
    <bean id="dataSource" class="org.springframework.jdbc.datasource.
DriverManagerDataSource">
        <property name="driverClassName" value="${jdbc.driver}"></property>
        <property name="url" value="${jdbc.url}"></property>
        <property name="username" value="${jdbc.username}"></property>
        <property name="password" value="${jdbc.password}"></property>
    </bean>
    <bean id="jdbcTemplate" class="org.springframework.jdbc.core.
JdbcTemplate">
        <property name="dataSource" ref="dataSource"></property>
    </bean>
    <bean id="transactionManager" class="org.springframework.jdbc.
datasource.DataSourceTransactionManager">
        <property name="dataSource" ref="dataSource"/>
```

```
        </bean>
        <tx:annotation-driven transaction-manager="transactionManager"/>
</beans>
```

3. 创建实体类

实体类 Student.java 文件与第 8 章综合案例中的 Student.java 文件内容一致，这里不再赘述。

4. 新建 DAO 层接口及其实现类

接口文件 StudentDao.java 及其实现类 StudentDaoImpl.java 与第 8 章综合案例中的文件内容一致，这里不再赘述。

5. 新建 Service 层接口及其实现类

接口文件 StudentService.java 及其实现类 StudentServiceImpl.java 文件与第 8 章综合案例中的文件内容一致，这里不再赘述。

6. 编写测试类

修改 com.imut.test 包下的测试类 StudentTest.java，测试不使用事务插入不符合规则的记录以及测试使用事务插入不符合规则的记录的功能。

程序 StudentTest.java 的代码如下。

```
@RunWith(SpringJUnit4ClassRunner.class)
@ContextConfiguration(locations="/applicationContext.xml")
public class StudentTest {
    @Autowired
    private StudentService studentService;
    //测试不使用事务插入不符合规则的记录,第一条记录插入,第二条没插入
    @Test
    public void addStudentTest() {
        Student student1=new Student();
        student1.setSno("20212020");
        student1.setSname("张三");
        studentService.addStudent(student1);
        Student student2=new Student();
        student2.setSno("2021202000000000222222222222222222");
        student2.setSname("张四");
        studentService.addStudent(student2);
    }
    //测试使用事务插入不符合规则的记录,第二条记录插不进去,所以第一条记录的插入也回
    //滚了
    @Test
    @Transactional(propagation=Propagation.REQUIRED, rollbackForClassName="
Exception")
    public void addStudentTxTest() {
        Student student1=new Student();
        student1.setSno("20212022");
        student1.setSname("王三");
```

```
        studentService.addStudent(student1);
        Student student2=new Student();
        student2.setSno("20212022222222222222222222222222");
        student2.setSname("王四");
        studentService.addStudent(student2);
    }
}
```

10.5 习题

1. 选择题

(1) 关于 Spring 管理数据库事务,下列选项错误的是(　　)。

A. 为了便于事务管理,Spring 提供了自己的解决方案

B. Spring 的编程式事务管理完全避免了事务管理和数据访问的耦合

C. Spring 的声明式事务管理需要基于 Spring AOP 实现

D. Spring 的声明式事务管理降低了事务管理和数据访问的耦合

(2) 在 TransactionDefinition 接口提供的方法中,用于返回事务传播行为的是(　　)。

A. getPropagationBehavior()　　B. getIsolationLevel()

C. getTimeout()　　D. isReadOnly()

(3) 以下关于@Transactional 注解可配置的参数信息及描述正确的是(　　)。

A. value 用于指定需要使用的事务管理器,默认为""

B. read-only 用于指定事务是否只读,默认为 true

C. isolation 用于指定事务的隔离级别,默认为 Isolation.READ_COMMITTED

D. propagation 用于指定事务的传播行为,默认为 Propagation.SUPPORTS

(4) 在下列隔离级别中,允许可重复读并且可以避免脏读、幻读的是(　　)。

A. ISOLATION_READ_UNCOMMITTED

B. ISOLATION_READ_COMMITTED

C. ISOLATION_REPEATABLE_READ

D. ISOLATION_SERIALIZABLE

(5) 以下有关事务管理方式的相关说法,错误的是(　　)。

A. Spring 中的事务管理分为两种方式,一种是编程式事务管理;另一种是声明式事务管理

B. 编程式事务管理:指通过 AOP 技术实现的事务管理,就是通过编写代码实现的事务管理,包括定义事务的开始、正常执行后的事务提交和异常时的事务回滚

C. 声明式事务管理:指将事务管理作为一个“切面”代码单独编写,然后通过 AOP 技术将事务管理的“切面”代码植入到业务目标类中

D. 声明式事务管理最大的优点在于开发人员无须通过编程的方式来管理事务,只要在配置文件中进行相关的事务规则声明,就可以将事务规则应用到业务逻辑中

2. 填空题

(1) 事务的四大特性分别是________、________、________、________。

(2) 在 Spring 提供的接口中，________用于定义事务的属性。

(3) 在 Spring 提供的接口中，________用于界定事务的状态。

(4) 在 Spring 提供的接口中，________用于根据属性管理事务。

3. 简答题

(1) 简述 Spring 提供的用于事务管理的核心接口。

(2) 简述如何使用 Annotation 方式进行声明式事务管理。

4. 编程题

现对新闻表 tb_news 的查询、添加、修改和删除操作，具体要求如下。

(1) 实现根据新闻标题名称模糊查询新闻信息列表的操作。

(2) 实现新闻信息的添加、修改和删除操作，使用 Spring 框架的事务切面实现声明式事务管理。

(3) tb_news 的数据表，具体表结构如表 10-9 所示。

表 10-9 tb_news 数据表

字段名称	数据类型	备注说明
id	INT	新闻编号，主键，自增长
title	VARCHAR(20)	新闻标题
content	VARCHAR(100)	新闻内容

第3部分

Spring MVC 篇

第11章 Spring MVC基础

Spring 提供了完备的表述层解决方案，即 Spring MVC。作为 Spring 的子项目，Spring MVC 是一个轻量级的 Web 开发框架，它基于 MVC 设计模式。在表述层框架多次更迭后，业界普遍将 Spring MVC 作为 Java 项目开发的首选方案。

本章首先介绍 Spring MVC 的基本概念和框架组件，然后对比 Spring MVC 与 Struts 2，概括 Spring MVC 的工作流程。接着讲述使用 Spring MVC 所需的准备，包括依赖、API 和配置文件。最后通过示例程序展示 Spring MVC 框架。

11.1 Spring MVC 概述

Spring MVC 是一款基于 MVC 架构模式的轻量级 Web 框架。它的目标是将 Web 开发模块化，解耦整体架构，简化开发流程。Spring MVC 采用请求驱动的方式，即使用请求—应答模型。它遵循 MVC 规范，将开发分为数据模型层(Model)、响应视图层(View)和控制层(Controller)，以便开发者设计出结构规范的 Web 层。

11.1.1 Spring MVC 简介

Spring Web MVC 是一种基于 MVC 设计模式的轻量级 Web 框架，采用请求—响应模型。MVC 代表 Model、View 和 Controller，这三部分按功能划分了 Web 应用程序中的资源。

(1) View(视图)：用户进行操作的可视化界面，可以是 HTML、JSP、XML 等。用户可以在视图上看到服务端传来的数据，或者在视图上录入数据，以便传递给服务端处理。

(2) Model(模型)：用于处理业务逻辑、封装、传输业务数据等。

(3) Controller(控制器)：程序的调度中心，控制程序的流转，接收客户端的请求，根据请求的类型判断调用哪个服务端程序来进行处理，处理完毕后把获得的模型数据显示到视图,返回给用户。

MVC 工作流程如图 11-1 所示:用户通过视图发送请求,Controller 接收请求,调用 Model 层处理请求,将结果返回给 Controller,Controller 找到相应的 View 视图,渲染数据后响应给用户。

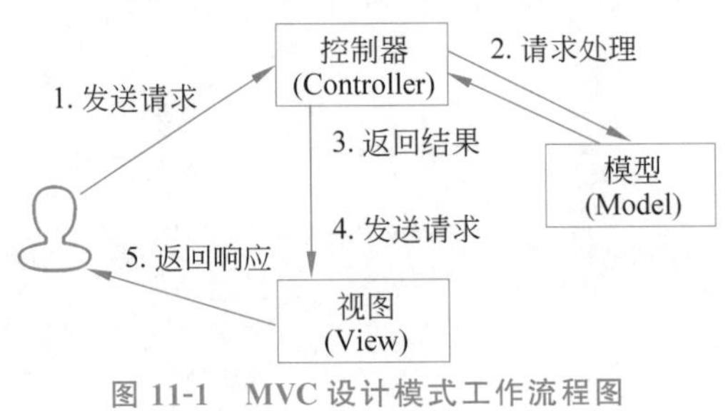

图 11-1 MVC 设计模式工作流程图

11.1.2 Spring MVC 的核心组件

Spring MVC 是一个基于组件式的 Web 框架,各个功能由独立的组件负责,它们相互协调,以完成工作流程。这种设计的好处在于各组件分工明确、协同工作,使请求处理和响应工作更加高效。每个组件都是独立的扩展点,使开发者可以轻松扩展功能而不必担心其他组件,非常灵活。核心组件的功能如下。

(1) 前端控制器(DispatcherServlet)是 Spring MVC 框架的核心组件,充当了控制器(controller)的角色。它是整个流程的控制中心,负责调用处理器映射器、处理器适配器、视图解析器等组件来处理用户请求。通过前端控制器,可以降低组件之间的耦合度,实现灵活的请求处理。

(2) 处理器映射器(HandlerMapping)的任务是根据用户请求找到适当的处理器(Handler),用于将用户请求映射到可处理该请求的处理器,并将其封装为处理器执行链传递给前端控制器。Spring MVC 提供多种映射方式,包括配置文件、接口实现、注解等方式。

(3) 处理器适配器(HandlerAdapter):根据处理器映射器找到的处理器(Handler)信息,按照特定规则执行相关的处理器。

(4) 处理器(Handler):实际处理请求的对象,接收前端控制器分发的请求,处理请求完成后将执行的结果(ModelAndView)返回给前端控制器。

(5) 视图解析器(ViewResolver):进行视图解析,将处理结果生成 View 视图。View Resolver 将逻辑视图名解析为物理视图名,即具体的页面地址,再生成 View 视图对象,最后将处理结果通过页面形式展示给用户。

(6) 视图(View):View 是一个接口,它的作用是渲染数据,将模型(Model)数据通过页面展示给用户。

11.1.3 Spring MVC 与 Struts 2 的区别

Spring MVC 和 Struts 2 都是用于构建 Java Web 应用程序的 MVC 框架,但它们在核心控制器、请求处理方式、拦截器机制、配置文件和学习难度等方面存在一些不同。

1. 核心控制器

Spring MVC 的核心控制器是前端控制器，它接收所有 HTTP 请求，并将请求分派给相应的处理器进行处理。

Struts 2 的核心控制器是 Filter，它采用基于类的方式来处理请求，通过一系列拦截器来处理请求和响应。

2. 请求处理方式

Spring MVC 基于方法进行请求处理，将请求映射到控制器类中的方法，使用方法参数来接收请求数据，然后通过方法的返回值返回响应数据。

Struts 2 基于类进行请求处理，将请求映射到 Action 类，使用 Action 类的属性来接收请求数据，并通过 Action 类的返回值返回响应数据。

3. 拦截器机制

Spring MVC 使用独立的 AOP 方式来实现拦截器，可以通过注解或 XML 配置来定义拦截器和切点。

Struts 2 使用自己的拦截器机制，需要在配置文件中定义拦截器和拦截器栈，并通过拦截器链来处理请求和响应。

4. 配置文件

Spring MVC 的配置文件相对简单和清晰，可以使用 XML 或注解进行配置。

Struts 2 的配置文件相对烦琐，需要定义大量的 Action、Result、Interceptor 等，并使用 XML 文件进行配置。

5. 学习难度

Spring MVC 的学习难度相对较低，更容易上手，主要使用常见的 Servlet 和 JSP 技术，提供了清晰的配置方式。

Struts 2 的学习难度相对较高，因为它采用了自己的标签库和配置文件格式，需要一定时间来掌握。

总的来说，Spring MVC 和 Struts 2 都是优秀的 MVC 框架，适用于不同的项目需求。Spring MVC 作为 Spring 框架的一部分，具有更大的灵活性、集成性和扩展性，适用于复杂的应用程序和大规模的团队开发，同时提供了丰富的生态系统支持，能够轻松实现与其他 Spring 模块的无缝衔接。

11.2 Spring MVC 的工作流程

在 Spring MVC 中，通过前端控制器 DispatcherServlet 统一接收客户端请求，DispatcherServlet 根据配置的映射规则将请求分派到不同业务处理的控制器中，各业务控制器调用相应的业务逻辑模型，完成处理后返回模型数据，前端控制器将模型数据交由视图解析器获得最终的视图进行请求响应。标准的请求和处理流程如图 11-2 所示。

Spring MVC 的运行原理和流程如图 11-2 所示。Spring MVC 的工作流程如下。

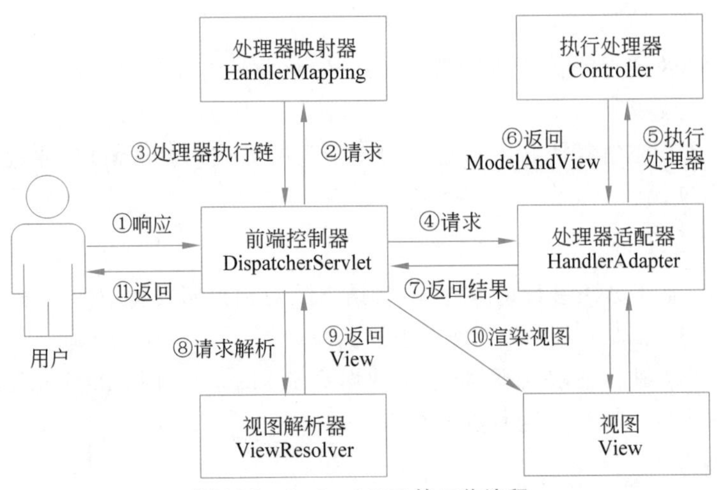

图 11-2　Spring MVC 的工作流程

(1) 用户向服务端提交请求,请求会被前端控制器(DispatcherServlet)拦截。

(2) 前端控制器(DispatcherServle)收到请求后,调用处理器映射器(HandlerMapping)。

(3) 处理器映射器(HandlerMapping)会根据请求,找到处理该请求的具体的处理器,生成处理器对象及处理器(Handler)和拦截器,并将其一并返回给前端控制器(DispatcherServlet)。

(4) 前端控制器(DispatcherServlet)请求处理器适配器(HandlerAdapter)执行处理。

(5) 处理器适配器(HandlerAdapter)调用执行处理器(Controller)。

(6) 处理器(Controller)将处理结果封装到一个对象(ModelAndView)中,并将其返回给处理器适配器(HandlerAdapter)。

(7) 处理器适配器(HandlerAdapter)接收后,将结果返回给前端控制器(DispatcherServlet)。

(8) 前端控制器(DispatcherServlet)接收到对象(ModelAndView)后,调用视图解析器(ViewResolver)对视图进行解析。

(9) 视图解析器(ViewResolver)根据视图对象(View)信息匹配到相应的视图结果,并反馈给前端控制器(DispatcherServlet)。

(10) 前端控制器(DispatcherServlet)调用视图对象(View)进行视图渲染(即将模型中的数据填充至视图图中),生成最终视图。

(11) 前端控制器(DispatcherServlet)向用户返回请求结果。

11.3　Spring MVC 使用前的准备

使用 Spring MVC 框架之前,除了必备的 Java Web 开发环境 JDK、Tomcat 以及 IDEA 外,还需要从 Spring MVC 的依赖、配置文件等方面考虑 Spring MVC 开发环境需要的依赖包。

11.3.1 Spring MVC 的依赖

Spring MVC 工程主要所需依赖如表 11-1 所示。其中核心“spring-webmvc-x. x. xRELEASE”是 Spring MVC 实现 MVC 结构的重要依赖，其余依赖与 Spring 的日志管理、AOP 及 Bean 的管理以及上下文管理有着密切的联系。

表 11-1 Spring、Spring MVC 相关 JAR

依赖 JAR 包名称	作　　用
spring-core	Spring 的核心包，实现 IOC(控制反转)和 DI(依赖注入)
spring-beans	提供了 BeanFactory，Spring 通过这个模块管理 Bean
spring-aop	Spring 的 AOP，提供 AOP 的实现
spring-aspects	提供了与 AspectJ 框架的整合
spring-expression	提供 Spring 表达式语言依赖
spring-context	提供 Spring 框架的上下文管理功能
spring-web	用于处理 Web 的基础模块
spring-webmvc	包含 Spring MVC 框架相关的所有类

如下所示，将可能用到的依赖在 pom.xml 文件中引入。

```
<dependency>
    <groupId>org.springframework</groupId>
    <artifactId>spring-webmvc</artifactId>
    <version>6.0.11</version>
</dependency>
<dependency>
    <groupId>org.springframework</groupId>
    <artifactId>spring-core</artifactId>
    <version>6.0.11</version>
</dependency>
<dependency>
    <groupId>org.springframework</groupId>
    <artifactId>spring-beans</artifactId>
    <version>6.0.11</version>
</dependency>
<dependency>
    <groupId>org.springframework</groupId>
    <artifactId>spring-aop</artifactId>
    <version>6.0.11</version>
</dependency>
<dependency>
    <groupId>org.springframework</groupId>
    <artifactId>spring-aspects</artifactId>
    <version>6.0.11</version>
</dependency>
```

```
<dependency>
    <groupId>org.springframework</groupId>
    <artifactId>spring-expression</artifactId>
    <version>6.0.11</version>
</dependency>
<dependency>
    <groupId>org.springframework</groupId>
    <artifactId>spring-context</artifactId>
    <version>6.0.11</version>
</dependency>
<dependency>
    <groupId>org.springframework</groupId>
    <artifactId>spring-web</artifactId>
    <version>6.0.11</version>
</dependency>
```

11.3.2 Spring MVC 配置方式

Spring MVC 目前有两种配置方式：基于注解和基于 XML 配置文件。

基于注解的配置方式是通过在类、方法和属性上添加注解来实现配置。例如，使用@Controller 注解标记控制器类，使用@RequestMapping 注解指定请求映射路径和处理方法。还有@Autowired、@Service、@Repository 等注解用于依赖注入、服务层和数据访问层的配置。基于注解的配置方式简洁明了，配置信息直接在代码中可见，减少配置错误的发生。

XML 配置文件方式通过定义 Bean 在 XML 文件中进行配置。在 Spring MVC.xml 文件中，可以定义控制器、服务、数据访问等各种 Bean，使用 XML 元素和属性配置它们之间的依赖和属性。这种方式的优势在于可以清晰地管理整个应用的配置信息，便于维护。另外，XML 配置文件还可以通过 Spring 的命名空间来简化配置，提高配置效率。

在实际开发中，通常根据项目需求和团队偏好选择配置方式。对于小型和简单应用，基于注解的配置简单快速；对于大型和复杂应用，XML 配置文件提供更灵活和可扩展的选项，更易于维护。

11.3.3 基于 XML 配置文件

Spring MVC 通常会在两个 XML 文件中进行配置，下面对两个配置文件分别讲解。

(1) 在 web.xml 文件中配置 DispatcherServlet。

DispatcherServlet 是 Spring MVC 用于前端请求分发的核心 Servlet，它处理所有前端请求。首先，在 web.xml 文件中配置一个 servlet 节点，使用<servlet-name>标签自定义名称，在<servlet-class>标签中指定 Servlet 的类为 org. springframework. web. servlet. DispatcherServlet。

使用<init-param>标签设置 Spring MVC 配置文件的位置和名称，通过<load-on-startup>标签来指定 DispatcherServlet 的初始化顺序。通常，将<load-on-startup>的值设置为 1，以确保 Spring 配置的 Bean 在容器启动时实例化。

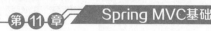

使用 contextConfigLocation 参数来指定 Spring 的配置文件的位置，通常使用 classpath：前缀表示从类路径中查找配置文件。

最后，需要配置哪些 URL 请求将由 DispatcherServlet 处理。这是通过在 web.xml 中添加 <servlet-mapping>节点实现的。在<servlet-name>标签中使用与 DispatcherServlet 相同的名字，然后通过<url-pattern>标签配置要处理的 URL 模式。例如，<url-pattern>/</url-pattern>表示处理所有请求，具体配置如下。

```
<!-- 配置 Spring MVC 的前端控制器-->
<servlet>
    <servlet-name>DispatcherServlet</servlet-name>
    <servlet-class>org.springframework.web.servlet.DispatcherServlet
</servlet-class>
    <!-- 通过初始化参数指定 Spring MVC 配置文件的位置和名称 -->
    <init-param>
        <!-- contextConfigLocation 为固定值 -->
        <param-name>contextConfigLocation</param-name>
        <param-value>classpath:Spring MVC.xml</param-value>
    </init-param>
    <load-on-startup>1</load-on-startup>
</servlet>
<servlet-mapping>
    <servlet-name>DispatcherServlet</servlet-name>
    <url-pattern>/</url-pattern>
</servlet-mapping>
```

(2) 在 Spring MVC.xml 文件中配置包扫描器、注解驱动和视图解析器。

我们已经在 web.xml 中指定了 Spring MVC 配置文件的位置和名称，只需要在相应位置创建即可。在 Spring MVC.xml 配置文件中使用<context:annotation-config/>或<context:component-scan/>标签开启组件注册(@Controller、@Service、@Component)或依赖注入(@Autowried)等注解功能。使用<mvc:annotation-driven/>标签全面开启 Spring MVC 相关注解功能，包括处理请求映射和参数转换相关注解。

InternalResourceViewResolver 是视图解析器，它会将视图名解析为 jsp 文件。属性 prefix 用于指定视图文件的前缀路径，suffix 属性指定视图文件的后缀。例如，以下配置理解为视图解析器会将控制器方法返回的字符串拼接为/WEB-INF/pages/xxx.jsp，并实现跳转。

```
<!-- 配置视图解析器 -->
<bean class="org.springframework.web.servlet.view.InternalResourceViewResolver">
    <!-- 配置视图前缀 -->
    <property name="prefix" value="/WEB-INF/pages/"></property>
    <!-- 配置视图后缀 -->
    <property name="suffix" value="jsp"></property>
</bean>
```

例 11-1：验证 Spring MVC 的环境搭建，程序 testController.java 如下。

```
@Controller
public class testController{
    @RequestMapping("/test")
    public String index() {
        //设置视图名称
        return "success";
    }
}
```

前段页面 index.jsp 如下所示。

```
<a href="test">访问 Spring MVC</a>
```

启动服务器，在浏览器地址栏输入 http://localhost:8080/ch11/index.jsp，成功跳转至 index.jsp 页面，如图 11-3 所示。单击“访问 Spring MVC”，请求跳转控制器 index 方法，通过视图解析器将 success 字符串拼接为/WEB-INF/pages/success.jsp 页面实现跳转。结果如图 11-4 所示。success(成功)代表 Spring MVC 框架搭建成功。

图 11-3　index.jsp 运行成功页面

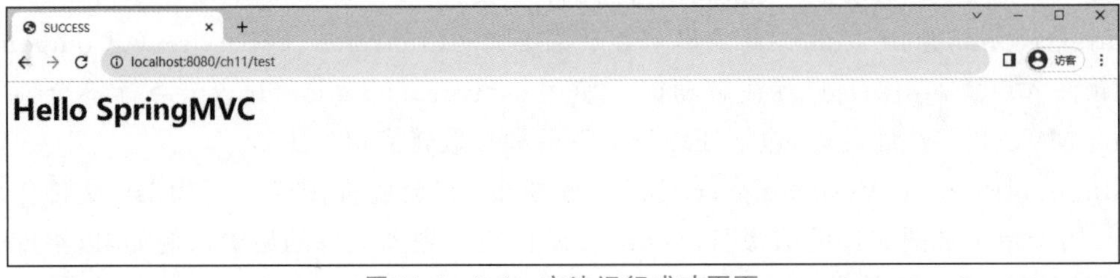

图 11-4　index 方法运行成功页面

11.4　综合案例

本章案例主要实现获取模型对象中字符串信息并在浏览器展示的功能。

11.4.1　案例设计

案例包括 Web 表现层，该层主要包括 Spring MVC 中的 Controller 类和 JSP 页面。Controller 类 HelloController.java 主要负责拦截用户的请求，向模型对象中添加一个字符串对象，然后返回给 JSP 页面进行展示。本章案例使用的主要文件如表 11-2 所示。

表 11-2　本章案例使用的主要文件

文　　件	所在包/路径	功　　能
web.xml	/WEB-INF/	web 配置文件
Spring MVC.xml	src/	Spring MVC 配置文件
HelloController.java	com.imut.controller	控制器类文件
index.jsp	/	首页
hello.jsp	/WEB-INF/pages/	显示信息页面

11.4.2　案例演示

在浏览器地址栏中输入 http://localhost:8080/ch11_Spring MVC/index.jsp，单击“HelloController”超链接，显示信息如图 11-5 所示。

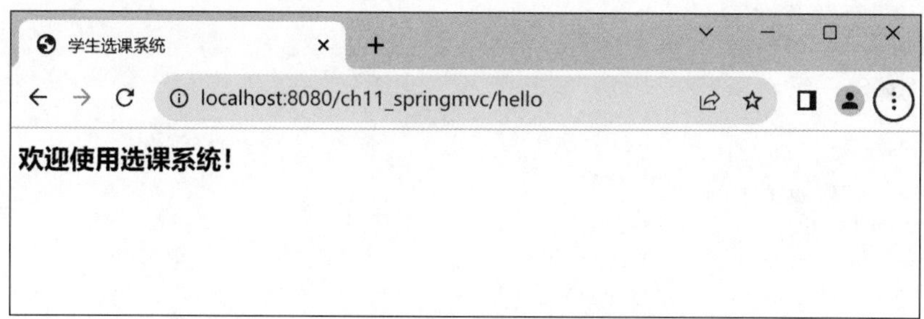

图 11-5　运行结果

11.4.3　代码实现

1. 创建项目，添加依赖

在 IDEA 中通过 Maven 创建 Web 工程 ch11_Spring MVC，在 pom.xml 文件中添加以下依赖：Spring 的 4 个基础包、aop 包、expression 包、web 包、webmvc 包以及日志包。代码如下。

```
<dependencies>
    <dependency>
      <groupId>org.springframework</groupId>
      <artifactId>spring-beans</artifactId>
      <version>5.2.5.RELEASE</version>
    </dependency>
    <dependency>
      <groupId>org.springframework</groupId>
      <artifactId>spring-context</artifactId>
      <version>5.2.5.RELEASE</version>
    </dependency>
    <dependency>
```

```
      <groupId>org.springframework</groupId>
      <artifactId>spring-core</artifactId>
      <version>5.2.5.RELEASE</version>
    </dependency>
    <dependency>
      <groupId>org.springframework</groupId>
      <artifactId>spring-expression</artifactId>
      <version>5.2.5.RELEASE</version>
    </dependency>
    <dependency>
      <groupId>org.springframework</groupId>
      <artifactId>spring-aop</artifactId>
      <version>5.2.5.RELEASE</version>
    </dependency>
    <dependency>
      <groupId>org.springframework</groupId>
      <artifactId>spring-web</artifactId>
      <version>5.2.5.RELEASE</version>
    </dependency>
    <dependency>
      <groupId>org.springframework</groupId>
      <artifactId>spring-webmvc</artifactId>
      <version>5.2.5.RELEASE</version>
    </dependency>
    <dependency>
      <groupId>javax.servlet</groupId>
      <artifactId>javax.servlet-api</artifactId>
      <version>3.1.0</version>
    </dependency>
    <dependency>
      <groupId>log4j</groupId>
      <artifactId>log4j</artifactId>
      <version>1.2.17</version>
    </dependency>
  </dependencies>
```

2. 编写配置文件

在 WEB-INF 下编写 web.xml 配置文件。在/src/main/resources 下编写 Spring MVC.xml 配置文件。

配置文件 web.xml 的代码如下。

```
<web-app xmlns:xsi="http://www.w3.org/2001/XMLSchema-instance" xmlns="http://java.sun.com/xml/ns/javaee" xsi:schemaLocation="http://java.sun.com/xml/ns/javaee http://java.sun.com/xml/ns/javaee/web-app_3_0.xsd" id="WebApp_ID" version="3.0">
    <display-name>ch11_Spring MVC</display-name>
    <welcome-file-list>
        <welcome-file>index.jsp</welcome-file>
```

```
    </welcome-file-list>
    <servlet>
          <servlet-name>DispatcherServlet</servlet-name>
          <servlet-class>org.springframework.web.servlet.DispatcherServlet
</servlet-class>
          <init-param>
              <param-name>contextConfigLocation</param-name>
              <param-value>classpath:Spring MVC.xml</param-value>
          </init-param>
          <load-on-startup>1</load-on-startup>
    </servlet>
    <servlet-mapping>
        <servlet-name>DispatcherServlet</servlet-name>
        <url-pattern>/</url-pattern>
    </servlet-mapping>
</web-app>
```

配置文件 Spring MVC.xml 的代码如下。

```
<?xml version="1.0" encoding="UTF-8"?>
<beans xmlns="http://www.springframework.org/schema/beans"
    xmlns:xsi="http://www.w3.org/2001/XMLSchema-instance"
    xsi:schemaLocation="
        http://www.springframework.org/schema/beans
        http://www.springframework.org/schema/beans/spring-beans.xsd">
    <bean id="simpleUrlHandlerMapping"
        class="org.springframework.web.servlet.handler.
SimpleUrlHandlerMapping">
        <property name="mappings">
            <props>
                <prop key="/hello">helloController</prop>
            </props>
        </property>
    </bean>
    <bean id="helloController" class="com.imut.controller.HelloController">
</bean>
    <bean class="org.springframework.web.servlet.view.
InternalResourceViewResolver">
        <property name="prefix" value="/WEB-INF/pages/"></property>
        <property name="suffix" value=".jsp"></property>
    </bean>
</beans>
```

3. 编写控制器类

在/src/main/java 下新建包 com.imut.controller，在该包下创建控制器类 HelloController.java。

程序 HelloController.java 的代码如下。

```
public class HelloController implements Controller {
    @Override
        public   ModelAndView   handleRequest   ( HttpServletRequest   request,
HttpServletResponse response) throws Exception {
        ModelAndView mv=new ModelAndView();
        mv.addObject("msg", "欢迎使用选课系统!");
        mv.setViewName("hello");
        return mv;
    }
}
```

4. 编写 JSP 页面文件

在根目录 webapp 下编写 index.jsp 文件。在 WEB-INF 下新建 pages 文件夹，在该文件夹下新建 hello.jsp 文件。由于篇幅原因，JSP 文件只放<body></body>之间的代码。

程序 index.jsp 的代码如下。

```
<body>
    <h3><a href="hello">HelloController</a></h3>
</body>
```

程序 hello.jsp 的代码如下。

```
<body>
    <h3>${msg }</h3>
</body>
```

11.5 习题

1. 选择题

(1) Spring MVC 是 Spring 提供的一个实现了(　　)设计模式的轻量级 Web 框架。

A. Web　　B. Web MVC　　C. Java　　D. 单例

(2) Spring MVC 中的前端控制器是指(　　)。

A. HandlerAdapter　　B. Handler

C. ViewReslover　　D. DispatcherServlet

(3) 用户通过浏览器向服务器发送请求时，负责拦截用户请求的是(　　)。

A. 前端控制器　　B. 处理器映射器

C. 处理器适配器　　D. 上述说法都不对

(4) 下列选项中，(　　)不是 Spring MVC 的核心组件。

A. DispatcherServlet　　B. SpringFactoriesLoader

C. HandlerMapping　　D. ModelAndView

(5) 下面关于 Spring MVC 的描述，正确的是(　　)。

A. 在 Spring MVC 中,可以配置多个 DispatcherServlet

B. DispatcherServlet 在 Spring MVC 中是核心 servlet,它负责接收请求,并将请求分发给适合的控制器

C. 要使 Spring MVC 可用,DispatcherServlet 需要在 web.xml 中配置

D. 上述说法都对

2. 填空题

(1) Spring MVC 中的核心组件为________、________、________、________、________、________。

(2) 属性________ 用于指定视图文件的路径,suffix 属性指定视图文件的________。

(3) ModelAndView 对象是 Spring MVC 中用于传递________和________信息的一种常用方式。

(4) Spring MVC 前端控制器是________。

(5) 在 Spring MVC 的配置文件中,可以配置处理器映射、处理器映射器、处理器适配器和________。

第12章 常用注解

Spring MVC 中有一些封装好的注解，使用这些注解可以简化开发过程，其他人阅读代码时也便于理解和管理。

本章主要介绍 Spring MVC 中几种常用的注解，包含@Controller、@RequestMapping、@PathVariable、@RequestParam、@Autowired、@ModelAttribute、@ResponseBody、@RequestBody 等。

12.1 @Controller

@Controller 是 Spring MVC 中的注解，用于标记一个类为 Controller。被标记为@Controller 的类会被 Spring 扫描机制自动注册为 Spring 应用程序上下文中的 Bean。这些 Controller 类是 Spring MVC 的核心处理器，负责处理 DispatcherServlet 分发的请求。

控制器的主要作用是处理请求，将用户请求数据经过业务处理层处理，调用 Service 层，最后将结果封装为一个 Model，返回给对应的 View 前端展示。在 Spring MVC 中，只需使用@Controller 注解标记控制器类，无须继承特定类或实现特定接口。使用@RequestMapping 和@RequestParam 等注解定义 URL 请求和 Controller 方法之间的映射，使 Controller 可以被外部访问。此外，Controller 方法参数可以轻松获取 HttpServletRequest 等 HTTP 相关对象。

需要注意的是，仅仅使用@Controller 注解标记一个类，并不能确保它被 Spring MVC 识别为控制器类。为了让 Spring 能够扫描和管理这个控制器类，有以下两种方式可以实现。

(1) 在 Spring MVC 的配置文件中定义 MyController 的 Bean 对象。

(2) 在 Spring MVC 的配置文件中告诉 Spring 该到哪里去找标记为@Controller 的 Controller 控制器。

例 12-1：演示两种方式的定义。

```
<!--方式一-->
<bean class="com.host.app.web.controller.MyController"/>
<!--方式二-->
< context:component-scan base-package = "basePackage" />
```

方式一是对 Controller 的 Bean 对象进行定义，让 Spring 直接找到这个控制器类的对象。方式二是配置包扫描器，扫描 Controller，将其生命周期纳入 Spring 的管理中。在<component-scan/>元素中指定控制类的基本包，指定时要确保所有控制器类都在基本包下，并且不要指定太广泛的基本包，太广泛的基本包会使 Spring MVC 扫描无关的包。

这两种定义方式任选其一，即可让 Spring 认识这个控制器类，并且是在 Spring MVC 的配置文件中完成的，配置文件后缀名为.xml。

例 12-2：演示@Controller 注解的使用，新建一个控制器类 ControllerUse。

```
@Controller
public class ControllerUse {
    @RequestMapping("/controller")
    public String controller() {
        return "success";
    }
}
```

在代码第 4 行中，@Controller 的作用是将下面的类 ControllerUse 标记为控制器类，如果不使用@Controller 注解，ControllerUse 只是一个普通的类。@RequestMapping("/controller")是为方法提供路径，便于直接通过路径使用。在实际代码的实现中，使用@Component 注解也可以起到和@Controller 相同的作用。代码运行结果如图 12-1 所示。根据 controller 方法，运行结果为返回 success 页面。

图 12-1　例 12-2 运行结果示例图

除了@Controller 注解，还有其他类级别的注解，如@Service、@Component 和@Repository，它们的作用类似，都是将被注解的类看作组件，并在基于注解的配置和类路径扫描时进行实例化。@Service 注解通常用于标记业务逻辑服务层的类，表示该类提供业务逻辑的实现。@Repository 注解通常用于标记数据库访问层的类，表示该类提供对数据库的访问操作。@Component 注解是一个通用的注解，用于将普通的 POJO 类实例化到 Spring 容器中，相当于在配置文件中使用<bean id="" class=""/>进行配置。

12.2 @RequestMapping

@RequestMapping 是一个用于处理请求地址映射的注解，可用于类或方法。当用于类上时，它将建立请求 URL 和控制器类之间的映射关系，成为父路径。当用于方法上时，它将建立请求 URL 和控制器方法之间的映射关系。在类级别上使用@RequestMapping，可以将特定请求或请求模式映射到整个控制器类上。此外，还可以在方法级别再添加注解，进一步指定处理方法的映射关系。@RequestMapping 注解用于定义请求的 URL 地址，当请求匹配该地址时，对应的控制器方法将被调用。控制器方法处理请求后返回视图，这个视图会被视图解析器解析。

@RequestMapping 注解有 6 个属性，下面分别说明。

(1) value：指定请求的实际地址，指定的地址可以是 URI Template 模式，如果是多个地址，用{}来指定就可以。

value 的 URL 值可以为以下 3 类：普通的具体值、含有某变量的一类值、含有正则表达式的一类值。

(2) method：指定请求的 method 类型，即 GET、POST、PUT、DELETE 等。

(3) consumes：指定处理请求的提交内容类型(Content-Type)，如 Json、text、html 等。

(4) produces：指定返回的内容类型，仅当 request 请求头中的 Accept 类型中包含该指定类型时才返回。

(5) params：指定 request 请求中必须包含某些参数值时，才让该方法处理请求。

(6) headers：指定 request 请求中必须包含某些指定的 header 值时，才让该方法处理请求。

例 12-3 是一个简单的@RequestMapping 注解的使用。

例 12-3：演示@RequestMapping 注解的简单使用。

```
@Controller
@RequestMapping("/index1" )
public class ControllerRequestMappingUse {
    @RequestMapping("/requestmapping")
    public String login() {
        return "success";
    }
}
```

例 12-3 中将@RequestMapping 注解在 ControllerRequestMappingUse 类上和 login 方法上，即请求的类的 URL 是/index1，请求的 login 方法的 URL 是/requestmapping。使用 ControllerRequestMappingUse 类直接请求/index1 即可。但是，使用的 login 方法在类之下，类本身有路径标注，所以 login 方法的请求路径要在本身类的路径基础上再加上/requestmapping，即请求路径为/index1/ requestmapping 才可正常使用。代码运行结果如图 12-2 所示。根据 login 方法，运行结果为返回 success 页面。

图 12-2 例 12-3 运行结果示例

在例 12-4 实现@RequestMapping 注解在不同请求方式下的使用。

例 12-4：演示@RequestMapping 注解两种请求方式(GET、POST)的使用。

```
@Controller
public class RequestMappingUse {
@RequestMapping(value = "/GEThello", method = RequestMethod.GET)
    public String hello(){
        return "hello";
    }
    @GetMapping("/GEThello2")
    public String hello2(){
        return "hello";
    }
    @RequestMapping(value = "/POSThello", method = RequestMethod.POST)
    public String world() {
        return "hello";
    }
    @PostMapping("/POSThello2")
    public String world2() {
        return "hello";
    }
}
```

在例 12-4 中，@RequestMapping(value = "/GEThello", method = RequestMethod.GET)表示该接口只能被 GET 请求访问到，用其他请求方式访问不可行，会报错。@RequestMapping(value = "/POSThello", method = RequestMethod.POST)表示该接口只能被 POST 请求访问到，用其他请求方式访问不可行，会报错。使用 POST 方法时，前端的方法要注明是 POST 方法，以免前端默认使用 GET 方法，导致运行报错。代码运行结果如图 12-3 和图 12-4 所示。

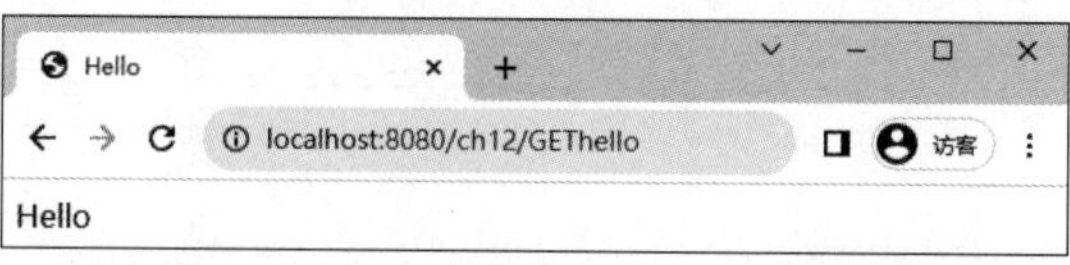

图 12-3 例 12-4 中 GET 方法运行结果示例

图 12-4 例 12-4 中 POST 方法运行结果示例

其中@RequestMapping(value = "/GEThello", method = RequestMethod.GET)等同于@GetMapping("/GEThello ");@RequestMapping(value = "/POSThello", method = RequestMethod.POST)等同于@PostMapping("/POSThello ")。根据 hello 方法和 world 方法,运行结果均为返回 hello 页面,如图 12-5 和图 12-6 所示。

图 12-5　例 12-4 中@GetMapping 运行结果示例

图 12-6　例 12-4 中@PostMapping 运行结果示例

@RequestMapping 还可以映射多个地址。其将多个请求映射到一个方法上,只需要给 value 指定一个包含多个路径的列表,例 12-5 是@RequestMapping 对多个地址映射的代码示例。

例 12-5:演示@RequestMapping 注解对多个地址映射的功能。

```
@Controller
public class RequestMappingAddressesUse {
    @RequestMapping({"/index","/hello"})
    public String hello(){
        return "hello";
    }
}
```

在例 12-5 中,用@RequestMapping 给 hello 方法指定了两个路径/index 和/hello,用这两个中的任意一个都可以访问到 Hello 方法。代码运行结果如图 12-7 和图 12-8 所示。两个路径都是根据 Hello 方法运行,运行结果均为返回 Hello 页面。

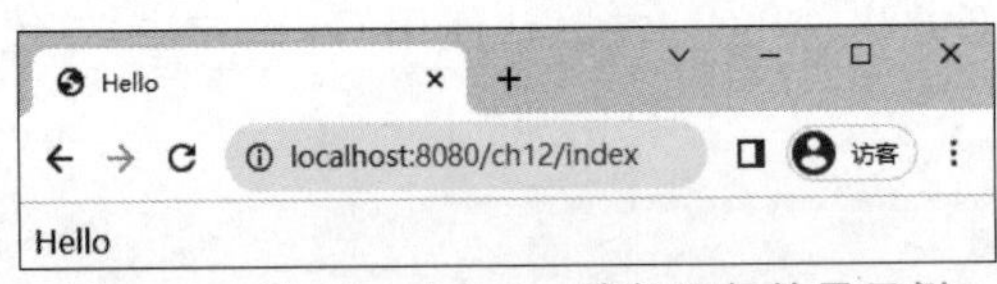

图 12-7　例 12-5 中 index 路径运行结果示例

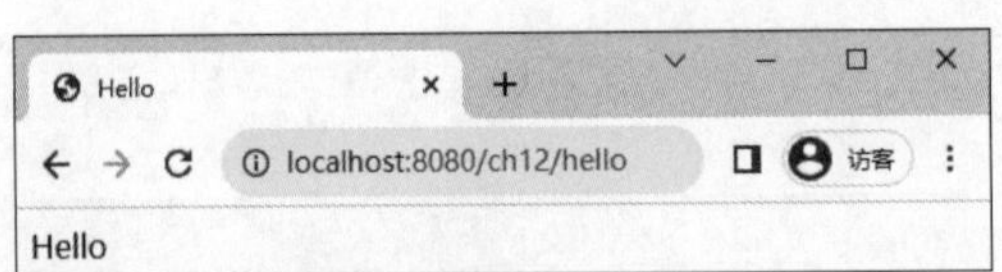

图 12-8　例 12-5 中 hello 路径运行结果示例

12.3 @PathVariable 和@RequestParam

@PathVariable 注解用于将请求 URL 中的模板变量映射到功能处理方法的参数上，即取出 URL 模板中的变量作为参数，通俗的解释就是获取请求 URL 中的动态参数。

@PathVariable 注解只支持一个属性 value，类型为 String。注解中在类型后要注明绑定的名称，如果省略，则默认绑定同名参数。以下为伪代码的示范。

```
@RequestMapping(value="/pathVariableTest/{userId}")
public void pathVariableTest(@PathVariable Integer userId)
```

@PathVariable 注解在参数为"/pathVariableTest/{userId}"的对应@RequestMapping 注解出现后使用。例 12-6 是@PathVariable 注解的使用方法演示。

例 12-6：演示@PathVariable 注解的具体使用。

```
@Controller
public class PathVariableUse {
    @ RequestMapping ( value ="/user/{ userId }/roles/{ roleId }", method =
RequestMethod.GET)
    public String getLogin(@PathVariable("userId") String userId,@PathVariable
("roleId") String roleId){
        System.out.println("User Id : " + userId);
        System.out.println("Role Id : " + roleId);
        return "hello";
    }
}
```

在例 12-6 中演示了两次@PathVariable 注解的使用，都是在 getLogin 方法中使用的，目的是将@RequestMapping（value ="/user/{ userId }/roles/{ roleId }", method = RequestMethod.GET）中请求的 URL 上的两个变量{userId}和{roleId}分别绑定到 getLogin 方法中的参数 userId 和 roleId 上，在方法中获取请求 URL 中的参数。代码运行结果如图 12-9 和图 12-10 所示。根据 getLogin 方法，运行结果为返回 Hello 页面。请求时传输了 userId 和 roleId 分别为 001 和 123456，因此输出结果为 User Id ：001 和 Role Id：123456。

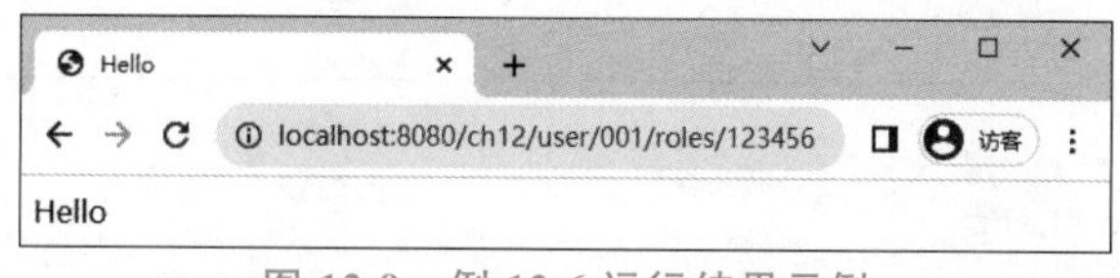

图 12-9 例 12-6 运行结果示例

@RequestParam 注解是把请求中的指定名称的参数传递给控制器中的形参去赋值，是 Spring MVC 中接收普通参数的注解，主要用于在 Spring MVC 后台控制层获取参数，类似 request.getParameter("name")方法。在实际开发中经常使用，其常用属性包括以下几个。

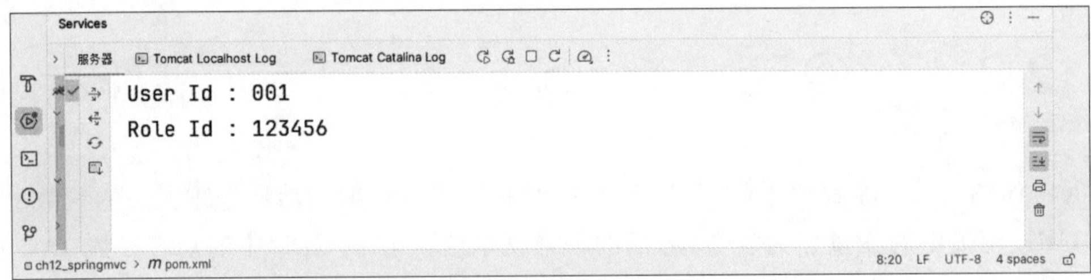

图 12-10 例 12-6 输出结果示例

(1) defaultValue：用于设置默认值。

(2) value：请求参数中的名称。

(3) required：请求参数中是否必须提供此参数，其默认值是 true，表示必须提供，如果不提供，系统就会报错。当设定为 false 时，不提供此参数不会报错，会赋值 null。

例 12-7：演示@RequestParam 注解的具体使用。

```
@Controller
public class RequestParamUse {
    @RequestMapping(value="/user/{userId}/username", method = RequestMethod.
GET)
    public String RequestParam(@RequestParam(value = "username", required =
false) String username){
        System.out.println("User name : " + username);
        return "success";
    }
}
```

在例 12-7 中，@RequestParam 要得到的是请求中名称为"username"的参数，将获取到的参数赋值给 username。因为使用@RequestParam 时 required 设置为 false，所以如果在请求时未提供"username"参数，不会报错，而会给 username 赋值 null。代码运行结果如图 12-11 和图 12-12 所示。根据 RequestParam 方法，运行结果为返回 success 页面。请求时传输了 username=abc，因此为 username 赋值后输出结果为 User name ：abc。

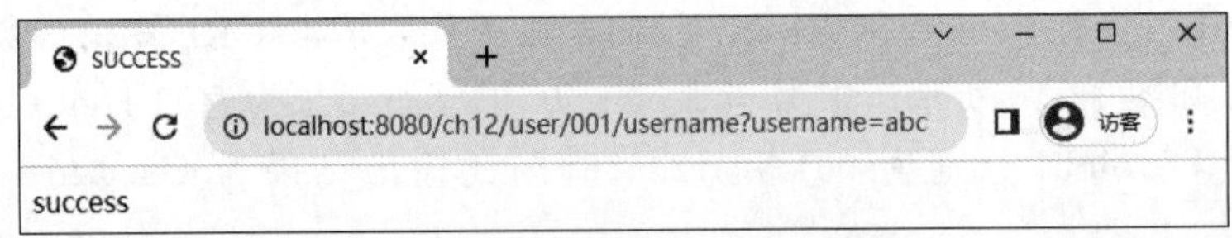

图 12-11 例 12-7 运行结果示例

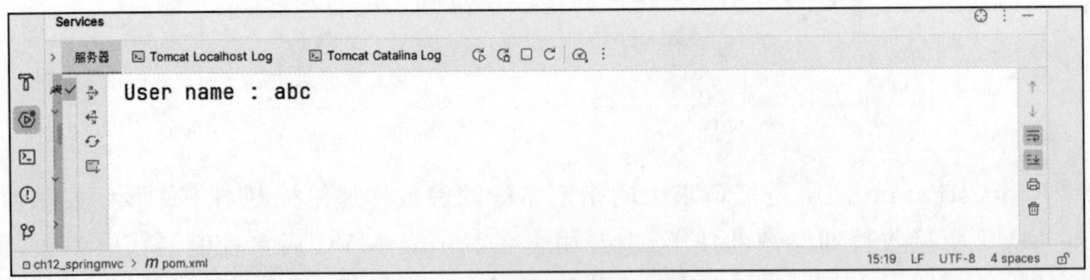

图 12-12 例 12-7 输出结果示例

12.4 @Autowired

@Autowired 注解可以对类成员变量、方法及构造函数进行标注，完成自动装配的工作。使用@Autowired 可以消除 set、get 方法，@Autowired 使 Spring 可以自动将 Bean 里面引用对象的 setter、getter 方法省略，自动实现 set、get。

在正常情况下，在 userService 里面需要做一个 userDao 的 setter、getter 方法，但如果使用了注解@Autowired，便不再需要写 setter、getter 方法，只需要在 UserService 的实现类中声明即可，之所以不用写，关键在于 Java 的反射机制。

Java 的反射机制允许在运行时获取类的属性和方法信息，并且可以动态调用对象的方法和属性。它提供了程序在运行时获取自身信息的能力。通过反射机制，只需提供类的名称，就可以获取类的所有信息。Java 反射机制在服务器程序和中间件程序中被广泛使用，特别是在服务器端根据客户请求动态调用对象的特定方法时。

例 12-8 和例 12-9 分别是通常写法的演示代码与使用注解@Autowired 后简化的演示代码。

例 12-8：演示未使用@Autowired 注解的通常写法。

```
@Controller
public class AutowiredUnUse {
    private UserService userService;
    public void setUserService(UserService injectedService) {
        this.userService = injectedService;
    }
}
```

不使用@Autowired 注解时，要在 userService 中写 setter 方法，如果需要 getter 方法也要写，较为烦琐。

因为使用了@Autowired 注解，上面的代码可以简化为例 12-9 所示。

例 12-9：演示使用@Autowired 注解的简化写法。

```
@Controller
public class AutowiredUse {
    @Autowired
    private UserService userService;
}
```

使用@Autowired 注解时，不需要写 setter、getter 方法，直接声明即可使用 set 和 get 两种功能。代码明显得到简化。

@Autowired 注解有两种修饰方式，一种是用在字段上的，另一种是用在方法上的，两种的演示代码如例 12-10 所示。

例 12-10：演示在两种不同位置上使用@Autowired 注解的代码。

```
public class AutowiredBothUse {
    //1 用在字段上
    @Autowired
    private UserDao userDao;
    //2 用在方法上
    @Autowired
    public void setUserDao(UserDao userDao) {
        this.userDao = userDao;
    }
}
```

其中,1 是将注解用在 UserDao 字段上;2 是将注解用在 setUserDao 方法上。

@Autowired 注解是按照类型(byType)装配依赖对象,默认情况下要求依赖对象必须存在。如果使用时允许 null 值,可以把它的 required 属性设置为 false。如果想使用按照名称(byName)装配,可以结合@Qualifier 注解一起使用。

例 12-11:演示按照名称(byName)装配使用@Autowired 注解的代码。

```
public class AutowiredbyNameUse {
    @Autowired
    @Qualifier("userDao")
    private UserDao userDao;
}
```

加上@Qualifier 注解后可以按照 byName 的方式对 name 为"userDao"的字段使用@Autowired 注解。

还需要了解的是,@Resource 注解和@Autowired 注解类似,也用来声明需要自动装配的 Bean。

@Resource 和@Autowired 的共同点是都可以写在字段和 setter 方法上。二者如果都写在字段上,就不需要再写 setter 方法。而二者之间的不同点,@Resource 并不是 Spring 的注解,它是 Java 的注解,由 J2EE 提供,包是 javax.annotation.Resource,需要先导入对应包才可使用,但是 Spring 也支持该注解的注入;@Autowired 则是 Spring 提供的注解,需要导入包 org.springframework.beans.factory.annotation.Autowired。而且@Resource 默认按照 byName 注入,@Autowired 默认按照 byType 注入。

@Resource 有两个重要的属性:name 和 type。Spring 将@Resource 注解的 name 属性解析为 Bean 的名字,将 type 属性解析为 Bean 的类型。所以,使用 name 属性时,使用 byName 注入策略;使用 type 属性时,则使用 byType 注入策略。如果既不制定 name 也不制定 type 属性,这时将通过 Java 的反射机制默认使用 byName 注入策略。

@Resource 注解也有两种修饰方式,一种是用在字段上的,另一种是用在方法上的,两种的演示代码如例 12-12 所示。

例 12-12:演示在两种不同位置上使用@Resource 注解的代码。

```
public class ResourceBothUse {
    //下面两种@Resource 只要使用一种即可
```

```
    //1 用在字段上
    @Resource(name="userDao")
    private UserDao userDao;
    //2 用在方法上
    @Resource(name="userDao")
    public void setUserDao(UserDao userDao) {
        this.userDao = userDao;
    }
}
```

其中,1 是将注解用在 UserDao 字段上;2 是将注解用在 setUserDao 方法上。

需要注意的是,最好将@Autowired 和@Resource 用在方法上,这样更符合面向对象的思想,即通过 set、get 操作属性,而不是直接操作属性。

12.5 @ModelAttribute

@ModelAttribute 是 Spring MVC 4.3 版本之后新增的注解,主要用于将数据添加到模型对象中,以便在视图页面展示时使用。根据注解的位置和与其他注解的组合使用,其含义会有所不同。

@ModelAttribute 用于修饰参数和方法。修饰参数时,将数据赋值给对象并添加到 ModelMap 中。修饰方法时,可以有返回值,返回值会自动添加到 ModelMap 中。编写控制器代码时,要注意多个 URL 映射的情况,建议将保存方法独立成一个控制器。

例 12-13:演示@ModelAttribute 注解修饰方法时无返回值的使用。

```
@Controller
public class ModelAttributeVoidUse {
    @ModelAttribute
    public void populateModel() {
        System.out.println("success");
    }
    @RequestMapping(value = "/ModelAttributeVoidUse")
    public String helloWorld() {
        return "helloWorld";
    }
}
```

在例 12-13 中,@ModelAttribute 用在 populateModel 方法上,该方法无返回值也可正常使用。在整个 ModelAttributeVoidUse 类中,访问控制器方法 helloWorld 时,会首先调用@ModelAttribute 注解的 populateModel 方法。代码运行结果如图 12-13 和图 12-14 所示。首先调用了@ModelAttribute 注解的 populateModel 方法,因此根据 populateModel 方法中的代码输出了 success。然后根据 helloWorld 方法,运行结果为返回 hello world 页面。

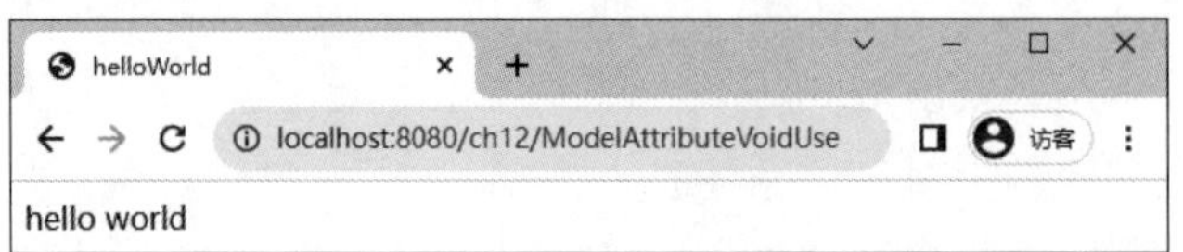

图 12-13　例 12-13 运行结果示例

图 12-14　例 12-13 输出结果示例

例 12-14：演示@ModelAttribute 注解修饰方法时返回具体类的使用。

```
@Controller
public class ModelAttributeClassUse {
    @ModelAttribute
    public User populateModel() {
        User user=new User();
        user.setAccount("ray");
        return user;
    }
    @RequestMapping(value = "/ModelAttribute")
    public String helloWorld(Model map) {
        System.out.println(map);
        return "success";
    }
}
```

当用户请求/ModelAttribute 时，首先访问@ModelAttribute 注解的 populateModel 方法，返回 User 对象，helloWorld 方法中声明了 Model map，向 Model 中的 map 放数据，数据为 populateModel 方法返回的名为 user 的 User 对象。代码运行结果如图 12-15 和图 12-16 所示。首先调用了@ModelAttribute 注解的 populateModel 方法，根据 populateModel 方法中的代码创建 User 对象 user，对其使用 setAccount 方法，将对象中的 account 赋值为 ray，返回 user。然后根据 helloWorld 方法输出 Model 中的 map 查看数据，并返回 success 页面。

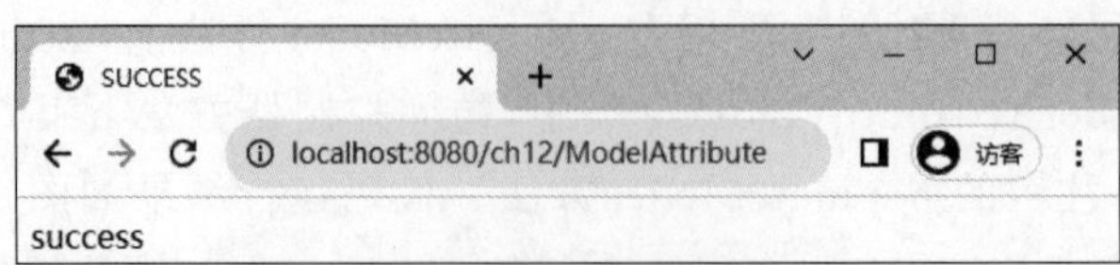

图 12-15　例 12-14 运行结果示例

```
{user=User [account=ray, name=null, getAccount()=ray, getName()=null, getClass()=class com.imut.pojo.User, hashCode()=667775806, toString()=com.imut.pojo.User@27cd733e]}
```

图 12-16　例 12-14 部分输出结果示例

@ModelAttribute 的 value 属性用于获取数据的 key。key 可以是 POJO 的属性名称，也可以是 Map 结构的 key。POJO(Plain Ordinary Java Object)即简单的 Java 对象，实际就是普通的 JavaBeans，是为了避免和 EJB 混淆所创造的简称。

例 12-15：演示@ModelAttribute 注解修饰方法时返回具体类时设置 value 的方法。

```
@Controller
public class ModelAttributeClassValueUse {
    @ModelAttribute(value="User1")
    public User populateModel() {
        User user=new User();
        user.setAccount("ray");
        return user;
    }
    @RequestMapping(value = "/ModelAttributeClassValueUse")
    public String helloWorld(Model map) {
        System.out.println(map);
        return "success";
    }
}
```

当用户请求/ModelAttributeClassValueUse 时，首先访问@ModelAttribute 注解的 populateModel 方法，返回 User 对象。@ModelAttribute(value="User1")指定名称时，意味着名称为“User1”的对象得到返回的 user 对象的信息。代码运行结果如图 12-17 和图 12-18 所示。首先调用了@ModelAttribute 注解的 populateModel 方法，根据 populateModel 方法中的代码创建 User 对象 user，对其使用 setAccount 方法，将对象中的 account 赋值为 ray，给名称为“User1”的对象返回 user。然后根据 helloWorld 方法输出 Model 中的 map 查看数据，并返回 success 页面。

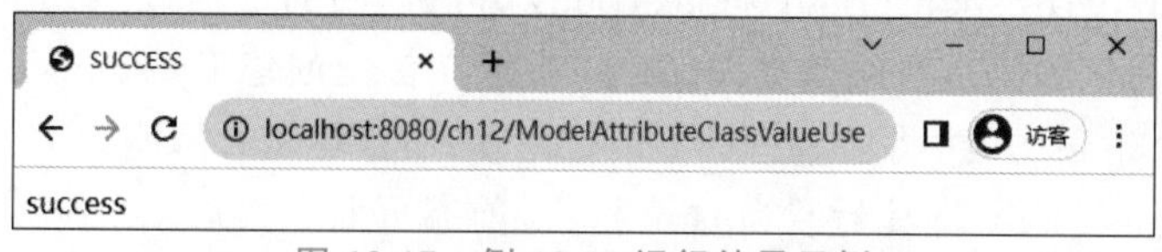

图 12-17　例 12-15 运行结果示例

@ModelAttribute 也可以在当表单提交的数据不是完整的实体类数据时，保证没有提交数据的字段使用数据库对象原来的数据。

```
{User1=User [account=ray, name=null, getAccount()=ray, getName()
=null, getClass()=class com.imut.pojo.User, hashCode()=1291511360,
toString()=com.imut.pojo.User@4cfae640]}
```

图 12-18　例 12-15 部分输出结果示例

12.6　@ResponseBody

@ResponseBody 注解通常使用在控制层的方法上，用于将 Controller 的方法返回的对象通过 Spring MVC 提供的 HttpMessageConverter 接口转换为指定格式的数据，如：json、xml 等，然后通过 Response 响应给客户端，写入到 Response 对象的 body 数据区。

@ResponseBody 注解用于控制层方法，将返回的对象通过 HttpMessageConverter 接口转换为指定格式的数据，如 JSON、XML 等，并写入 Response 的 body 数据区，响应给客户端。需要注意的是，返回对象时使用 UTF-8 编码，返回字符串可能导致乱码。可以通过@RequestMapping 注解的 produces 属性手动指定编码格式，如@RequestMapping(value="/query", produces="text/html;charset=utf-8")，前面是请求的路径，后面是编码格式。

例 12-16：演示@ResponseBody 注解的使用。

```
@Controller
public class ResponseBodyUse {
    @RequestMapping("/ResponseBodyUse")
    @ResponseBody
    public Person testBody(){
        Person person = new Person();
        person.setAge(10);
        person.setName("ly");
        return person;
    }
}
```

在例 12-16 中，@ResponseBody 将 testBody 方法返回的数据以 json 格式输出到前端，供前端解析后使用。代码运行结果如图 12-19 所示。首先创建 Person 对象 person，对其使用 setAge 和 setName 方法，将对象中的 name 赋值为 ly，将 age 赋值为 10，直接给前端返回 person。@ResponseBody 将其封装为 json 格式，前端页面直接展示 json 格式数据。

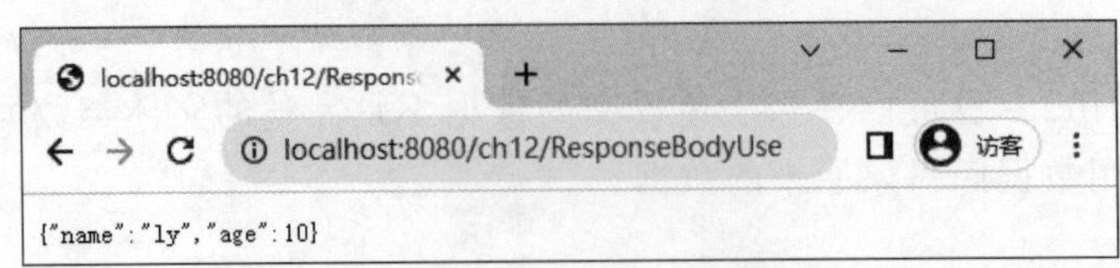

图 12-19　例 12-16 运行结果示例

12.7 @RequestBody

@RequestBody 注解用于接收前端传递给后端的请求体中的 JSON 字符串数据。通常情况下，使用 POST 请求方式进行提交。

@RequestBody 的属性 required 用于指示请求体是否必须存在，默认值为 true。当 required 为 true，GET 请求方式会报错，应使用 POST 请求。当 required 为 false，GET 请求时，请求体内容为 null，但不会报错。

例 12-17：演示@RequestBody 注解的使用。

```
@Controller
public class RequestBodyUse {
    @RequestMapping(value = "/RequestBodyUse")
    public String testBody (@RequestBody String name) {
        System.out.println(name);
        return "success";
    }
}
```

例 12-17 中@RequestBody 注解使用在 name 上，使用后获取前端的 name 数据。因为其在使用时未声明 required 属性，因此当前情况下是使用的默认值 true，前端请求必须使用 POST 方法，使用 GET 会发生报错。代码运行结果如图 12-20 和图 12-21 所示。前端根据/RequestBodyUse 路径和 POST 方法发出请求，并且前端页面上要有名称为 name 的数据。根据 testBody 方法中的@RequestBody 注解获取前端数据 name 为 001，根据方法输出获取的 name，并返回 success 页面。

图 12-20　例 12-17 运行结果示例

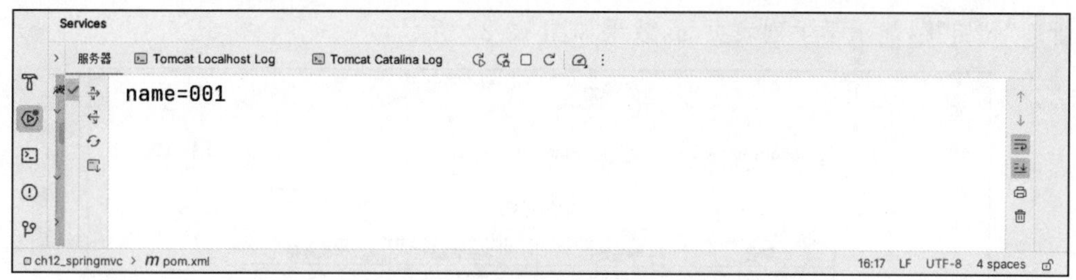

图 12-21　例 12-17 输出结果示例

@ModelAttribute 也可以在当表单提交数据不是完整的实体类数据时，保证没有提交数据的字段使用数据库对象原来的数据。

如果@RequestBody 作用在方法上，则表示该方法的返回结果是直接按写入 Http responsebody 中，是一种一般在异步获取数据时使用的注解。

在后端的同一个接收方法里，@RequestBody 与@RequestParam 可以同时使用。在一个请求中，@RequestBody 最多只能有一个，而@RequestParam 可以有多个。

12.8 综合案例

本章案例主要实现获取模型对象中 List 集合中所有学生的信息，并在浏览器显示的功能。

12.8.1 案例设计

案例包括持久层和 Web 表现层。持久层由实体类 Student.java 组成。Web 表现层主要包括 Spring MVC 中的 Controller 类和 JSP 页面。Controller 类 StudentController.java 主要负责拦截用户的请求，向模型对象中添加一个包含 Student 对象的 List 集合，然后返回给 JSP 页面进行展示。JSP 页面 studentList.jsp 展示 List 集合中的所有学生信息。本章案例使用的主要文件如表 12-1 所示。

表 12-1　本章案例使用的主要文件

文　件	所在包/路径	功　能
web.xml	/WEB-INF/	Web 配置文件
Spring MVC.xml	src/	Spring MVC 配置文件
Student.java	com.imut.pojo	封装学生信息的类
StudentController.java	com.imut.controller	控制器类
index.jsp	/	首页
studentList.jsp	/WEB-INF/pages/	显示信息的页面

12.8.2 案例演示

在浏览器地址栏中输入 http://localhost:8080/ch12_Spring MVC/index.jsp，单击“学生管理”超链接，显示学生列表页面，效果如图 12-22 所示。

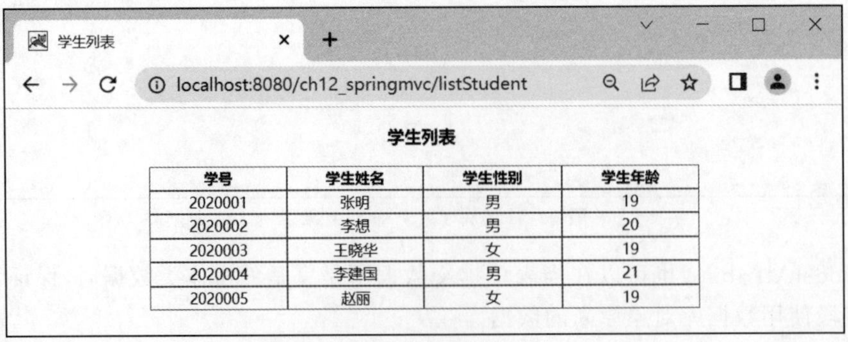

学生列表

localhost:8080/ch12_springmvc/listStudent

学生列表

学号	学生姓名	学生性别	学生年龄
2020001	张明	男	19
2020002	李想	男	20
2020003	王晓华	女	19
2020004	李建国	男	21
2020005	赵丽	女	19

图 12-22　学生列表页面

12.8.3 代码实现

1. 创建项目，添加依赖

在 IDEA 中使用 Maven，创建 Web 动态工程 ch12_Spring MVC，在 pom.xml 文件中添加以下依赖：Spring 的 4 个基础包、aop 包、expression 包、web 包、webmvc 包、日志包以及 jstl 包，代码如下。

```
<dependencies>
    <dependency>
      <groupId>org.springframework</groupId>
      <artifactId>spring-beans</artifactId>
      <version>5.2.5.RELEASE</version>
    </dependency>
    <dependency>
      <groupId>org.springframework</groupId>
      <artifactId>spring-context</artifactId>
      <version>5.2.5.RELEASE</version>
    </dependency>
    <dependency>
      <groupId>org.springframework</groupId>
      <artifactId>spring-core</artifactId>
      <version>5.2.5.RELEASE</version>
    </dependency>
    <dependency>
      <groupId>org.springframework</groupId>
      <artifactId>spring-expression</artifactId>
      <version>5.2.5.RELEASE</version>
    </dependency>
    <dependency>
      <groupId>org.springframework</groupId>
      <artifactId>spring-aop</artifactId>
      <version>5.2.5.RELEASE</version>
    </dependency>
    <dependency>
      <groupId>org.springframework</groupId>
      <artifactId>spring-web</artifactId>
      <version>5.2.5.RELEASE</version>
    </dependency>
    <dependency>
      <groupId>org.springframework</groupId>
      <artifactId>spring-webmvc</artifactId>
      <version>5.2.5.RELEASE</version>
    </dependency>
    <dependency>
      <groupId>javax.servlet</groupId>
      <artifactId>javax.servlet-api</artifactId>
      <version>3.1.0</version>
    </dependency>
```

```
    <dependency>
      <groupId>javax.servlet</groupId>
      <artifactId>jstl</artifactId>
      <version>1.2</version>
      <scope>compile</scope>
    </dependency>
    <dependency>
      <groupId>log4j</groupId>
      <artifactId>log4j</artifactId>
      <version>1.2.17</version>
    </dependency>
</dependencies>
```

2. 编写配置文件

在 WEB-INF 下编写 web.xml 配置文件。在/src/main/resources 下编写 Spring MVC.xml 配置文件。web.xml 文件与第 11 章综合案例的 web.xml 文件内容相同，这里不再赘述。

配置文件 Spring MVC.xml 的代码如下。

```
<?xml version="1.0" encoding="UTF-8"?>
<beans xmlns="http://www.springframework.org/schema/beans"
    xmlns:xsi="http://www.w3.org/2001/XMLSchema-instance"
    xmlns:context="http://www.springframework.org/schema/context"
    xmlns:mvc="http://www.springframework.org/schema/mvc"
    xsi:schemaLocation="
        http://www.springframework.org/schema/beans
        http://www.springframework.org/schema/beans/spring-beans.xsd
        http://www.springframework.org/schema/context
        http://www.springframework.org/schema/context/spring-context-4.0.xsd
        http://www.springframework.org/schema/mvc
        http://www.springframework.org/schema/mvc/spring-mvc-4.0.xsd">
    <mvc:annotation-drive/>
    <context:component-scan base-package="com.imut.controller"/>
    <bean class="org.springframework.web.servlet.view.
InternalResourceViewResolver">
    <property name="prefix" value="/WEB-INF/pages/"></property>
    <property name="suffix" value=".jsp"></property>
    </bean>
</beans>
```

3. 创建实体类

在/src/main/java 下新建包 com.imut.pojo，在该包下创建实体类 Student.java。

程序 Student.java 的代码如下。

```
public class Student implements Serializable{
    private int sid;
    private String sno;
```

```
    private String sname;
    private String sgender;
    private int sage;
    //此处省略构造方法、getXxx()、setXxx()和 toString()方法
}
```

4. 编写控制器类

在/src/main/java 目录下新建包 com. imut. controller，在该包下创建控制器类 StudentController.java。

程序 StudentController.java 的代码如下。

```
@Controller
public class StudentController {
    @RequestMapping("/listStudent")
    public ModelAndView findAllStudent() {
        List<Student> list=new ArrayList<Student>();
        list.add(new Student(1,"2020001","张明","男",19));
        list.add(new Student(2,"2020002","李想","男",20));
        list.add(new Student(3,"2020003","王晓华","女",19));
        list.add(new Student(4,"2020004","李建国","男",21));
        list.add(new Student(5,"2020005","赵丽","女",19));
        ModelAndView mv=new ModelAndView();
        mv.addObject("list", list);
        mv.setViewName("studentList");
        return mv;
    }
}
```

5. 编写 JSP 页面文件

在根目录 webapp 下编写 index.jsp 文件。在 WEB-INF 下新建 pages 文件夹，在该文件夹下新建 studentList.jsp 文件。由于篇幅原因，JSP 文件只放<body></body>之间的代码。

程序 index.jsp 的代码如下。

```
<body>
    <h3><a href="listStudent">学生管理</a></h3>
</body>
```

程序 studentList.jsp 代码如下：

```
<body>
<div class="mydiv">
    <h3>学生列表</h3>
    <table id="mytb" style="width:600px;border-spacing: 0px 0px;">
        <thead>
            <tr>
```

```
                <th>学号</th>
                <th>学生姓名</th>
                <th>学生性别</th>
                <th>学生年龄</th>
            </tr>
        </thead>
        <tbody>
            <c:forEach items="${list}" var="student">
                <tr>
                    <td>${student.sno }</td>
                    <td>${student.sname }</td>
                    <td>${student.sgender }</td>
                    <td>${student.sage }</td>
                </tr>
            </c:forEach>
        </tbody>
    </table>
    </div>
</body>
```

12.9 习题

1. 选择题

(1) 关于@Controller 注解类，以下说法错误的是(　　)。

A. @Controller 注解是@Component 注解的一种类型

B. @Controller 注解有资格在 Spring MVC 中处理请求

C. @Controller 与@RestController 可以互换，类里面的方法不需要额外的代码修改

D. 可以通过组件扫描发现

(2) 关于@RestController 控制器，以下说法正确的是(　　)。

A. 是 Spring 的内容注解　　B. 是 Spring 的路径注解

C. 是 Spring 的控制器注解　　D. @Controller 注解更强大

(3) 关于@RequestMapping 注解类，以下说法正确的是(　　)。

A. 该注解是用于处理请求地址映射的注解，只能用在类上

B. 该注解是用于处理请求地址映射的注解，只能用在方法上

C. 该注解是用于处理请求地址映射的注解，可用于类或方法上

D. 该注解用于接收基本数据类型参数

(4) 在 Spring MVC 中，通过(　　)注解可以将 URL 请求与业务方法进行映射。

A. @Controller　　B. @RequestMapping

C. @Component　　D. @Repository

(5) 关于@Autowired 和@Resource，以下说法正确的是(　　)。

A. @Autowired 是 JSR250 提供的，@Resuource 是由 Spring 提供的

B. @Autowired 默认采用 byName 方式装配，@Resuorce 默认按照 byType 方式装配

C. @Autowired 和@Resource 完全相同

D. @Autowired 可以用在属性字段上，也可以用在 Setter 方法上，用在属性字段上时可以省略该字段的 Setter 方法

(6) 关于@RequestBody，以下说法正确的是(　　)。

A. 可用于类或方法上

B. 用于类上，则表示类中的所有响应请求参数都是 json

C. 注解实现接收 http 请求的 json 数据，将 json 转换为 java 对象

D. 注解实现将 conreoller 方法返回对象转化为 json 对象响应给客户

(7) 下列关于 Spring MVC 注解的叙述中，正确的是(　　)(多选)。

A. Spring MVC 中@RequestMapping 注解的作用是：可以对类的成员变量、方法及构造函数进行标注，完成对象自动装配的工作

B. @Override 注解可帮助检查某方法名是否与父类中方法相同，或在重写父类的 onCreate 时系统可以帮你检查方法的正确性

C. DispatcherServlet 是 Spring MVC 的一个核心控制器，可通过 HandlerMapping 将请求映射到处理器

D. Spring MVC 只可以通过注解的方式注入

2. 填空题

(1) 在 Spring MVC 中使用________，也可以起到@Controller 的作用。

(2) @RequestMapping 是用于________的注解，能用在________上。

(3) @Autowired 是按照________进行注入，@Resource 是按照________进行注入。

(4) Spring MVC 中，在方法上加注解________可以将方法的返回值传到页面。

(5) @ResponseBody 注解可以将后端 Java 对象转换成________进行响应。

3. 简答题

(1) 请简述@Controller 注解的使用步骤。

(2) 请简述简单数据类型中的@RequestParam 注解及其属性作用。

(3) 请简述 JSON 数据交互使用的两个注解的作用。

第13章 Spring MVC数据处理

在 Spring MVC 中,处理表单提交的请求参数时,不论是 int 类型还是 String 类型,处理器方法都能自动接收相应的数据类型。这得益于 Spring MVC 框架内置的类型转换器,它能够自动将 String 类型的数据转换成对应的数据类型。本章主要讲解 Spring MVC 中常用的数据处理注解和相关代码。

13.1 数据转换

在 Spring MVC 中,通过注解,控制器(Controller)能够获取各种类型的参数,这得益于 Spring MVC 的消息转换机制。然而,在 Spring MVC 的执行过程中,涉及 HandlerAdapter 的角色,它并非与 Controller 相同。处理器(Handler)在控制器功能的基础上添加了一层封装,其主要作用是在 HTTP 请求到达控制器之前,对 HTTP 的各种消息进行处理。

当一个请求到达 DispatcherServlet 时,首先需要找到相应的 HandlerMapping,然后根据 HandlerMapping 找到相应的 HandlerAdapter,来执行处理器。处理器调用控制器之前,需要先获取 HTTP 发送过来的信息,并将其转换为控制器需要的各种不同类型的参数。这就是各种注解能够获取丰富类型参数的原因。在消息转换方面,Spring MVC 使用 HTTP 消息转换器(HttpMessageConverter)进行转换,其中 String 类型和文件类型的转换比较简单。此外,Spring 4 引入了转换器和格式化器,使得通过注解和参数类型能够将 HTTP 发送的各种消息转换为控制器所需的各类参数。

处理器完成参数转换后,对数据进行校验,并执行自定义开发的控制器。此时,控制器已获得各种 Java 数据类型的支持,控制器中的处理方法能够完成业务逻辑,最终返回结果。处理器还会根据返回结果的类型查找对应的 HttpMessageConverter 的实现类,然后调用相应实现类的方法,将控制器返回的结果进行 HTTP 转换。前提是能够找到对应的转换器,这样,处理器的任务就完成了。图 13-1 为 Spring MVC 消息转换流程图。

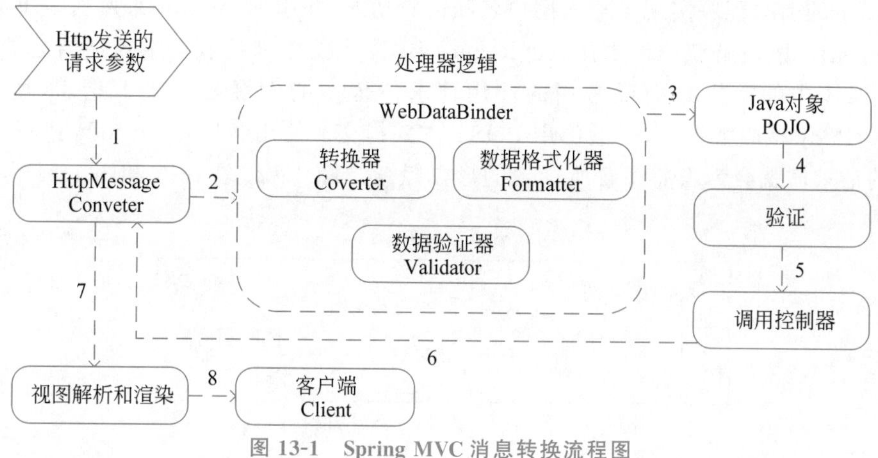

图 13-1 Spring MVC 消息转换流程图

13.1.1 HttpMessageConveter

HttpMessageConverter，也称为报文信息转换器，具有将请求报文转换为 Java 对象以及将 Java 对象转换为响应报文的功能。它在 Spring MVC 中提供了两个注解和两个类型。

@RequestBody：用于将请求报文的内容转换为 Java 对象。

@ResponseBody：用于将 Java 对象转换为响应报文的内容。

此外，还有两个相关的类型：

RequestEntity：用于封装 HTTP 请求报文和请求头信息。

ResponseEntity：用于封装 HTTP 响应报文和响应头信息。

这些注解和类型在 Spring MVC 中非常有用，它们允许灵活地处理 HTTP 请求和响应，以实现数据的转换和交互。

13.1.2 @RequestBody

例 13-1：演示前端传递给后端的 json 字符串，前端程序 user.jsp 如下。

```
<form th:action="@{/testRequestBody}" method="post">
    <input type="text" name="username">
    <input type="text" name="password">
    <input type="submit" name="测试@RequestBody">
</form>
```

后端接收前端表单传来的数据，后端程序 UserController.java 如下。

```
@RequestMapping(value = "/testRequestBody", method = "RequestMethod.POST")
public String testRequestBody(@RequestBody String RequestBody) {
        System.out.println("requestBody"+RequestBody);
        return"success";
}
```

例 13-1 的前端程序实现了输入用户名和密码功能，并添加 submit 来发送表单数据，测试@RequestBody 注解获取请求体。@RequestBody 主要用来接收前端传递给后端的 json 数据(请求体中的数据)，而最常用的使用请求体传参的无疑是 POST 请求，所以使用@RequestBody 接收数据时，一般都用 POST 方式提交。使用@RequestBody 进行标识，当前请求的请求体就会为当前注解所标识的形参赋值。图 13-2、图 13-3 和图 13-4 为演示效果图。

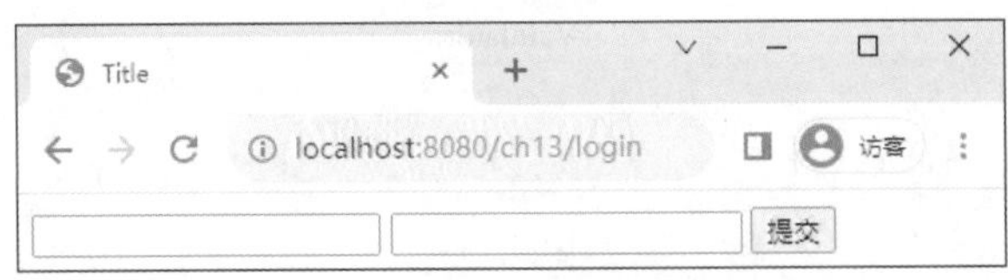

图 13-2　前端提交页面

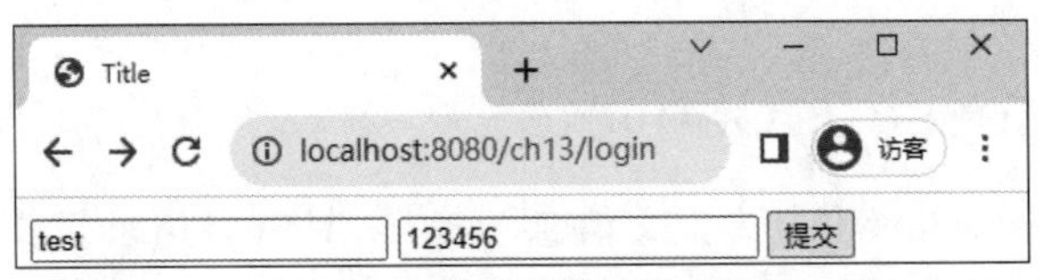

图 13-3　输入用户名和密码

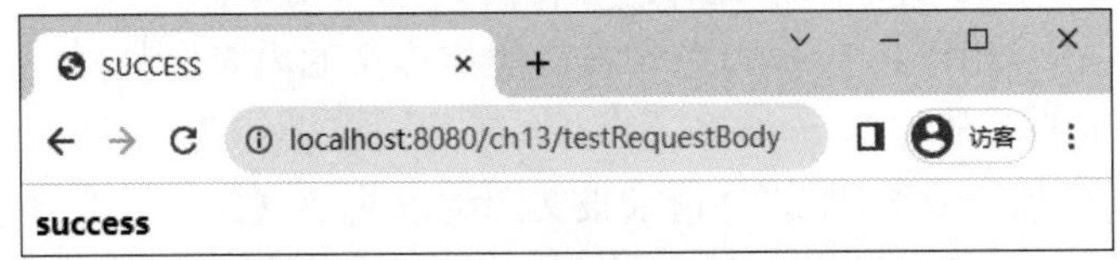

图 13-4　成功接收并跳转页面

13.1.3　@ResponseBody

例 13-2：例 13-1 演示了一个前端页面(user.jsp)和后端控制器(UserController.java)，用于展示如何将前端传递的 JSON 字符串发送到后端。以下例子演示将后端控制器返回值直接作为响应体返回给前端。

后端控制器(UserController.java)。

```
@RequestMapping("/testResponseBody")
@ResponseBody
public String testResponseBody() {
    return "success";
}
```

这里更新了@RequestMapping 的路径定义方式，使其更加清晰。这个控制器方法使用@ResponseBody 注解，将方法的返回值直接作为响应体返回给前端。这样，用户界面将直接收到方法返回的字符串信息，如图 13-5 所示。

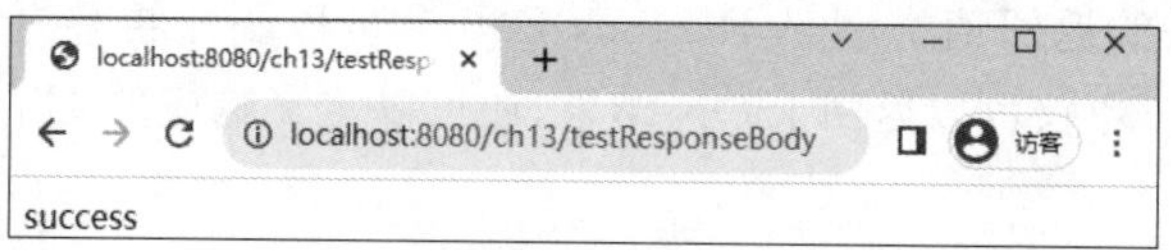

图 13-5　成功接收并跳转页面

13.1.4　ResquestEntity

例 13-3：演示 getHeaders()获取请求头信息，getBody()获取请求体信息，后端程序 UserController.java 如下。

```
@RequestMapping("testRequestEntity")
public String testRequestEntity(RequestEntity<String> requestEntity) {
    System.out.println("requestHeader"+requestEntity.getHeaders());
    System.out.println("requestBody"+requestEntity.getBody());
    return "success";
}
```

RequestEntity 是封装请求报文的一种类型，需要在控制器方法设置该类型的形参，当前请求的报文就会赋值给该形参，可以通过 getHeaders()获取请求头信息，通过 getBody()获取请求体信息。显示结果如图 13-6 所示。

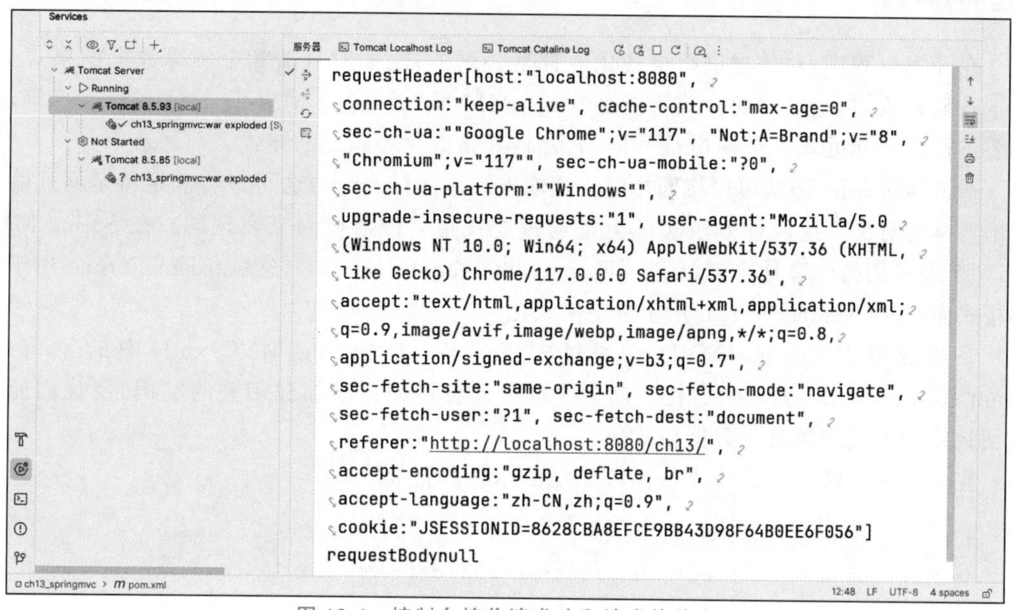

图 13-6　控制台接收请求头和请求体信息

13.1.5　ResponseEntity

ResponseEntity 用于指定控制器方法的返回值类型，该返回值即是将要发送给浏览器的响应报文。它的优先级高于 @ResponseBody 注解，只有在不使用 ResponseEntity 的情况下才会检查是否存在 @ResponseBody 注解。

如果控制器方法的返回值类型是 ResponseEntity，可以不用再显式添加@ResponseBody 注解。ResponseEntity 不仅可以返回 JSON 结果，还可以定义响应的 HTTP 状态码、响应体和响应头信息。

以下是一个示例。

```
@RequestMapping("/testResponseEntity")
public ResponseEntity testResponseEntity() {
    //使用方式一
    //ResponseEntity responseEntity = new ResponseEntity(new User("IMUT", 77),
HttpStatus.OK);
    //使用方式二
    return ResponseEntity.ok(new User("IMUT", 71));
}
```

ResponseEntity 类继承自 HttpEntity，具有 3 个重要属性：HttpStatus 代表响应状态码、Body 代表响应体、HttpHeader 代表响应头信息。这允许你更精确地定义控制器方法的响应报文。

13.2 数据格式化

在开发过程中，对数据进行格式化通常是必需的，比如对日期和金额等数据的格式化。对于日期来说，传递的日期格式可能是“yyyy-MM-dd”或“yyyy-MM-dd HH:mm:ss” 等，都需要进行格式化处理。金额也是一样，需要将其格式化为固定的样式。

通常通过 print 方法可以按照指定格式输出结果，而 parse 方法可以将满足特定格式的字符串转换为对象。这两种方法实际上是通过委托给转换器机制来实现的。在 Spring MVC 中，两个最常用的注解是 @DateTimeFormat 和@NumberFormat。@DateTimeFormat 用于日期格式化，而@NumberFormat 用于数字格式化。

系统启动时，Spring MVC 会通过配置文件（如 Spring MVC.xml）中的 <mvc:annotation-driven> 完成初始化。这个注解驱动会帮助自动加载相关的实例，使我们能够方便地使用上述注解进行数据格式化。

在 Spring MVC.xml 中添加注解驱动的示例配置如下。

```
<!-- 开启 MVC 的注解驱动 -->
<mvc:annotation-driven></mvc:annotation-driven>
```

通过这样的配置，Spring MVC 将能够自动处理相关的注解，简化开发者在数据格式化方面的工作。

13.2.1 @DateTimeFormat

@DateTimeFormat 注解可对 java.util.Date、java.util.Calendar、java.long.Long 时间类型进行标注。DateTimeFormat 有以下 3 种属性，详情如表 13-1 所示。

表 13-1 DateTimeFormat 的 3 种属性

属性	说　明
Pattern	类型为字符串。指定解析/格式化字段数据的模式，如“yyyy-MM-dd hh:mm:ss”
Iso	类型为 DateTimeFormat.ISO。指定解析/格式化字段数据的 ISO 模式，包括 4 种：ISO.NONE 表示不使用；ISO. DATE(yyyy-MM-dd)；ISO. TIME(hh: mm: ss. SSSZ)；ISO. DATE_TIEM (yyyy-MM-dd hh:mm:ss.SSSZ)三种为默认格式
Style	字符串类型。通过样式指定日期时间的格式，由两位字符组成，第一位表示日期的格式，第二位表示时间的格式：S：短日期/时间格式；M：中日期/时间格式；L：长日期/时间格式；F：完整日期/时间格式；-：忽略日期或时间格式

13.2.2 @NumberFormat

@NumberFormat 可对类似数字类型的属性进行标注，它拥有两个互斥的属性，如表 13-2 所示。

表 13-2 NumberFormat 的两种互斥属性

属性	说　明
Style	类型为 NumberFormat.Style。用于指定样式类型，包括 3 种：Style.NUMBER(正常数字类型)、Style.CURRENCY(货币类型)、Style.PERCENT(百分数类型)
Patterh	类型为 String，自定义样式，如 patterh="＃，＃＃＃"

例 13-4：演示日期和金钱接收格式，创建表单分别有日期和金钱属性。前端程序 infoConver.jsp 代码如下。

```
<form th:action="@{/formatter}">
    日期:<input type="text" name="date"> <a>日期的格式为 XXXX-XX-XX</a>
    <br>
    金额:<input type="text" name="money">
<br>
    <input type="submit" name="测试 format">
</form>
```

后端程序 UserController.java 如下。

```
@RequestMapping("formatter")
public String formatter(@RequestParam("date")@DateTimeFormat(pattern = "yyyy-
MM-dd") Date date, @RequestParam("money")@NumberFormat(pattern = "#,###.##")
Double money) {
    System.out.println("Date:"+date);
    System.out.println("Money:"+money);
    return "success";
}
```

控制台输出结果为 Date：Mon Apr 11 00:00:00 CST 2022；Money：1123.44，这说明前端输入的格式后台可以解析出来并输出。显示结果如图 13-7 所示。提交完成后控制台显示前端输入的日期和金额，如图 13-8 所示；当输入的格式错误，则会有错误 400 提示，如图 13-9 所示。

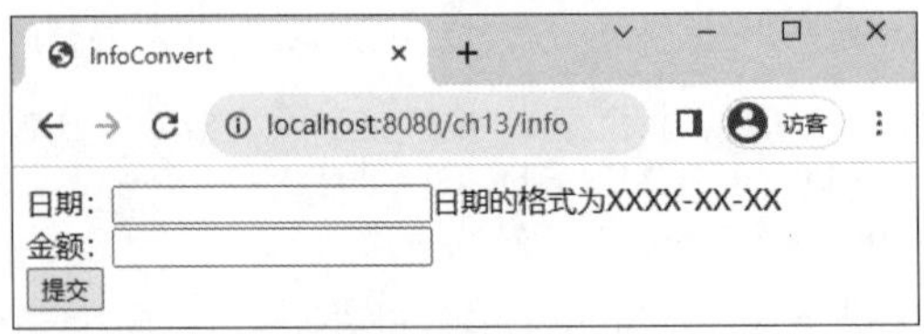

图 13-7　输入页面

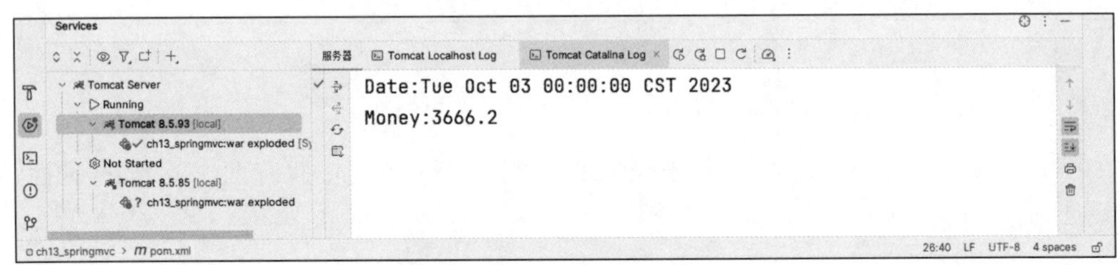

图 13-8　后台解析数据格式

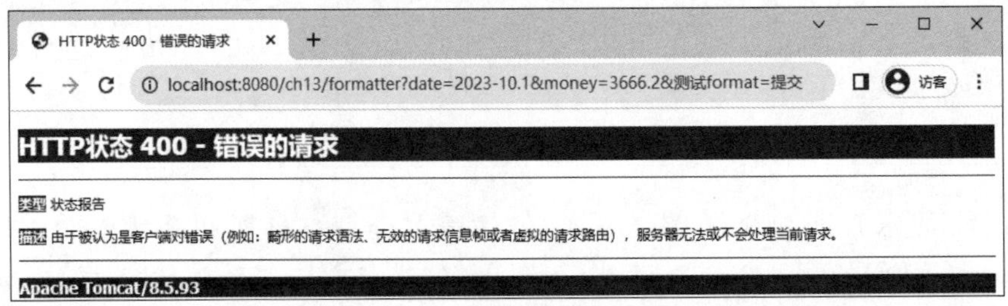

图 13-9　输入格式错误提示页面

13.3　数据校验

13.3.1　数据校验的需求

在实际开发中，数据的验证是非常关键的步骤。通常，可以使用注解进行数据验证，以检查数据的正确性，并将错误信息返回给前端。然而，对于一些逻辑上的验证规则，例如年薪等于月薪乘以 12，注解验证可能不太适用。在这种情况下，可以使用 Spring 提供的验证器(Validator)来进行验证。需要注意的是，所有的验证器都需要先进行注册，但 Spring MVC 会自动加载这些验证器。

13.3.2　常用的数据验证规则

Spring 提供了通过@Valid 注解启用的 Bean 数据合法性校验框架，基于 JSR 303 规范的注解，可用于标记 Bean 属性的验证规则，确保数据的有效性。这些注解定义在 javax.validation.constraints 中，包括@NotNull、@Max 等，常用的验证规则见表 13-3。

表 13-3 数据验证规则

注　　解	验证规则
@AssertTrue	验证注解必须为 true
@AssertFalse	验证注解必须为 false
@Digits(integer=,fraction=)	验证注解的元素必须为一个数字,其值必须在可接受的范围内
@DecimalMax(value)	验证注解的元素必须为一个数字,其值必须小于或等于指定的最大值
@DecimalMin(value)	验证注解的元素必须为一个数字,其值必须大于或等于指定的最小值
@Future	验证注解必须为日期,检查给定的日期是否比现在晚
@Max(value)	验证注解的元素必须为一个数字,其值必须小于或等于指定的最大值
@Min(value)	验证注解的元素必须为一个数字,其值必须大于或等于指定的最小值
@Null	验证注解元素必须为空
@NotNull	验证注解元素必须不为空
@Past(java.util.Date/Calendar)	验证注解元素必须是过去的日期,检查标注对象中的值表示的日期要比当前早
@Pattern(regex= ,flag=)	验证注解元素是否符合正则表达式,检查该字符串是否能在 match 指定的情况下被 regex 定义的正则表达式匹配
@Size(min=,max=)	验证注解元素必须限制在指定的范围(数据类型:String、Collection、Map、Arrays)
@Valid	递归地对关联对象进行校验,如果关联对象是个集合或者数组,那么对其中的元素进行递归校验,如果是 map,则对其中的部分值进行校验
@Email	验证注解元素必须是电子邮箱地址
@Length(min=,max=)	验证注解元素必须是字符串,并在指定的长度范围内
@NotBlank	验证注解元素必须是字符串且不能为空,忽略空格
@NotEmpty	验证注解元素必须不为空(数据:String、Collectoin、Map、Arrays)

13.4 域对象共享数据

视图是向用户展示业务处理结果的方式。通常,业务处理方法会返回数据,它们将在页面上显示给用户。在之前的 Spring MVC 开发流程中提到,控制器获取数据后,将这些数据添加到模型对象中,然后将视图名称传递给视图解析器。解析器会找到最终的视图,将模型数据渲染到视图中,最终呈现给用户。Spring MVC 支持多种方式输出模型数据,包括 ModelAndView、Map、Model、ModelMap 等。

13.4.1 ModelAndView

例 13-5:使用 ModelAndView 向 request 域对象共享静态数据,前端程序 DynamicSuccess.

jsp 如下所示。显示结果如图 13-10 所示。

```
<body>
    <h4>success</h4>
    <h5>数据传输如下:</h5>
    ${testScope}
</body>
```

后端程序 UserController.java 如下。

```
@RequestMapping("/testModelAndView")
public ModelAndView testModelAndViewStatic(){
    ModelAndView mav = new ModelAndView();
    mav.addObject("testScope", "hello,ModelAndView");
    mav.setViewName("DynamicSuccess ");
    return mav;
}
```

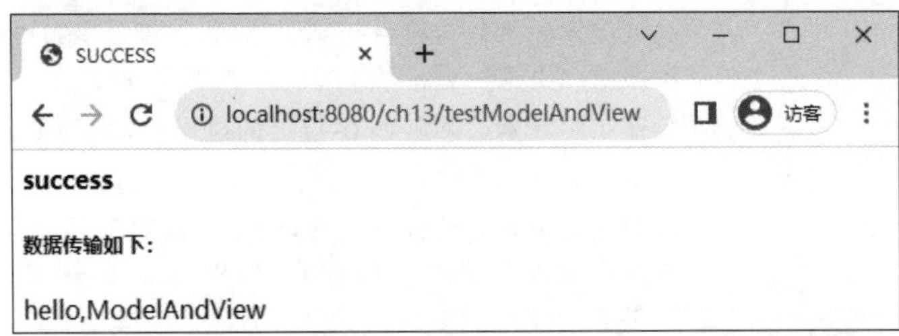

图 13-10 ModelAndView 向 request 域对象共享静态数据

例 13-6：使用 ModelAndView 向 request 域对象共享动态数据，后端程序 UserController.java 如下。

```
//创建学生列表
private List<Student> list = new ArrayList<Student>();
@RequestMapping("/listStudent")
public ModelAndView findAllStudent() {
    ModelAndView mv = new ModelAndView();
    mv.addObject("list", list);
    mv.setViewName("StudentList");
    return mv;
}
```

当控制器处理方法的返回值类型为 ModelAndView 时，可以通过该对象方便地添加模型数据。ModelAndView 不仅包含视图信息，还包含模型数据信息。模型数据可以通过 ModelAndView 的方法设置，而视图信息也可以通过相应的方法配置，使开发更加简便，添加模型数据的方式如表 13-4 所示，设置视图方式如表 13-5 所示。

表 13-4 添加模型数据的方式

方　法	说　明
ModelAndView addObject(String attributeName, Object attributeValue)	设置属性名和属性值
ModelAndView addAllObjects(Map<String,?> modelMap)	如果有多个属性需要传递,可以先保存在一个 Map 集合中,然后直接传递一个 Map 集合,达到实现传递多个属性的效果

表 13-5 设置视图方式

方　法	说　明
void setViewName(String viewName)	传递一个视图名称,DispatcherServlet 会根据该视图名称结合视图解析器进行解析
void setView(View view)	传递一个 View 类型的对象

13.4.2 Model

例 13-7:使用 Model 向 request 域对象共享静态数据,前端 DynamicSuccess.jsp 页面代码如下。显示结果如图 13-11 所示。

```
<body>
    <h4>success</h4>
    <h5>model 数据传输如下:</h5>
    ${testScope}
</body>
```

后端程序 UserController.java 如下。

```
@RequestMapping("/testModel")
public String testModel(Model mdoel) {
    mdoel.addAttribute("testScope", "hello,Model");
    return "DynamicSuccess";
}
```

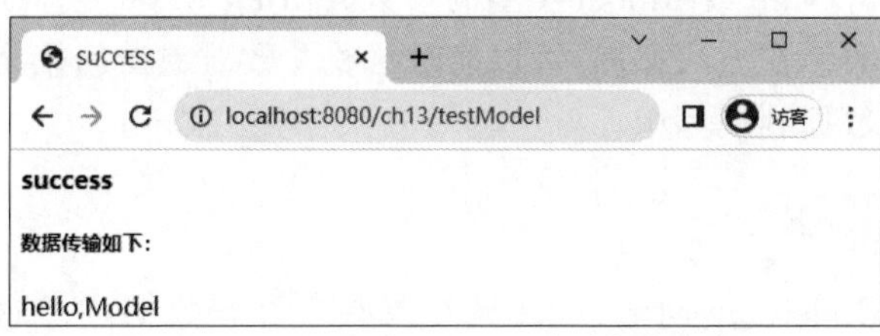

图 13-11 Model 向 request 域对象共享数据

例 13-8:使用 Model 向 request 域对象共享动态数据,后端程序 UserController.java 如下。

```
private List<Student> list = new ArrayList<Student>();
@RequestMapping("/listStudent")
```

```
public String testModelDynamic(Model model) {
    model.addAttribute("list", list);
    return "StudentList";
}
```

13.4.3 Map

例 13-9：使用 Map 向 request 域对象共享静态数据，前端 DynamicSuccess.jsp 页面代码如下，显示结果如图 13-12 所示。

```
<body>
    <h4>success</h4>
    <h5>model 数据传输如下:</h5>
    ${testScope}
</body>
```

后端程序 UserController.java 如下。

```
@RequestMapping("/testMap")
public String testMap(Map<String, Object> map){
    map.put("testScope", "hello,Map");
    return "DynamicSuccess";
}
```

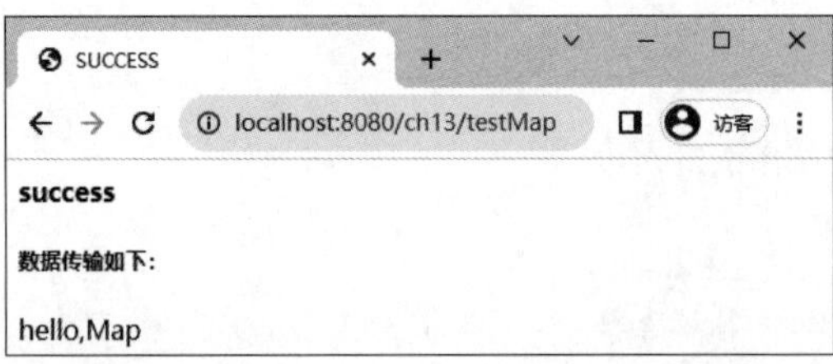

图 13-12　Map 向 request 域对象共享数据

在调用控制器方法之前，Spring MVC 会创建一个隐含的模型对象用于存储模型数据。如果方法的参数中包含 Map 或 Model 类型的参数，Spring MVC 会将这个隐含的模型对象的引用传递给这些参数。在方法体内，可以通过这些参数对象访问模型数据中的所有属性，也可以添加新的属性数据到模型中。

13.4.4 ModelMap

例 13-10：使用 Modelmap 向 request 域对象共享静态数据，后端程序 UserController.java 如下。显示结果如图 13-13 所示。

```
@RequestMapping("/testModelMap")
public String testModelMap(ModelMap modelmap) {
    modelmap.addAttribute("testScope", "hello,ModelMap");
    return "DynamicSuccess";
}
```

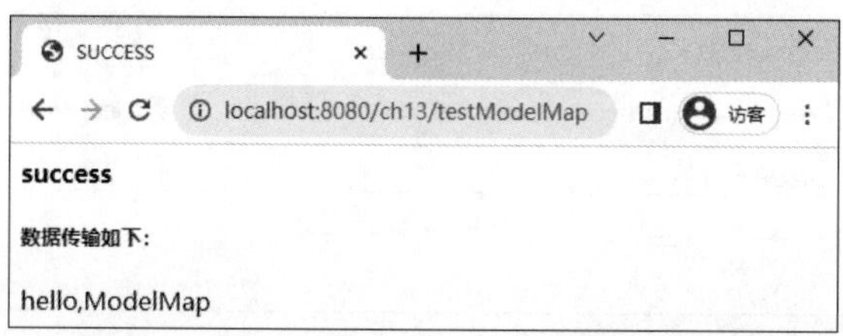

图 13-13　ModelMap 向 request 域对象共享数据

例 13-11：使用 ModelMap 向 request 域对象共享动态数据，后端程序 UserController.java 如下。

```
@RequestMapping("/listStudent")
public String testModelDynamic(ModelMap modelMap) {
    modelMap.addAttribute("list", list);
    return "StudentList";
}
```

Model、ModelMap 和 Map 参数实际都是 BindingAwareModelMap 类型的参数。Spring MVC 支持多种视图类型。当发起请求后，Spring MVC 控制器将数据绑定到数据模型中，然后视图可以使用数据模型中的信息进行展示。默认情况下，Spring MVC 支持转发视图和重定向视图。以下简要介绍这两种视图。

13.5　Spring MVC 的视图

13.5.1　转发视图

例 13-12：使用 forward：转发视图，后端程序 Student.java 如下。显示结果如图 13-14 所示。提交完成后控制台提示已转发视图，如图 13-15 所示。

```
@Controller
@RequestMapping("/student")
public class Student {
    @RequestMapping("/studentIndex")
    public String studengIndex() {
        return "success";
    }
}
```

后端程序 UserController.java 如下。

```
@RequestMapping("/testForward")
public String testForward() {
    System.out.println("演示转发");
    return "forward:student/studentIndex ";
}
```

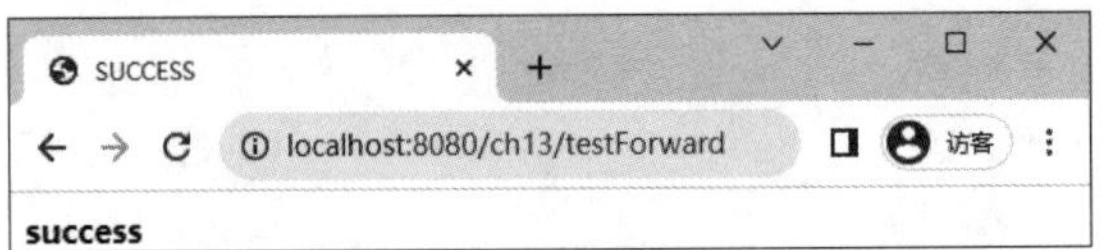

图 13-14　转发视图

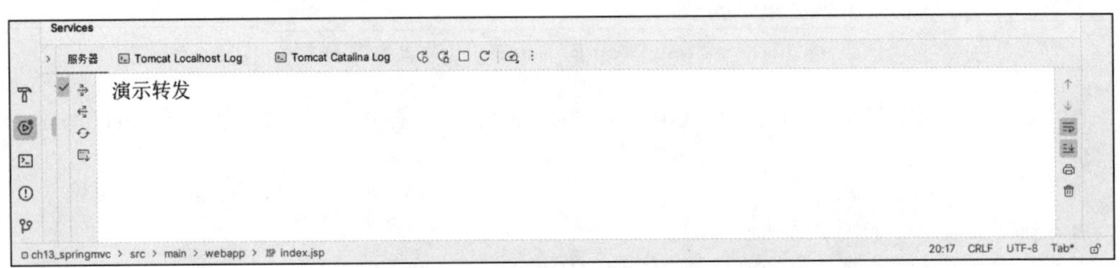

图 13-15　控制台提示已转发视图

在 Spring MVC 中,默认的转发视图是 InternalResourceView。Spring MVC 创建转发视图的情况是当控制器方法中设置的视图名称以“forward:”为前缀时。在这种情况下,视图名称不会经过 Spring MVC 配置文件中配置的视图解析器解析,而是会将前缀“forward:”去掉,然后将剩余部分作为最终路径,通过转发的方式实现跳转。

13.5.2　重定向视图

例 13-13:使用“redirect:”重定向视图后端程序 UserController.java 如下。显示结果如图 13-16 所示。提交完成后控制台提示已重定向视图,如图 13-17 所示。

```
@RequestMapping("/testRedirect")
public String testRedirect() {
    System.out.println("演示重定向");
    return "redirect:studentIndex";
}
@RequestMapping("/studentIndex")
public String studengIndex() {
    return "success";
}
```

图 13-16　重定向视图

图 13-17　控制台提示已经重定向视图

在 Spring MVC 中，默认的重定向视图是 RedirectView。当控制器方法中设置的视图名称以“redirect:”为前缀时，将创建 RedirectView 视图。在这种情况下，视图名称不会受到 Spring MVC 配置文件中配置的视图解析器的影响。相反，它会将前缀“redirect:”去掉，然后使用剩余部分作为最终路径，通过重定向的方式实现跳转。

13.5.3 转发与重定向

例 13-14：测试域对象共享动态数据与 Spring MVC 的视图，后端代码如下，显示结果如图 13-18 和图 13-19 所示。

```
private List<Student> list = new ArrayList<Student>();
@RequestMapping("/listStudent")
public ModelAndView findAllStudent() {
    ModelAndView mv = new ModelAndView();
    mv.addObject("list", list);
    mv.setViewName("StudentList"); return mv;
}
@RequestMapping("/toAddStudent")
public String toAddStudent() {
    return "addStudent";
    //addStudent 与前端页面<form action="addStudent" method="post">form 里
action 保持一致。
}
@RequestMapping("/addStudent")
public String add(Student student) throws Exception {
    list.add(student);//运用 list 自带的 add 方法添加
    return "redirect:/listStudent";//注意:重定向的地址为@RequestMapping 里的
//listStudent,因为需要更新列表里的学生属性,才能获取已经添加的学生信息
}
```

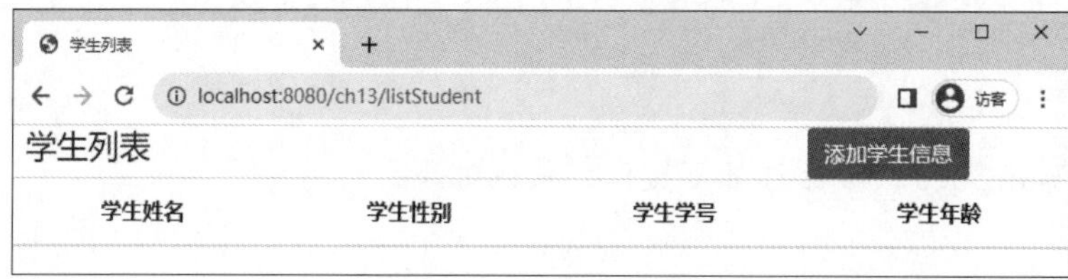

图 13-18 listStudent 界面

图 13-19 接收完学生信息的界面

13.5.4 利用转发与重定向测试后端数据传至前端

例 13-15：利用 ModelAndView 测试后端数据通过重定向与转发至前端，代码如下。显示结果如图 13-20 和图 13-21 所示。

在 index.jsp 页面输入需要传输的信息，触发超链接携带姓名和年龄信息，并请求转发 myInfo.do。

```
<h3>
<a href="${pageContext.request.contextPath}/myInfo.do? name=张三 &age=25">访问服务,进行数据携带跳转-测试后端数据传至前端</a>
</h3>
```

前端接收页面 main.jsp 如下。

```
<body>
<h2>main...显示数据</h2>
    requestUsers:${requestUser}<br>
    sessionUsers:${sessionUser}<br>
    modelUsers:${modelUser}<br>
    mapUsers:${mapUser}<br>
    modelMapUsers:${modelMapUser}<br>
    从 index.jsp 页超链接获取的姓名${param.name}
    从 index.jsp 页超链接获取的年龄${param.age}
</body>
```

编写 User 实体。代码如下。

```
public class User {
    private String name;                          //姓名
    private int age;                              //年龄
    public User() {
    }
    //此处省略构造方法、getter()、setter()等方法
}
```

在 UserController.java 添加如下接口。代码如下。

```
/**
 * 所有形参都是 Spring MVC 默认内置的对象,不需要自己 new,直接使用即可
 */
@RequestMapping("/myInfo.do")
public String data(HttpServletRequest request,
        HttpServletResponse response,      //响应对象无法携带参数
        HttpSession session,
        Model model,
        Map map,
        ModelMap modelMap) {
            User u = new User("张三,24);
```

```
            request.setAttribute("requestUser",u);
            session.setAttribute("sessionUser",u);
            model.addAttribute("modelUser",u);
            map.put("mapUser",u);
            modelMap.addAttribute("modelMapUser",u);
            return "main";                       //请求转发方式跳转页面
    }
    @RequestMapping("/UserIndex")
    public String studengIndex2() {
    return "main";
    }
```

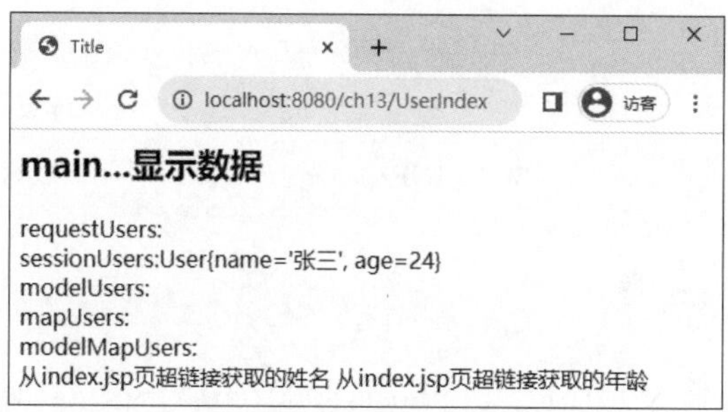

图 13-20　重定向成功页面

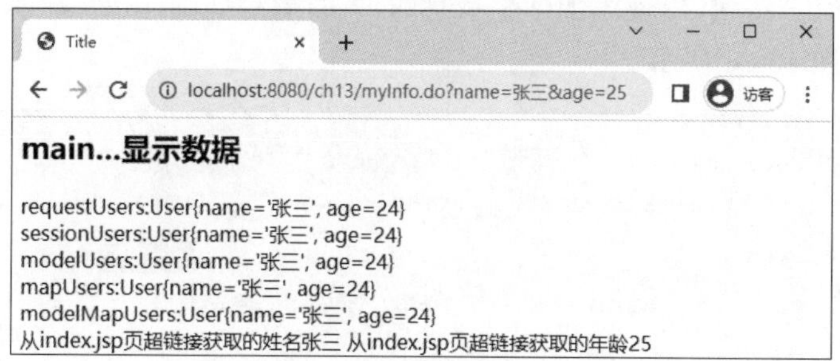

图 13-21　转发成功页面

重定向通常只能通过会话(session)方式传递数据,而其他方式无法携带数据进行跳转。从浏览器的角度来看,重定向是生成数据,然后将数据传递到主页面。与之不同,转发可以直接通过超链接等方式获取数据,因此可以更容易地传递数据给目标页面。

13.6 综合案例

本章案例主要实现添加学生信息的功能。由于本章不涉及数据库的操作,所以只实现由页面提交数据,经由 Controller 处理后向 List 集合添加学生信息的功能。

13.6.1 案例设计

案例包括持久层和 Web 表现层。持久层包含实体类 Student.java。Web 表现层主要包括 Spring MVC 中的 Controller 类和 JSP 页面。Controller 类 StudentController.java 主要负责拦截用户的请求，并将请求的处理结果返回给 JSP 页面进行展示。JSP 页面 studentList.jsp 负责向 List 中添加学生信息以及展示 List 集合中的所有学生信息。本章案例使用的主要文件如表 13-6 所示。

表 13-6　本章案例使用的主要文件

文　　件	所在包/路径	功　　能
web.xml	/WEB-INF/	web 配置文件
Spring MVC.xml	src/	Spring MVC 配置文件
Student.java	com.imut.pojo	封装学生信息的类
StudentController.java	com.imut.controller	控制器类
index.jsp	/	首页
studentList.jsp	/WEB-INF/pages/	显示信息页面

13.6.2　案例演示

在浏览器地址栏中输入 http://localhost:8080/ch13_Spring MVC/index.jsp，单击“学生管理”超链接，显示学生列表页面，由于当前没有学生信息，所以显示效果如图 13-22 所示。单击“添加学生”按钮，出现添加学生信息的页面，效果如图 13-23 所示。添加了 5 位学生的列表页面如图 13-24 所示。

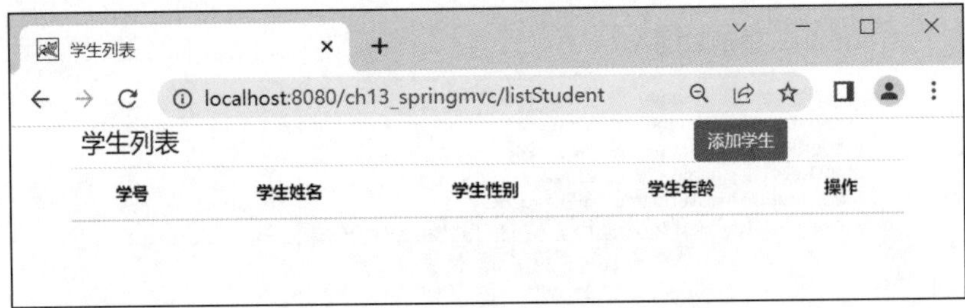

图 13-22　学生列表页面

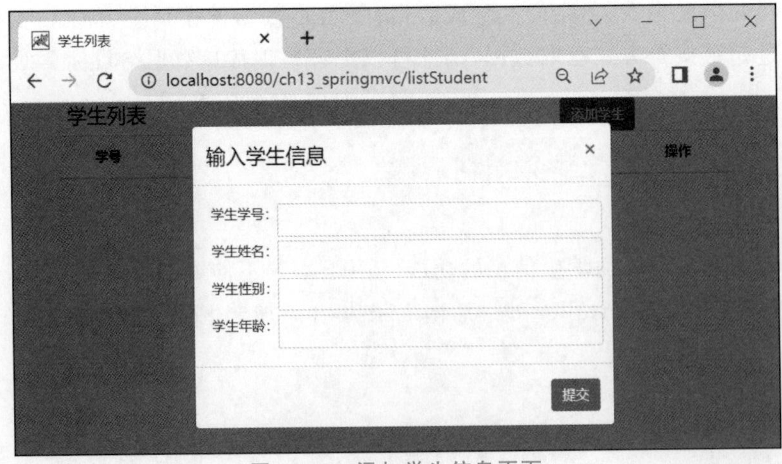

图 13-23　添加学生信息页面

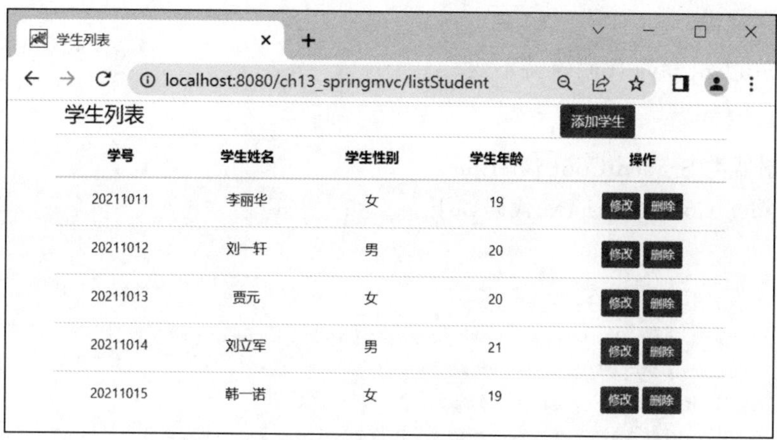

图 13-24　添加学生信息后的学生列表页面

13.6.3　代码实现

1. 创建项目

复制 Web 动态工程 ch12_Spring MVC 为 ch13_Spring MVC。

2. 编写配置文件

修改 web.xml 配置文件，增加对编码过滤器的配置。修改 Spring MVC.xml 配置文件，增加对静态资源访问的配置。

配置文件 web.xml 增加的代码如下。

```
<filter>
    <filter-name>CharacterEncodingFilter</filter-name>
    <filter-class>org.springframework.web.filter.CharacterEncodingFilter
</filter-class>
    <init-param>
        <param-name>encoding</param-name>
        <param-value>UTF-8</param-value>
    </init-param>
    <init-param>
        <param-name>forceEncoding</param-name>
        <param-value>true</param-value>
    </init-param>
</filter>
<filter-mapping>
    <filter-name>CharacterEncodingFilter</filter-name>
    <url-pattern>/*</url-pattern>
</filter-mapping>
```

配置文件 Spring MVC.xml 增加的代码如下。

```
<mvc:resources location="/js/" mapping="/js/**"></mvc:resources>
```

3. 创建实体类

实体类 Student.java 代码保持不变。

4. 编写控制器类

修改控制器类 StudentController.java。

程序 StudentController.java 代码如下。

```
@Controller
public class StudentController {
    private  List<Student> list = new ArrayList<Student>();
    @RequestMapping("/listStudent")
    public ModelAndView findAllStudent() {
        ModelAndView mv = new  ModelAndView();
        mv.addObject("list",list);
        mv.setViewName("studentList");
        return mv;
    }
    @RequestMapping("/toAddStudent")
    public String toAddStudent() {
        return "addStudent";
    }
    @RequestMapping("/addStudent")
    public String add(Student student) throws Exception {
        list.add(student);
        return "redirect:/listStudent";
    }
}
```

5. 编写 JSP 页面文件

修改 studentList.jsp 文件，增加添加学生的界面代码。由于篇幅原因，JSP 文件只放<body></body>之间的代码。

程序 studentList.jsp 的代码如下。

```
<body style="width: 800px;margin:0px auto;">
    <div style="margin-left:10px;width:150px;float: left;">
        <h4>学生列表</h4>
    </div>
    <div style="width:200px;float:right;">
        <button type="button" class="btn btn-primary" data-toggle="modal" data
-target="#myModal">添加学生</button>
    </div>
    <div class="modal fade" id="myModal" tabindex="-1" role="dialog" aria-
labelledby="myModalLabel">
        <div class="modal-dialog">
            <div class="modal-content">
                <div class="modal-header">
                    <h4 class="modal-title">输入学生信息</h4>
```

```
                <button data-dismiss="modal" class="close" type="button">
                    <span aria-hidden="true">×</span><span class="sr-
only">Close</span>
                </button>
            </div>
            <form action="addStudent" method="post">
                <div class="modal-body">
                    <div class="input-group" style="margin: 5px">
                        <span class="input-group-addon">学生学号:</span>
                        <input type="text" class="form-control" name="sno">
                    </div>
                    <div class="input-group" style="margin: 5px">
                        <span class="input-group-addon">学生姓名:</span>
                        <input type="text" class="form-control" name=
"sname">
                    </div>
                    <div class="input-group"style="margin: 5px">
                        <span class="input-group-addon">学生性别:</span>
                        <input type="text" class="form-control" name=
"sgender">
                    </div>
                    <div class="input-group"style="margin: 5px">
                        <span class="input-group-addon">学生年龄:</span>
                        <input type="text" class="form-control" name="sage">
                    </div>
                </div>
                <div class="modal-footer">
                    <button class="btn btn-primary" type="submit">提交
</button>
                </div>
            </form>
        </div>
    </div>
</div>
<table id="mytb" class="table table-hover">
    <thead>
        <tr>
            <th>学号</th>
            <th>学生姓名</th>
            <th>学生性别</th>
            <th>学生年龄</th>
            <th>操作</th>
        </tr>
    </thead>
    <tbody>
        <c:forEach items="${list}" var="student">
            <tr>
                <td>${student.sno }</td>
                <td>${student.sname }</td>
                <td>${student.sgender }</td>
```

```
                <td>${student.sage }</td>
                <td>
                    <a href="#">
                        <button type="button" class="btn btn-primary btn-sm">
修改</button>
                    </a>
                    <a href="#">
                         <button type="button" class="btn btn-primary btn-
sm" onclick="return confirm('确认要删除吗?')">删除</button>
                    </a>
                </td>
            </tr>
        </c:forEach>
    </tbody>
  </table>
</body>
```

13.7 习题

1. 选择题

(1) 在 Spring MVC 中,(　　)类是负责处理 HTTP 请求和响应的。

A. DispatcherServlet　　B. RequestHandler

C. HttpServlet　　D. SpringController

(2) 某个类要返回 JSON 对象,下面(　　)注解可以满足条件。

A. @ResponseBody　　B. @RequestMapping

C. @SessionAttributes　　D. @Controller

(3) 返回一个 JSON 对象,这个类应该标有(　　)注解。

A. @ResponseBody　　B. @SessionAttributes

C. @Controller　　D. @RequestMapping

(4) 在 Spring MVC 中,下面说法不正确的是(　　)。

A. Spring MVC 框架是基于 Model2 实现的技术框架

B. Spring MVC 框架由 ModelAndView 和 controller 组成

C. Spring MVC 框架中 Controller 的返回值一般只是一个 View Name

D. Spring MVC 框架中 DispatcherServlet 的作用是将请求发送到一个 Spring MVC 控制器

(5) 以下不是 Spring MVC 核心组件的是(　　)。

A. DispatcherServlet　　B. JVM

C. Controller　　D. ModelAndView

2. 简答题

(1) 请简述静态资源的几种配置方法。

(2) 在 Spring MVC 中,有几种响应数据到视图的方法?

(3) Spring MVC 是怎样设定重定向和转发的?

(4) Spring MVC 是用什么对象从后端向前端传递数据的? Spring MVC 中有个类,把视图和数据都合并的一起的,叫什么?

3. 程序设计

(1) 创建一个项目,内有学生类,实现学生信息的数据验证功能。

(2) 创建一个项目,运用数据验证注解,对注册功能进行验证。

第14章 Spring MVC拦截器和异常处理

Spring MVC 拦截器是对控制器方法的预处理和后处理，实际上是 AOP 思想的具体实现。通过实现 HandlerInterceptor 接口，Spring MVC 允许对请求和响应进行拦截和处理。这有助于将请求处理与拦截逻辑分开，实现清晰和可维护的代码结构。

另外，Spring MVC 异常处理机制将各种异常从处理过程中解耦，确保了处理过程的单一性，并实现了异常信息的一致处理和维护。这种异常处理方式可以在 Controller 类层级中进行单独操作，也可以在整个应用中进行全局处理，提供了更灵活的异常管理和处理选项。

14.1 拦截器

Spring MVC 拦截器用于拦截请求处理，常见应用包括登录验证、权限验证和日志记录。与 Servlet 在 web.xml 中配置＜url-pattern＞拦截方式相比，两者有以下两个区别。

① 使用范围：Spring MVC 拦截器仅在使用 Spring MVC 框架的工程中可用，而 Servlet 配置方式可以在任何 Java Web 工程中使用。

② 拦截范围：Spring MVC 拦截器仅拦截控制器方法，而 Servlet 可以配置＜url-pattern＞/ *＜/url-pattern＞来拦截请求的静态和动态资源，包括 jsp、html、css、image、js 等各种资源。

14.1.1 自定义拦截器

自定义拦截器的步骤如下。

(1) 创建拦截器类，实现 HandlerInterceptor 接口。

HandlerInterceptor 接口包含 3 个方法：preHandle()、postHandle()和 afterCompletion()。可以根据需要选择是否实现这些方法，具体用途如下。

preHandle()：在请求处理方法执行之前调用。常用于登录验证、权限验证等场景。

postHandle()：在请求处理方法执行之后调用。如果返回的 ModelAndView 对象不为 null,它就在视图渲染之前执行。

afterCompletion()：在视图渲染之后调用,通常用于资源释放或日志记录等情况。

(2) 配置拦截器类。

在 Spring MVC 的配置文件中,需要将自定义拦截器类配置为拦截器,并定义拦截规则。这可以通过 XML 配置或 Java 配置来完成。

(3) 测试拦截器的拦截效果。

可以运行应用程序并发送请求来测试自定义拦截器的拦截效果。根据实际需求和拦截器的逻辑确保拦截器按照预期进行拦截和处理请求。

例 14-1：演示通过上述方式实现拦截器的创建。MyIntercepter 拦截器类代码如下。

(1) 创建拦截器类。

```
public class MyIntercepter implements HandlerInterceptor {
    @Override
    public boolean preHandle(HttpServletRequest request, HttpServletResponse
response, Object handler) throws Exception {
        System.out.println("preHandle...");
        //返回 true 表示不对该请求进行拦截,false 则表示拦截
        return false;
    }
    @Override
    public void postHandle(HttpServletRequest request, HttpServletResponse
response, Object handler, ModelAndView modelAndView) throws Exception {
        System.out.println("postHandle...");
        //此处可编写请求方法执行完成后要执行的操作
    }
    @Override
    public void afterCompletion(HttpServletRequest request,
HttpServletResponse response, Object handler, Exception ex) throws Exception {
        System.out.println("afterCompletion...");
    }
}
```

为了方便查看上述 3 个方法的执行过程,已在上述代码中使用 System.out.println 输出相应的字符串。由于自定义拦截器除实现 HandlerInterceptor 接口外,还需要对该拦截器类进行配置,所以具体的输出效果将在步骤(3)中展示。

(2) 对拦截器类进行配置。

拦截器的配置有多种方式,其中常用的方式如下。

使用<mvc:interceptors>配置：这种方式会拦截所有请求,即所有请求都会经过拦截器。示例配置如下。

```
<mvc:interceptors>
    <bean class="com.imut.intercepter.MyInterceptor"/>
</mvc:interceptors>
```

使用<mvc:interceptor>进行配置：这种方式允许选择哪些请求会经过拦截器，同时可以指定哪些请求不会经过拦截器。这对于特定需求非常有用，如验证用户登录等操作请求时，这些请求不会经过拦截器。示例配置如下。

```
<mvc:interceptors>
    <mvc:interceptor>
        <mvc:mapping path="/**"/>
        <mvc:exclude-mapping path="/login" />
        <bean class="com.imut.intercepter.MyInterceptor"/>
    </mvc:interceptor>
</mvc:interceptors>
```

(3) 测试拦截器的拦截效果。

完成(1)和(2)的配置之后，可以测试自定义拦截器的有效性。运行应用程序，并发送请求来验证拦截器是否按照配置预期地拦截和处理请求。

例 14-2：自定义拦截器的简单使用。intercepterController.java 的代码如下。

```
public class intercepterController {
    @RequestMapping("/loginTest")
    public ModelAndView loginTest(){
        System.out.println("loginTest start...");
        ModelAndView modelAndView = new ModelAndView();
        modelAndView.addObject("userName","IMUT");
        modelAndView.setViewName("success");
        System.out.println("loginTest end...");
        return modelAndView;
    }
}
```

为了更好地体会到拦截器的作用，在 MyIntercepter 拦截器类中添加一些判断操作，来判断前端请求中有无 param＝yes 参数，进而有选择地对前端请求进行拦截。当采用<mvc:interceptor>来进行自定义拦截器配置时，利用前端请求 http://localhost:8080/loginTest 来访问 loginTest()方法，控制台仅输出 preHandle...。这是因为访问的请求为/loginTest，而不是/login，不是/login 的请求都会经过拦截器，又因为步骤(1)中的自定义拦截器会对所有经过的请求进行拦截(return false)，所以控制台有且仅有唯一输出。

例 14-3：自定义拦截器中的判断操作。自定义拦截器 MyIntercepter 的代码如下。

```
public class MyIntercepter implements HandlerInterceptor {
    @Override
    public boolean preHandle(HttpServletRequest request, HttpServletResponse
response, Object handler) throws Exception {
        System.out.println("preHandle...");
        String param = request.getParameter("param");
        if ("yes".equals(param)){
            return true;
        }else{
```

```
            request.getRequestDispatcher("/WEB-INF/pages/error.jsp").forward
(request,response);
            return false;
        }
    }
    @Override
    public void postHandle(HttpServletRequest request, HttpServletResponse
response, Object handler, ModelAndView modelAndView) throws Exception {
        System.out.println("postHandle...");
        //此处可编写请求方法执行完成后要执行的操作
    }
    @Override
    public void afterCompletion(HttpServletRequest request,
HttpServletResponse response, Object handler, Exception ex) throws Exception {
        System.out.println("afterCompletion...");
        //此处可编写视图渲染完成后要执行的操作
    }
}
```

当利用前端请求 http://localhost:8080/loginTest?param=yes 来访问上述 Controller 中的代码时,因为该请求符合拦截器中的判断条件,所以控制台的输出结果如下所示,可结合步骤(1)中各方法执行的顺序进行更好的理解。

preHandle...
loginTest start...
loginTest end...
postHandle...
afterCompletion...

若当前端请求中没有 param 参数或者 param 参数不为 yes 时,会在控制台中仅输出 preHandle...,并跳转到/WEB-INF/pages/error.jsp 页面(进行一些自定义的信息展示),除此之外不会再进行其他的操作处理。拦截器的执行顺序如图 14-1 所示。因此,可以将该原理应用到实际开发中,如判断用户是否登录:在用户登录时可以将用户的数据写入到 session 中,当进行其他前端请求时,可以利用拦截器来判断 session 中有无该用户的数据,进而判断出该用户有没有登录。

14.1.2 拦截器作用范围

在自定义拦截器中,介绍了如何通过 Spring MVC 配置文件进行拦截器的配置。其中,第二种配置方法是使用 mvc:interceptor 元素,可以有选择地拦截前端请求。在这种方法中,可以配置特定的请求,使其不被拦截。举例来说,可以使用<mvc:exclude-mapping>元素来定义哪些请求不需要被拦截,比如:

```
<mvc:exclude-mapping path="/login" />
```

上述配置表示不对/login 路径的请求进行拦截,允许这些请求绕过拦截器执行。这种配置方式更灵活地控制哪些请求需要经过拦截器处理,哪些请求不需要拦截。

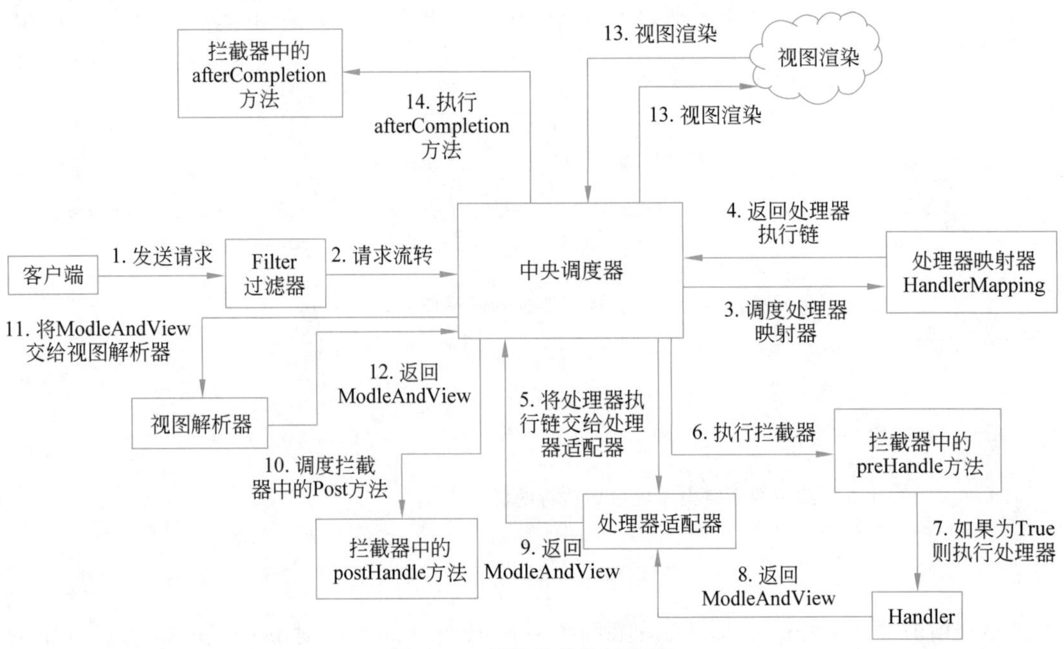

图 14-1　拦截器的执行顺序

例 14-4：拦截器的作用范围示例。intercepterController.java 的代码如下。

```
public class intercepterController {
    @RequestMapping("/login")
    public ModelAndView login(){
        System.out.println("login start...");
        ModelAndView modelAndView = new ModelAndView();
        modelAndView.addObject("userName","IMUT");
        modelAndView.setViewName("success");
        System.out.println("login end...");
        return modelAndView;
    }
}
```

控制台输出结果如下：

```
login start...
login end...
```

在这个例子中，当使用前端请求 http://localhost:8080/login 访问一个名为 login()的方法时，自定义拦截器中添加的 preHandle...、postHandle...、afterCompletion...语句没有被输出。这表明该请求没有被拦截器拦截，也说明了拦截器的作用范围完全由拦截器的配置决定。

当没有使用<mvc:exclude-mapping path=" " />进行配置时，拦截器会拦截所有的请求。但是，当使用<mvc:exclude-mapping path="/login" />进行配置时，拦截器会解除对于/login 请求的拦截。

这种配置方式可以根据具体需求选择性地拦截请求，提高应用程序的灵活性和性能。通过合理配置拦截器的作用范围，可以确保只有需要特殊处理的请求才会被拦截器拦截，而其他请求可以自由通过，从而提高应用程序的响应速度和效率。

14.1.3 拦截器执行顺序

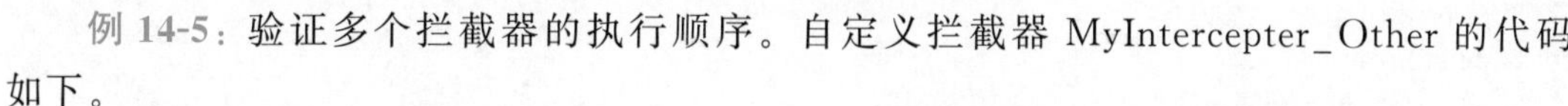

对于多个拦截器，拦截器的执行顺序和其配置顺序相关。

例 14-5：验证多个拦截器的执行顺序。自定义拦截器 MyIntercepter_Other 的代码如下。

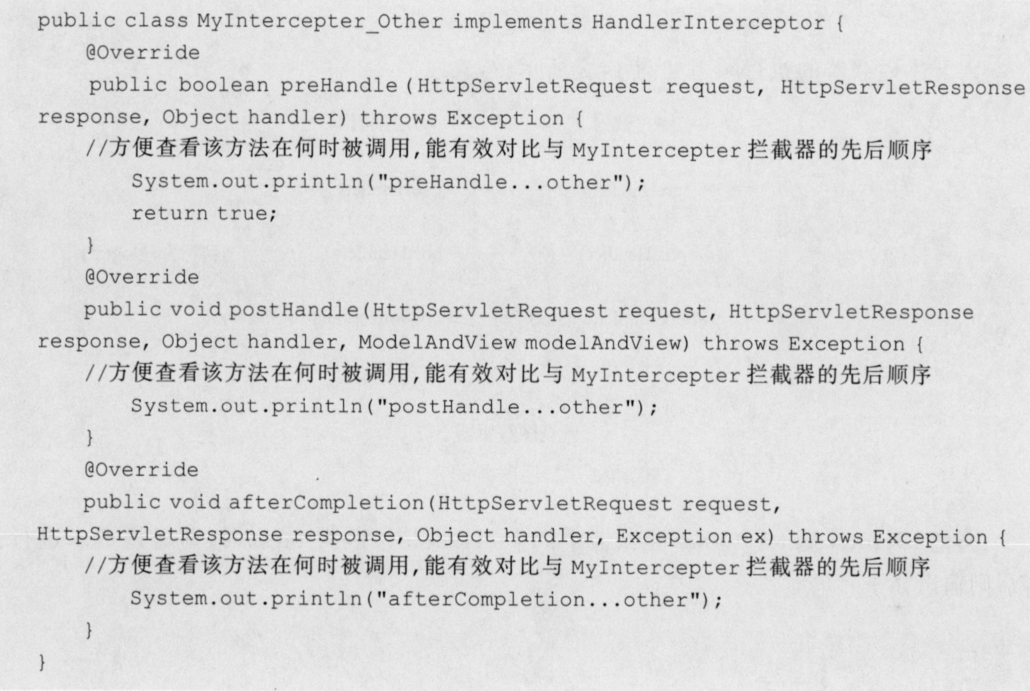

```
public class MyIntercepter_Other implements HandlerInterceptor {
    @Override
    public boolean preHandle(HttpServletRequest request, HttpServletResponse
response, Object handler) throws Exception {
    //方便查看该方法在何时被调用，能有效对比与 MyIntercepter 拦截器的先后顺序
        System.out.println("preHandle...other");
        return true;
    }
    @Override
    public void postHandle(HttpServletRequest request, HttpServletResponse
response, Object handler, ModelAndView modelAndView) throws Exception {
    //方便查看该方法在何时被调用，能有效对比与 MyIntercepter 拦截器的先后顺序
        System.out.println("postHandle...other");
    }
    @Override
    public void afterCompletion(HttpServletRequest request,
HttpServletResponse response, Object handler, Exception ex) throws Exception {
    //方便查看该方法在何时被调用，能有效对比与 MyIntercepter 拦截器的先后顺序
        System.out.println("afterCompletion...other");
    }
}
```

在 Spring MVC 配置文件中有如下配置。

```
<mvc:interceptors>
    <mvc:interceptor>
      <mvc:mapping path="/**"/>
       <mvc:exclude-mapping path="/login"/>
       <bean class="com.imut.intercepter.MyIntercepter"/>
    </mvc:interceptor>
    <mvc:interceptor>
      <!-- 对哪些资源进行拦截  -->
      <mvc:mapping path="/**"/>
      <!--对哪些资源不拦截-->
      <mvc:exclude-mapping path="/login"/>
      <!--与例 14-2、14-3 有关，防止该拦截器的输出 -->
      <mvc:exclude-mapping path="/loginTest"/>
      <bean class="com.imut.intercepter.MyIntercepter_Other"/>
      </mvc:interceptor>
</mvc:interceptors>
```

当利用前端请求 http://localhost:8080/loginTestOther?param=yes 访问控制器中的 loginTestOther()时,控制台的输出如下:

```
preHandle...
preHandle...other
loginTest start...
loginTest end...
postHandle...other
postHandle...
afterCompletion...other
afterCompletion...
```

则上述拦截器的执行顺序如图 14-2 所示。

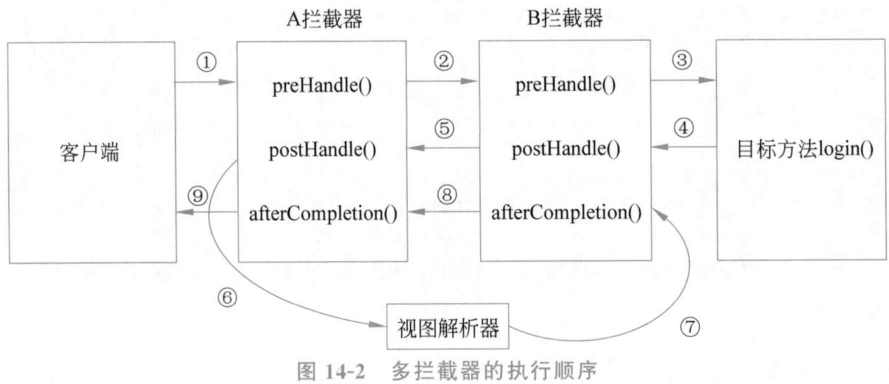

图 14-2 多拦截器的执行顺序

当把 MyIntercepter_Other 拦截器中的 preHandle()改为 false 时,即对该请求进行拦截,则输出如下:

```
preHandle...
preHandle...other
afterCompletion...
```

(1) 当每个拦截器的 preHandle()方法都返回 true,则 preHandle()会按照配置文件中的顺序执行,而 postHandle()和 afterComplation()会按照配置文件的反序执行。

(2) 当某个拦截器的 preHandle()返回了 false,则返回 false 的拦截器与它之前拦截器的 preHandle()方法都会执行,postHandle()都不会执行,返回 false 的拦截器之前拦截器的 afterComplation()会执行。

14.2 异常处理

在 Web 项目中,未处理的异常通常以状态码的形式呈现在前端页面上,如 403、404、500 等状态码,如图 14-3 所示。然而,这些状态码对普通用户来说并不直观和友好。为了提供更好的用户体验,Web 项目通常会通过 try-catch 方式捕捉这些异常,并将用户重定向到自定义的错误界面，以展示更友好的错误信息。针对这种情况，Spring MVC 提供了更便捷的方式来处理。

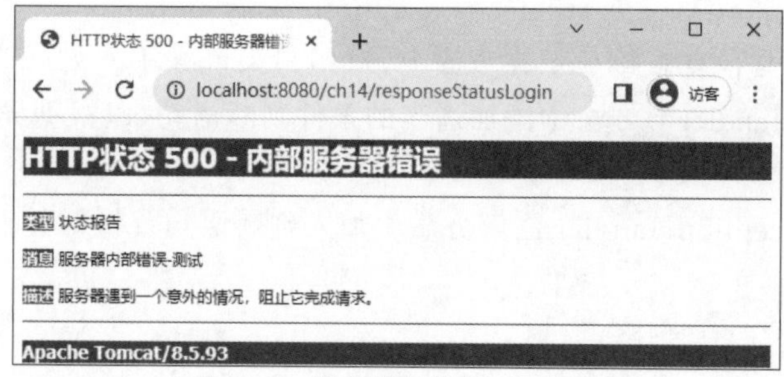

图 14-3　服务器出错的页面显示

14.2.1　ExceptionHandler 注解方式

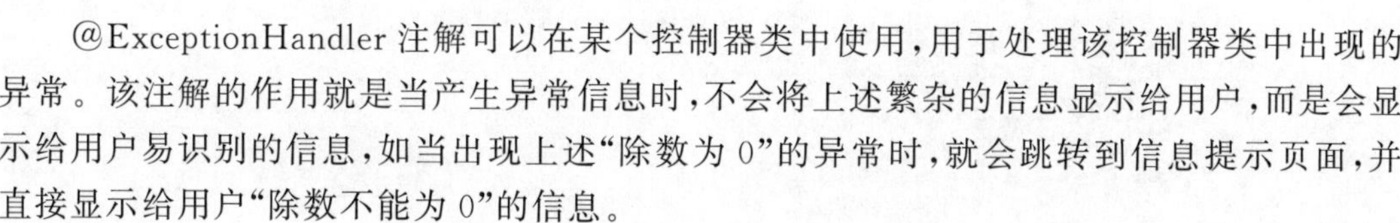

@ExceptionHandler 注解可以在某个控制器类中使用，用于处理该控制器类中出现的异常。该注解的作用就是当产生异常信息时，不会将上述繁杂的信息显示给用户，而是会显示给用户易识别的信息，如当出现上述“除数为 0”的异常时，就会跳转到信息提示页面，并直接显示给用户“除数不能为 0”的信息。

例 14-6：ExceptionHandler 注解方式的简单异常处理。exceptionController 中的 exceptionLogin()方法如下。

```
@RequestMapping("/exceptionLogin")
public ModelAndView exceptionLogin (){
    int i=1/0; //人为制造的异常
    System.out.println("exceptionLogin start...");
    ModelAndView modelAndView = new ModelAndView();
    modelAndView.addObject("userName","IMUT");
    modelAndView.setViewName("success");
    System.out.println("exceptionLogin end...");
    return modelAndView;
}
```

在该 Controller 类中添加异常处理方法，代码如下。

```
@ExceptionHandler()
public ModelAndView MyExceptionHandler(Exception e){
    ModelAndView modelAndView =new ModelAndView();
    //将异常信息添加到 ModelAndView 中
    modelAndView.addObject("errorString",e.toString());
    //出现异常跳转到/WEB-INF/pages/exception.jsp 的视图
    modelAndView.setViewName("exception");
    return modelAndView;
}
```

为了演示这个注解类的作用，在 Controller 类中的 exceptionLogin()方法中故意引发一个除零异常。当使用前端请求 http://localhost:8080/exceptionLogin 时，异常会被同一

Controller 类中的异常处理方法 MyExceptionHandler()捕获并处理。随后，页面会被重定向到/WEB-INF/pages/exception.jsp，并通过 ${errorString}输出异常信息，结果是页面上显示了异常信息，如 java.lang.ArithmeticException：/ by zero。

当然，将这个异常信息直接显示在页面上，对用户来说并不友好。为改善用户体验，可以在 MyExceptionHandler()方法中添加适当的条件判断语句，以根据异常类型向用户提供更友好的反馈信息。

例 14-7：在 ExceptionHandler 注解方法中加入判断语句。详细异常处理方法如下。

```
@ExceptionHandler()
public ModelAndView MyExceptionHandler(Exception e){
    ModelAndView modelAndView =new ModelAndView();
    if (e instanceof ArithmeticException)
        modelAndView.addObject("errorString","除数不能为 0");
    else
        modelAndView.addObject("errorString","普通异常");
        modelAndView.setViewName("exception");
        return modelAndView;
}
```

@ExceptionHandler 注解默认可以处理所有异常，但也可以指定要处理的异常类型，如@ExceptionHandler(value = {java.lang.NullPointerException.class})。

需要注意的是，这个注解只能处理当前 Controller 类中产生的异常，不能处理所有 Controller 类中的异常。如果希望对所有 Controller 类中的异常进行统一处理，需要在每个 Controller 类中定义异常处理方法。然而，更好的解决方法是使用全局异常处理方法。全局异常处理可以集中处理所有 Controller 类中的异常，从而避免在每个 Controller 类中都定义异常处理方法。这可以提高代码的可维护性，减少冗余的异常处理代码，同时建议采用全局异常处理的方式，以提高代码的效率。

14.2.2 ResponseStatusExceptionResolver

ResponseStatusExceptionResolver 解析器在 DispatcherServlet 默认启用，其主要职责是根据自定义异常上配置的注解@ResponseStatus，将这些自定义异常映射到设定的 HTTP 状态码中，让自定义的异常能通过状态码在前端页面进行信息的展示，这种展示相比 java.lang.ArithmeticException：/ by zero 也更加直观，如图 14-4 所示。

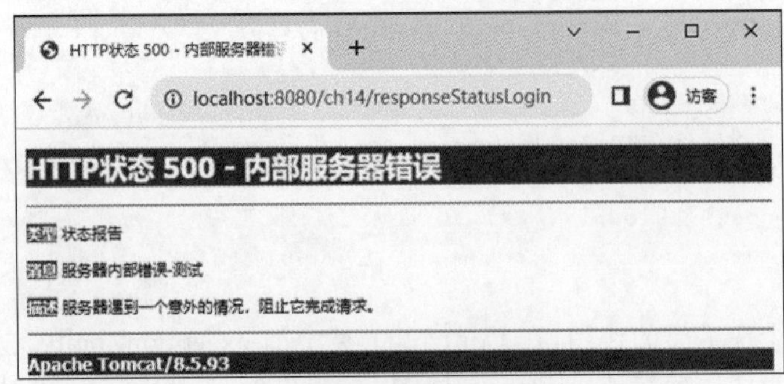

图 14-4 使用注解进行自定义异常显示

@ResponseStatus 注解有两个参数：value 和 reason。value 对应 HTTP 状态码，常见的有 HttpStatus.NOT_FOUND(404)、HttpStatus.INTERNAL_SERVER_ERROR(500)、HttpStatus.FORBIDDEN(403)，它指定了自定义异常的状态码信息。reason 用于提供异常的摘要和提示。

这个注解可用在自定义异常类和控制器方法上。但在控制器方法上使用时，无论方法内是否发生异常，最终页面输出都是@ResponseStatus 注解中的内容。

例 14-8：将@ResponseStatus 注解运用在 responseStatusLogin()方法中。exceptionController 中的 responseStatusLogin ()方法如下。

```
@RequestMapping("/responseStatusLogin ")
//value 的值为产生异常时的状态码类型,reason 的值需自定义,用于异常信息的展示
@ResponseStatus(value = HttpStatus.INTERNAL_SERVER_ERROR, reason = "服务器内部
错误-测试")
public ModelAndView responseStatusLogin (){
    System.out.println("responseStatusLogin start...");
    ModelAndView modelAndView = new ModelAndView();
    modelAndView.addObject("userName","IMUT");
    modelAndView.setViewName("success");
    System.out.println("responseStatusLogin end...");
    return modelAndView;
}
```

前端界面如图 14-5 所示。需要注意的是，虽然 responseStatusLogin()方法没有异常，但当访问 http://localhost:8080/responseStatusLogin 时，仍然会显示异常错误。尽管 responseStatusLogin()方法会继续执行(控制台中会输出"responseStatusLogin start..."和"responseStatusLogin end...")，但前端不会显示 success.jsp 页面。因此，最好将@ResponseStatus 注解与自定义异常一起使用。

在@ResponseStatus 注解中使用的 reason ＝"服务器内部错误-测试"会在页面的"消息"一栏中显示，而 value ＝ HttpStatus.INTERNAL_SERVER_ERROR 则显示为"HTTP 状态 500 -内部服务器错误"。

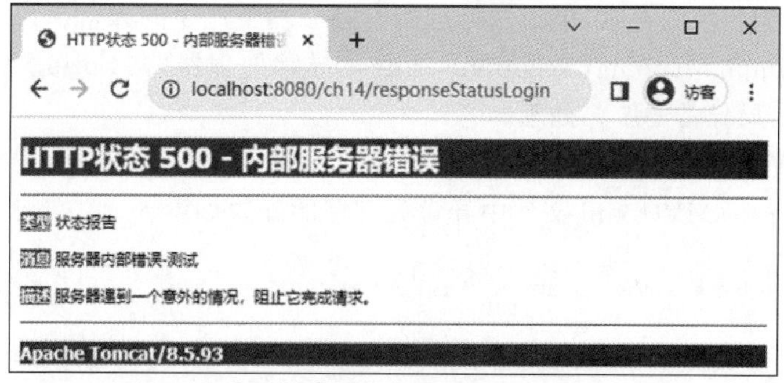

图 14-5 在目标方法上使用 ResponseStatus 注解

例 14-9：在自定义异常类上使用@ResponseStatus 注解。

(1) 首先需要根据现实需求定义一个异常类，并在异常类上使用@ResponseStatus 注解，当目标方法出现某异常时，就抛出我们定义的异常。异常类 TestException 的代码

如下。

```
@ResponseStatus(value = HttpStatus.INTERNAL_SERVER_ERROR,reason = "服务器异常")
public class TestException extends RuntimeException{ }
```

(2) 通过目标方法 testExceptionLogin ()抛出自定义异常。testExceptionLogin ()方法代码如下。

```
@RequestMapping("/testExceptionLogin")
public ModelAndView testExceptionLogin (){
    throw new TestException();
}
```

当使用 http://localhost:8080/testExceptionLogin 进行访问时,前端异常显示如图 14-6 所示。

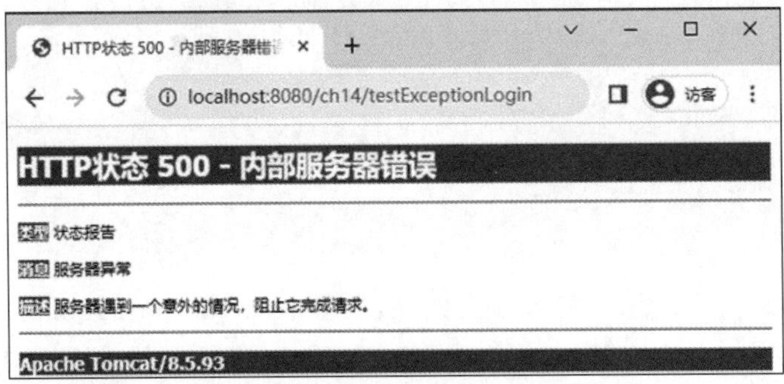

图 14-6 在自定义异常类中使用注解

当在目标方法和异常类上同时使用@ResponseStatus 注解时,若目标方法能够抛出自定义异常,则会展示自定义异常中的信息。

14.2.3 SimpleMappingExceptionResolver

以上两节讲解了两种基于注解的异常信息展示方法。@ExceptionHandler 注解用于处理当前控制器类产生的异常,但无法处理所有控制器类产生的异常。为了解决这个问题,可以配置一个 SimpleMappingExceptionResolver 异常处理器,在 Spring MVC 配置文件中设置它,以实现全局异常的统一处理。

例 14-10:使用 SimpleMappingExceptionResolver 类型的异常处理器来进行全局异常的统一处理。Spring MVC.xml 文件中异常处理器的配置如下。

```
<bean class="org.springframework.web.servlet.handler.
SimpleMappingExceptionResolver">
    <property name="defaultErrorView" value="defaultException"/>
    <property name="exceptionMappings">
        <map>
            <entry key="java.lang.ClassCastException" value="classCastException"/>
        </map>
```

```
    </property>
</bean>
```

在上述代码中，java.lang.ClassCastException 为类型转换异常，当出现类型转换异常时，该异常将由 SimpleMappingExceptionResolver 异常解析器捕获，并把这个异常映射到 classCastException.jsp 视图中进行自定义处理。defaultErrorView 为默认异常处理的视图，当存在其他异常时，该异常将映射到 defaultException.jsp 中。

当使用 http://localhost:8080/simpleMappingExceptionLogin 访问 simpleMappingExceptionLogin() 方法时，将产生类型转换异常。simpleMappingExceptionLogin()方法如下。

```
@RequestMapping("/simpleMappingExceptionLogin ")
@ResponseBody
public void simpleMappingExceptionLogin (){
    Object str = "IMUT";
    Integer num = (Integer) str;      //类型转换异常
}
```

类转换异常界面 classCastException.jsp 的代码如下。

```
<%@ page contentType="text/html;charset=UTF-8" language="java" %>
<html>
<head>
    <title>Title</title>
</head>
<body>
    <h1>类型转换异常错误!!</h1>
</body>
</html>
```

异常页面显示如图 14-7 所示。

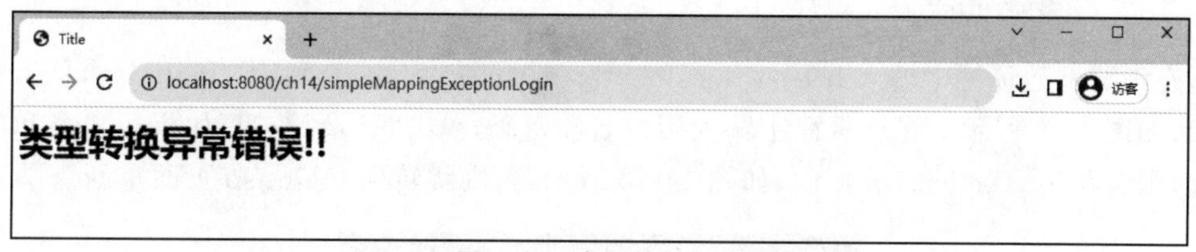

图 14-7　类型转换异常

与上述原理相似，若产生“除数为 0”异常时，由于没有在配置文件中配置该异常处理的视图，所以会在默认异常解析器中进行处理。结果如图 14-8 所示。

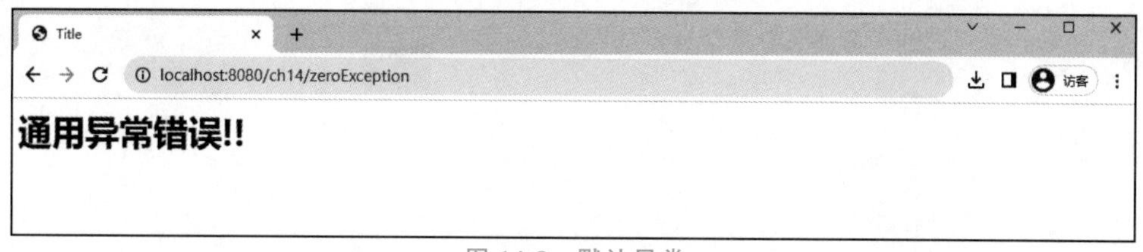

图 14-8　默认异常

14.3 综合案例

本章案例主要实现登录的功能。由于本章不涉及数据库操作,所以登录的账号如果与指定的值相同,则认为登录成功。

14.3.1 案例设计

登录案例包括持久层和 Web 表现层。持久层包含实体类 User.java。Web 表现层主要包括 Spring MVC 中的 Controller 类和 JSP 页面。Controller 类 UserController.java 负责拦截用户的登录请求,并将请求的处理结果返回给 JSP 页面进行展示。JSP 页面 login.jsp 是登录界面,index.jsp 是登录成功后的菜单页面。本章案例使用的主要文件如表 14-1 所示。

表 14-1 本章案例使用的主要文件

文　　件	所在包/路径	功　　能
web.xml	/WEB-INF/	Web 配置文件
Spring MVC.xml	src/	Spring MVC 配置文件
User.java	com.imut.pojo	封装用户信息的类
UserController.java	com.imut.controller	用户控制器类
LoginInterceptor.java	com.imut.interceptor	拦截器
login.jsp	/	登录页
index.jsp	/	菜单页面

14.3.2 案例演示

在浏览器地址栏中输入 http://localhost:8080/ch14_Spring MVC/login.jsp,显示登录页面,如图 14-9 所示。在该页面中输入用户名和密码,单击"登录"按钮,如果用户名和密码正确,则会跳转到 index.jsp 页面,如图 14-10 所示,否则跳转到 login.jsp 页面重新登录。

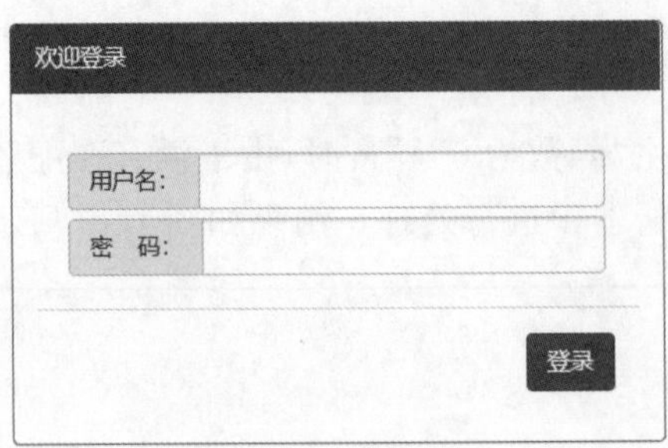

图 14-9 登录页面

图 14-10　登录成功后的页面

14.3.3　代码实现

1. 创建项目

复制 Web 动态工程 ch13_Spring MVC 为 ch14_Spring MVC。

2. 编写配置文件

web.xml 配置文件内容不改变。修改 Spring MVC.xml 配置文件，增加对拦截器的配置。

配置文件 Spring MVC.xml 增加的代码如下。

```
<!-- 配置权限拦截器 -->
<mvc:interceptors>
    <mvc:interceptor>
        <mvc:mapping path="/**"/>
        <mvc:exclude-mapping path="/login"/>
        <bean class="com.imut.interceptor.LoginInterceptor"/>
    </mvc:interceptor>
</mvc:interceptors>
```

3. 创建实体类

在 com.imut.pojo 包下新建实体类 User.java。程序 User.java 代码如下。

```
public class User {
    private int id;
    private String username;
    private String password;
    //此处省略 getXxx()和 setXxx()方法
}
```

4. 编写控制器类

修改控制器类 UserController.java。程序 UserController.java 的代码如下。

```
@Controller
public class UserController {
    @RequestMapping("/login")
```

```
    public String login(String username, String password,HttpSession session) {
        if("admin".equals(username)&&"admin123".equals(password)) {
            User user=new User();
            user.setUsername(username);
            user.setPassword(password);
            session.setAttribute("user", user);
            return "redirect:index.jsp";
        }else {
            return "redirect:login.jsp";
        }
    }
}
```

5. 编写拦截器类

在 com.imut 下新建包 interceptor，在该包下新建类 LoginInterceptor.java。

程序 LoginInterceptor.java 的代码如下。

```
public class LoginInterceptor implements HandlerInterceptor {
    public boolean preHandle(HttpServletRequest request, HttpServletResponse
response,Object handler) throws IOException {
        HttpSession session=request.getSession();
        User user=(User)session.getAttribute("user");
        if(user==null) {
            response.sendRedirect(request.getContextPath()+"/login.jsp");
            return false;
        }
        return true;
    }
}
```

6. 编写 JSP 页面文件

在根目录 webapp 下编写 login.jsp 文件，修改 index.jsp 文件。由于篇幅原因，JSP 文件只放<body></body>之间的代码。

程序 login.jsp 的代码如下。

```
<body style="width: 400px;margin:0px auto;">
    <div class="panel panel-primary">
        <div class="panel-heading">欢迎登录</div>
        <div class="panel-body">
            <form action="${pageContext.request.contextPath }/login" method=
"post">
                <div class="modal-body">
                    <div class="input-group" style="margin: 5px;" >
                        <span class="input-group-addon">用户名:</span>
                        <input type="text" class="form-control" name="username">
                    </div>
                    <div class="input-group" style="margin: 5px">
```

```
            <span class="input-group-addon">密        码:
</span>
                    <input type="password" class="form-control" name="password">
                    </div>
                </div>
                <div class="modal-footer">
                    <button class="btn btn-primary" type="submit">登录</button>
                </div>
            </form>
        </div>
    </div>
</body>
```

修改后的 index.jsp 代码如下。

```
<body  style="width: 300px;margin:0px auto;">
    <div>
        <ul class="nav nav-pills nav-stacked">
            <li class="active"><a href="listStudent">学生管理</a></li>
            <li class="active"><a href="listCourse">课程管理</a></li>
            <li class="active"><a href="listCcourse">选课管理</a></li>
        </ul>
    </div>
</body>
```

14.4 习题

1. 选择题

(1) 关于 Spring MVC 拦截器，下列说法错误的是(　　)。

A. 开发 Spring MVC 拦截器，需实现 WebMvcConfigurer 接口

B. preHandle 方法在 Controller 之前执行，若返回 false，则终止执行后续的请求

C. postHandle 方法在 Controller 之后、模板之前执行

D. afterCompletion 方法在模板之后执行

(2) Spring MVC 拦截器的 3 个方法的执行顺序是(　　)。

A. preHandle()、postHandle()、afterCompletion()

B. preHandle()、postHandle()、afterHandle()

C. preCompletion()、postHandle()、afterHandle()

D. preCompletion()、postHandle()、afterCompletion()

(3) 当请求不需要经过拦截器时，可在配置文件中进行(　　)的配置。

A. <mvc:include-mapping path=" " />

B. <mvc:exclude path=" " />

C. <mvc:exclude-mapping path=" " />

D. <mvc:include path=" " />

(4) 下列关于@ExceptionHandler 注解，描述正确的是(　　)。

A. 该注解的方法会对所有异常类型进行处理，不可以单独处理某个异常类型

B. @ExceptionHandler 为全局异常处理方法

C. 要求该注解的方法和出现问题的控制器必须在同一个类中才能生效

D. 以上描述都不对

(5) 下列关于 SimpleMappingExceptionResolver 异常处理器，解释错误的是(　　)。

A. 可实现全局异常的统一处理

B. <property name="defaultErrorView" value="defaultException"/>中 value 值可进行自定义

C. 在<property name="exceptionMappings"></property>中进行配置可实现某异常的单独处理

D. 以上说法都不对

2. 填空题

(1) Spring MVC 拦截器中的 HandlerInterceptor 接口中定义了 3 个方法，即________、________、________。

(2) 自定义 Spring MVC 拦截器的步骤可分为________、________以及测试拦截器的拦截效果。

(3) 多个拦截器的执行顺序：当每个拦截器的 preHandle()方法都返回 true，则________会按照配置文件中的顺序执行，而________和________会按照配置文件的反序执行。

3. 程序设计题

(1) 编写程序，发起前端请求时先判断用户是否登录。

(2) 编写程序，在该程序中先创建几个异常，之后利用全局异常处理器进行异常的处理。

第15章 Spring MVC 其他功能

本章重点介绍如何使用 Spring MVC 框架实现常见操作，包括文件上传、下载和国际化功能。将通过实际案例来深入理解这些基本操作。

15.1 Spring MVC 实现文件上传

数据上传是指客户端向服务器传递数据，其中客户端的请求都可以视为数据上传。而文件上传则是数据上传的一种特殊方式，即将客户端上传的文件存储到服务器上。

Spring MVC 提供了直接支持文件上传的功能，这是通过 MultipartResolver 接口实现的。该接口的作用是将上传请求包装成可以轻松处理文件的请求，以便进行操作。Spring MVC 提供了两个主要的实现类，分别是 StandardServletMultipartResolver 和 CommonsMultipartResolver。前者使用了 Java 官方的 Servlet 3.0 标准文件上传方式，但仅支持 Servlet 3.0 及更高版本。后者则使用了 Apache 发布的开源组件 Commons FileUpload，这是 Java 中较早并广泛使用的文件上传方式。无论使用哪种方式，都采用了一致的 API 来处理文件上传。本节将详细介绍使用 Apache Commons FileUpload 组件进行文件上传的方法。

15.1.1 环境配置

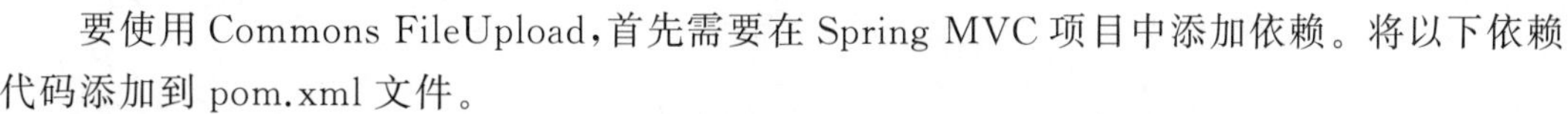

要使用 Commons FileUpload，首先需要在 Spring MVC 项目中添加依赖。将以下依赖代码添加到 pom.xml 文件。

```
<dependency>
    <groupId>commons-fileupload</groupId>
    <artifactId>commons-fileupload</artifactId>
    <version>1.5</version>
</dependency>
```

在默认情况下,Spring MVC 不会配置任何 MultipartResolver,因此无法处理文件上传。因此,需要在 Spring MVC 配置文件中配置 MultipartResolver 的实现类为 CommonsMultipartResolver。请注意,容器 bean 的 id 必须设置为 multipartResolver,否则文件上传会导致 400 错误。DispatcherServlet 初始化时会使用 initMultipartResolver()方法来初始化多部分解析器,它会通过 id 查找名为 multipartResolver 的 bean 容器。如果存在该 bean,请求处理会自动将 HttpServletRequest 转换为相应解析器的 MultipartHttpServletRequest 实现类型。

具体的配置如下。

```
<bean id="multipartResolver" class="org.springframework.web.multipart.
commons.CommonsMultipartResolver">
    <property name="defaultEncoding" value="UTF-8" />
</bean>
```

之前学到,在 JavaWeb 中,文件上传操作可以使用 DiskFileItemFactory 对象创建临时文件,并设置临时文件的大小限制,通过文件上传核心组件 ServletFileUpload 来限制上传文件的大小,并解决文件名乱码等问题。在 Spring MVC 中,可以通过配置 CommonsMultipartResolver 的属性来完成这些任务。表 15-1 列出了一些常用的属性。

表 15-1　CommonsMultipartResolver 的常用基本属性

属　性	说　明
defaultEncoding	默认字符编码
maxUploadSize	最大上传文件限度,以字节为单位
resolveLazily	推迟文件解析
maxInMemorySize	设置文件放入临时文件夹的最大限制

15.1.2　单文件上传

完成上述操作后,Spring MVC 框架就具备了文件上传能力。

例 15-1:演示 Spring MVC 单文件上传功能,前端程序 singleFileUpload.jsp 如下。

```
<h3>单文件上传</h3>
<form method="post" action="fileupload1" enctype="multipart/form-data">
    <ul>
        <li>图片:</li>
        <li><input type="file" name="photo" multiple></li>
    </ul>
    <ul><li><input type="submit" value="上传" /></li></ul>
</form>
```

示例 15-1 展示了一个用于文件上传的前端页面的实现。首先,页面需要在 form 表单中添加 enctype="multipart/form-data"属性。enctype 属性用于指定在将数据发送到服务器时浏览器应使用的编码类型。默认情况下,编码类型为 application/x-www-form-urlencoded,这种编码方式将表单域中的值处理为 URL 编码形式,即键值对形式,这是标准的编码格式。

然而,文件上传需要客户端表单提交为 multipart 请求,即包含多部分数据的请求。这种编码方式要求使用 multipart/form-data 多部分表单数据类型,它以二进制形式传输音频、视频、图片等多部分表单数据。

在 form 表单中指定了 method="post",POST 请求具有高安全性,并且可以传输大量数据,非常适合文件上传。<input>标签的 type 属性设置为 file,用户通过这个标签来选择要上传的文件。前端页面的单文件上传页面如图 15-1 所示。

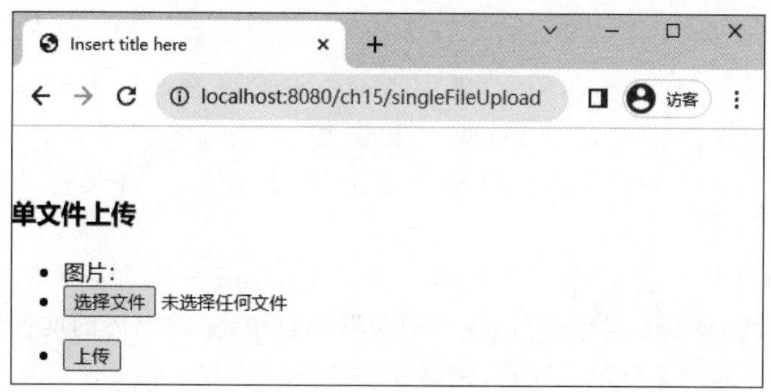

图 15-1　单文件上传页面

一旦配置好文件上传解析器,Spring MVC 会自动将上传的文件解析为 MultipartFile 对象。因此,控制器方法的参数应该是 MultipartFile 类型,而不是 File 类型,因为 MultipartFile 是专门用于处理文件上传的接口。可以使用以下方法来获取 MultipartFile 对象的相关信息,如表 15-2 所示。

表 15-2　获取目标文件对象信息的主要方法

方　法	返　回　值	说　明
getBytes()	Byte[]	返回文件内容,以字节数组形式
getName()	String	以多部分的形式返回参数的名称
getContentType()	String	获取文件 MIME 类型
getInputStream()	InputStream	获取文件输入流
getOriginalFilename()	String	返回客户端本地驱动器中的原始文件名
getSize()	long	返回文件大小,以字节为单位
isEmpty()	boolean	判断上传的文件是否为空
transferTo()	void	将上传的文件保存到目标目录下

在例 15-1 的基础上通过创建控制器方法实现文件上传的最后一步。后端程序 uploadController.java 如下所示。

```
@RequestMapping("/fileupload1")
    public ModelAndView upload1(@RequestParam MultipartFile photo, HttpSession
session) throws IOException {
    String imagesPath = session.getServletContext().getRealPath("images");
```

```
        String originalFilename = photo.getOriginalFilename();
        File file = new File(imagesPath, originalFilename);
        if (!file.exists()) {
            file.mkdir();
        }
        photo.transferTo(file);
        ModelAndView modelAndView = new ModelAndView();
        modelAndView.addObject("filename", originalFilename);
        modelAndView.setViewName("singleFileSuccess");
        return modelAndView;
    }
```

需要注意的是，在控制器方法中，确保 MultipartFile 参数的名称与 form 表单中的 <input type="file">标签的 name 属性相匹配。CommonsMultipartResolver 会根据 name 属性的值(如 name="photos")来提取上传的文件，并将其封装为 MultipartFile 对象。

此外，控制器方法的参数中可以包含一个 HttpSession 对象，用于获取当前项目的 session 对象。通过 session 对象，可以找到服务器的目录，并将上传的文件保存到新创建的 images 文件夹中。如果该目录不存在，则需要创建它。

在之前的文件处理方式中，通常会使用 IO 流来操作文件。首先，获取文件的输入流和输出流，然后使用 write 方法将文件内容写出，最后关闭流。然而，使用 MultipartFile 之后，不再需要手动进行 IO 操作来读取和写入文件。相反，可以使用 transferTo()方法，直接将上传的文件写入到目标目录中。这大大简化了 JavaWeb 中的文件上传操作。

例如，如果在文件上传页面上传名为 a.jpg 的图片文件，然后单击上传按钮，会跳转到文件上传成功的页面，如图 15-2 所示。前端可以通过访问控制器方法返回的 ModelAndView 对象中的文件名来查找并显示已经上传的 a.jpg 文件。这使得在前端页面上显示上传的文件非常容易。

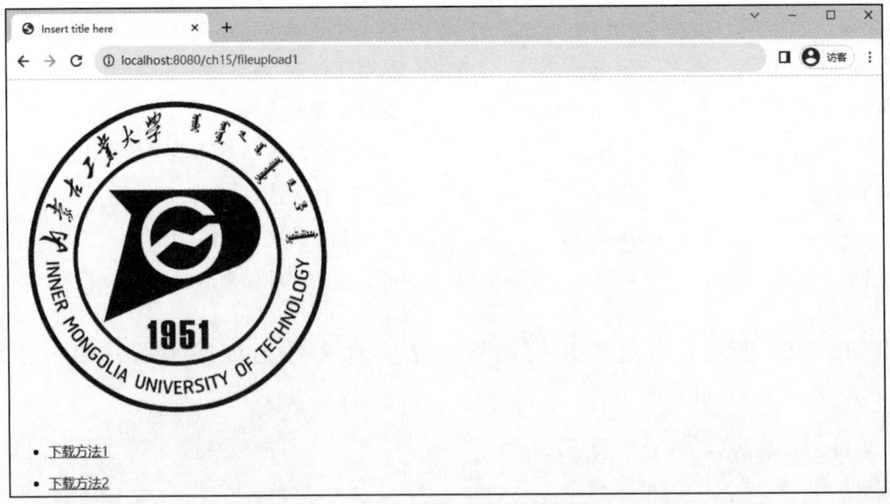

图 15-2 单文件上传成功页面

15.1.3 多文件上传

多文件上传是在单文件上传的基础上实现的，唯一不同之处是控制器方法需要使用数组或 List 来接收相关参数。不管是上传单个文件还是多个文件，都会被包装在 MultipartFile 对象中。控制器方法可以通过循环遍历 MultipartFile 对象中的多个文件，然后将它们保存到服务器目录，以完成多文件上传的操作。这简化了多文件上传的过程。

例 15-2：演示 Spring MVC 多文件上传功能，前端程序 multiFileUpload.jsp 如下。

```
<h3>多文件上传</h3>
    <form action="fileupload2" method="post" enctype="multipart/form-data">
    <ul>
        <li>图片 1:</li>
        <li><input type="file" name="photos" multiple></li>
        </ul>
        <ul>
            <li>图片 2:</li>
            <li><input type="file" name="photos" multiple></li>
        </ul>
        <ul>
            <li><input type="submit" value="上传" /></li>
    </ul>
</form>
```

多文件上传需要创建多个 type=file 的 input 标签，控制器方法为了能够循环遍历每个上传的文件，input 标签的 name 属性值需要一致。其他内容同单文件上传，多文件上传前端页面如图 15-3 所示。控制器中方法如 uploadController.java 所示。

```
@RequestMapping("/fileupload2")
public String testFileUpload(@RequestParam MultipartFile[] photos, HttpSession
session) throws IOException {
    //获取服务器中 images 目录的路径
    String imagesPath = session.getServletContext().getRealPath("images");
    File file = new File(imagesPath);
    //如果服务器中不存在 images 路径，则创建一个
    if (!file.exists()) {
        file.mkdir();
    }
    for (MultipartFile photo : photos) {
        if (!photo.isEmpty()) {
            String filename = photo.getOriginalFilename();
            String finalPath = imagesPath + File.separator + filename;
            photo.transferTo(new File(finalPath));
        }
    }
    return "multiFileSuccess";
}
```

控制器方法接收 MultipartFile[]数组类型参数，通过 for 循环遍历数组中的文件，在循

环中使用 transferTo()方法实现多文件上传，上传成功页面如图 15-4 所示。打开服务器目录，找到 images 文件夹，如图 15-5 所示，可以看到通过单文件方式上传的 a.jpg 和通过多文件上传的 b.jpg、c.jpg，说明文件均已上传至服务器。

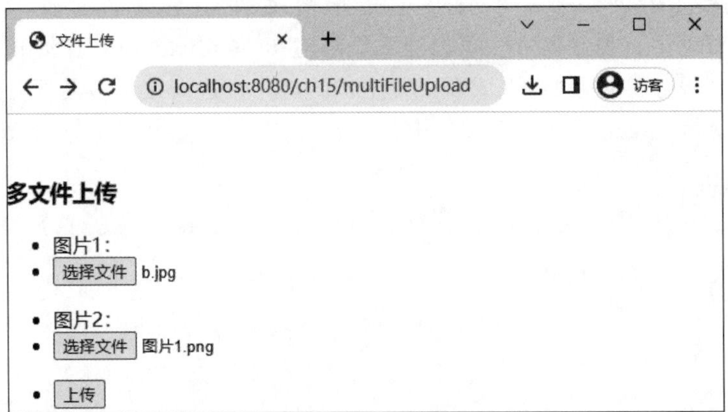

图 15-3　多文件上传页面

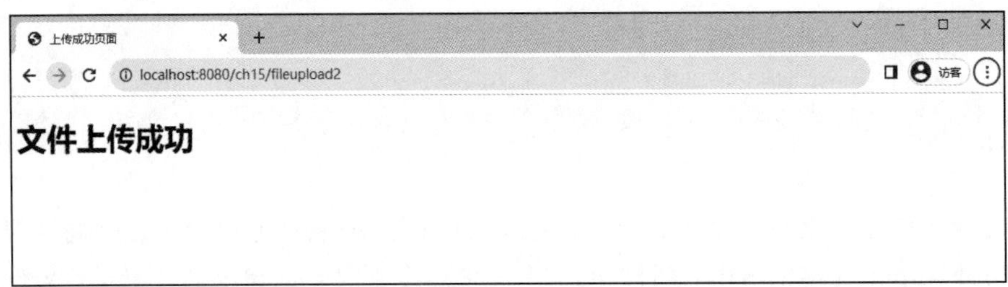

图 15-4　多文件上传成功页面

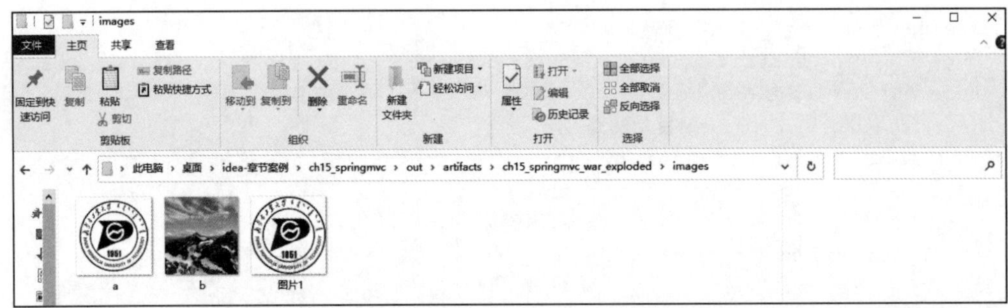

图 15-5　服务器目录 images 文件夹

15.2　Spring MVC 实现文件下载

数据下载是指客户端从服务器获取数据的过程，其中文件下载是数据下载的一种特例。通常，向服务器发送请求，大多数请求都是文件下载请求。通过这些请求，客户端可以从服务器下载文本、图片、声音、视频等文件。这些文件通常需要被客户端浏览器解析才能查看文件内容。

然而，这里提到的文件下载是指文件从服务器下载到浏览器后并不直接解析，而是以附件形式保存在客户端本地。

Spring MVC 提供了一种通过单击超链接下载按钮进行文件下载的方法，其中链接直接指向待下载文件的地址。但这种方法不够灵活，在企业开发中，通常需要前端获取文件名，然后将文件名传递给控制器类中的文件下载控制器方法，从而解决了文件下载地址的静态指定问题。

在本小节中，在单文件上传成功的基础上，通过例 15-1 中 ModelAndView 对象获取待下载的文件名，前端代码在 singleFileSuccess.jsp 中如下所示。

```
<img src="${basePath }images/${filename }" width="400px" /><br>
<ul>
    <li><a href="download/filedownload1? filename=${filename}">下载方法 1</a>
</li>
</ul>
<ul>
    <li><a href="download/filedownload2? filename=${filename}">下载方法 2</a>
</li>
</ul>
```

15.2.1 HttpServletResponse

例 15-3：演示一种 Spring MVC 的文件下载功能，控制器类 download.java 如下所示。

```
@RequestMapping("/filedownload1")
public void testFileDownload (String filename, HttpServletResponse respose,
HttpSession session) throws IOException {
respose.setHeader("Content-Disposition","attachment;filename="+filename);
    String imagesPath = session.getServletContext().getRealPath("/images");
    String file = imagesPath+"\\"+filename;
    IOUtils.copy(new FileInputStream(file), respose.getOutputStream());
}
```

其中，filename 为前端获取到的待下载文件名，通过设置 HttpServletResponse 响应头以附件形式进行下载，并将文件名拼接在服务器中文件真实路径后，最后使用 IOUtils.copy 方法，以流的形式输出服务器中指定的文件，直接减少了 byte 数组对象以及字节数组转换字符串这一大步骤。案例运行结果如图 15-6 所示。

15.2.2 ResponseEntity

Spring MVC 中，文件下载不仅限于例 15-3 演示的方式。可以使用 Spring MVC 提供的 ResponseEntity 实现文件下载功能。ResponseEntity 是一种泛型类型，可以将任何类型作为响应主体。该类继承自 HttpEntity，包含 3 个关键属性：Body(响应体)、HttpHeader(响应头)和 HttpStatus(响应状态码信息)。需要注意的是，ResponseEntity 与@ResponseBody 注解有所区别。如果只需返回 JSON 格式的数据，可以直接使用@ResponseBody 注解。而 ResponseEntity 用作方法的返回类型，不仅可以返回 JSON 格式的数据，还可以返回完整的响应报文等信息。

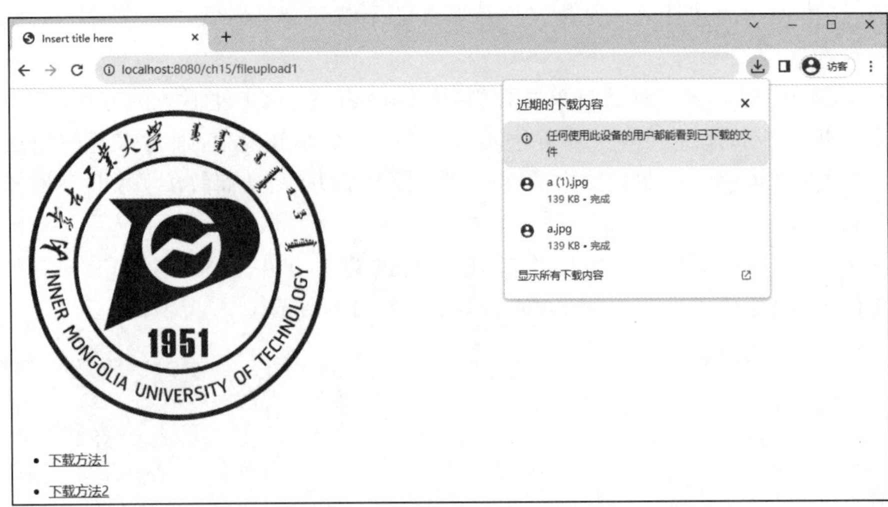

图 15-6　文件下载成功页面

因此，使用 ResponseEntity 返回类型相当于自定义一个响应报文，用于向浏览器发送响应，这是 Spring MVC 文件下载常用的方式。

例 15-4：演示使用 ResponseEntity 进行文件下载操作，控制器类 download.java 如下所示。

```
@RequestMapping("/filedownload2")
public ResponseEntity<byte[]> testResponseEntity(String filename, HttpSession
session) throws IOException {
    byte[] body=null;
    //获取服务器中文件的真实路径
    String imagesPath = session.getServletContext().getRealPath("/images");
    String file = imagesPath+"\\"+filename;
    InputStream inputStream = new FileInputStream(file);
    body=new byte[inputStream.available()];
    inputStream.read(body);
        MultiValueMap<String,String> httpHeaders = new HttpHeaders();
    httpHeaders.add("Content-Disposition","attachment;filename="+filename);
    HttpStatus statusCode=HttpStatus.OK;
    ResponseEntity<byte[]> responseEntity=new ResponseEntity<byte[]>(body,
headers,statusCode);
    return responseEntity;
}
```

若要在控制器方法中使用 ResponseEntity 来实现文件下载功能，需要遵循以下步骤。

① 返回类型必须为 ResponseEntity，表示自定义的响应报文。

② 获取服务器存储上传文件的目录，并定义文件名和编码格式。

③ 设置响应头、响应体和响应状态码。可以创建一个 HttpHeaders 对象来设置响应头信息，以键值对的形式存在。使用 setContentDispositionFormData 方法来定义文件下载方式和文件名，例如，使用 attachment 表示以附件形式下载，还可以指定下载文件的默认名称。通过 setContentType 方法将响应内容类型设置为 application/octet-stream，表示二进

制流数据。

④ 使用 FileUtils.readFileToByteArray()方法读取文件,并将其作为响应体。

⑤ 建议使用 HttpStatus.OK 来设置文件下载的 HTTP 状态码。

一旦 ResponseEntity 的属性都设置好,就可以将响应直接发送到浏览器。用户单击下载后,文件会以附件形式保存在本地。运行程序后的效果如图 15-6 所示。

15.3 国际化

实现软件系统国际化涉及两个重要概念。首先是国际化(i18n 的缩写),指的是将程序设计轻松翻译成多种语言,而无须做任何修改的过程。这使得应用系统或网站能够根据不同的语言或国家/地区显示相应的页面。其次是本地化(l10n 的缩写),指将国际化应用程序改造成支持特定语言区域的技术,使软件适用于特定语言或地区的过程。

15.3.1 语言区域

Java 中使用 java.util.Locale 类定义语言环境,即表示一个语言区域。由于语言本身并不能区分一个区域,例如说英语的国家很多,但是美国和英国所说的英语具有差异。所以必须指定语言区域。常见的语言和国家/地区说明如表 15-3 所示。

表 15-3 常见语言和国家/地区说明

语　　言	国家/地区	说　　明
en	US	美国英语
en	GB	英国英语
en	CA	加拿大英语
zh	CN	中国汉语
fr	FR	法国法语
de	DE	德国德语
ja	JP	日本日语
ko	KR	韩国韩语

Locale 提供了 3 个主要元件,language、country 分别代表对应语言和国家/地区,variant 是一个特定于供应商或者特定于浏览器的代号。例如,使用 Locale 构造函数构造一个表示英国所用英语的 Locale 对象,表示为:

```
Locale locale = new Locale("en", "GB");
```

除此之外,还可以使用 Locale 类提供的静态成员实例,直接返回特定国家/地区或语言的语言区域。遵循的规则是:Locale.国家/地区,代表使用某种语言的某个国家/地区;Locale.语言,代表某种语言(国家/地区不限定)。

15.3.2 国际化资源文件

Spring MVC 在国际化应用程序时需要满足以下两个条件。

(1) 资源文件配置：在国际化应用程序中，每一种语言环境都对应一个属性文件，其中内容以键值对的形式存储。每个属性文件的键必须是唯一的，用来表示特定语言区域的对象。属性文件的命名规范如下。

```
resourceBasename_language_country.properties
```

其中：

resourceBasename 是资源文件的基本名称，需要保持一致。

language 和 country 分别代表语言和国家/地区，它们之间使用下画线分隔。

例如，为了支持美国英语和中国汉语，需要配置两个属性文件。假设基准名为 message，那么属性文件的命名如下。

```
message_en_US.properties
message_zh_CN.properties
```

需要切换语言的属性文件中的键必须相同。例如，如果要实现登录页面的中英文切换，属性文件的内容分别如下。

美国英语属性文件如下。

```
resource.username = username
resource.password = password
```

中国汉语属性文件如下。

```
resource.username = \u7528\u6237\u540d
resource.password = \u5bc6\u7801
```

(2) 读取属性文件：Spring MVC 通过配置 ReloadableResourceBundleMessageSource 来加载属性文件。Spring MVC 启动时，会自动查找 id 为 messageSource 的 bean，如果存在的话，就会自动加载。这个 bean 会在 Spring MVC 响应过程中介入，用于解析国际化资源文件。以下是配置 messageSource bean，以读取属性文件的示例。

```
<bean id="messageSource" class="org.springframework.context.support.
ReloadableResourceBundleMessageSource">
    <property name="basename" value="classpath:message"></property>
</bean>
```

在这段配置代码中，basename 属性指定了要加载的资源文件的路径，这里是 classpath: message，表示容器会加载位于根目录下的基准名为 message 的资源文件。

通过这两个条件，Spring MVC 可以实现国际化应用程序，根据用户的语言环境动态切换并加载对应的属性文件，从而提供多语言支持。

15.3.3 语言区域选择

Spring MVC 提供语言区域解析器接口 LocaleResolver 获取 Locale 信息。Spring MVC 初始化过程中,DispatcherServlet 会解析一个 LocaleResolver 接口对象,可以通过它来决定用户区域,读出对应用户系统设定的语言或用户选择的语言。该接口常用实现类有以下 4 种。

AcceptLanguageLocaleResolver:根据请求头中名为 Accept-Language 的头来获取 Locale。

SessionLocaleResolver:使用 Session 传输语言环境,根据用户 Session 的属性获取 Locale。

CookieLocaleResolver:使用 Cookie 传送语言环境,根据前端请求的某个 Cookie 属性获取 Locale。

FixedLocaleResolver:忽略其他所有来源的 Locale 信息,而始终返回预先设定的 Locale。

(1) 使用 Accept-Language 参数。

Spring MVC 默认情况下通过读取用户浏览器的 Accept-Language 标题值来获取当前语言环境。AcceptHeaderLocaleResolver 根据请求头的 Accept-Language 属性值解析 Locale,这个属性在浏览器发送请求时会由浏览器自动带上。DispatcherServlet 初始化时会设置容器的语言环境解析器,在 Spring MVC 配置文件中配置容器 Bean 来实现。配置如下。

```
< bean id =" localeResolver" class =" org. springframework. web. servlet. i18n.
AcceptHeaderLocaleResolver">
<property name="defaultLocale" value="zh_CN"></property>
</bean>
```

如果浏览器的某个语言区域和 Spring MVC 应用程序支持的某个语言区域匹配,就会使用这个语言区域。如果请求头没有包含 Accept-Language,则返回属性为 defaultLocale 的值作为默认语言区域,一般我们选择中国汉语。若采用 Accept-Language 参数方式获取语言区域,则通过切换浏览器自身语言环境完成国际化语言环境的切换。

(2) 使用动态拦截器。

除了可以根据 Accept-Language 确定语言区域之外,开发中常用的国际化是通过设置拦截器实现的,根据某个特定请求自行设置 Locale。上述 3 种语言环境解析器中只有 Session 和 Cookie 支持对 Locale 的修改,并且需要结合拦截器使用。我们经常在网站上看到可以通过单击切换语言环境,这种方式就是通过自行设置 Locale 实现的。

首先配置 Locale 改变拦截器,根据拦截某个请求参数确定 Locale。在拦截器中添加 LocaleChangeInterceptor 的 Bean,拦截名为 locale 的请求参数,拦截器会根据 locale 的参数值来设置当前语言区域。拦截器的配置如下。

```
<mvc:interceptors>
<bean class="org.springframework.web.servlet.i18n.LocaleChangeInterceptor">
</bean>
</mvc:interceptors>
```

接下来对存储区域配置初始默认的语言环境，有 Session 和 Cookie 两种配置方式。因为 Cookie 语言环境解析器会优先从 Session 中获取对应属性值，所以 Session 语言环境解析器是比较常用的。SessionLocaleResolver 通过一个预定义会话名将区域化信息存储在会话中，使用会话中的属性解析 Locale，从而获取需要更改的语言区域。Session 语言环境解析器配置默认语言环境为中文，配置如下。

```
<bean id="localeResolver" class="org.springframework.web.servlet.i18n.
SessionLocaleResolver">
    <property name="defaultLocale" value="zh_CN"></property>
</bean>
```

15.3.4 国际化使用

例 15-5：演示采用动态拦截器的方式实现 Spring MVC 国际化，i18n.jsp 的前端代码如下所示。

```
<%@taglib prefix="spring" uri="http://www.springframework.org/tags"%>
<spring:message code="resource.username" />:<input type="text" name="username">
<br>
<spring:message code="resource.password" />:<input type="text" name="password">
<br>
<a href="i18n?locale=zh_CN">简体中文</a>|<a href="i18n?locale=en_US">
English</a>
```

Spring MVC 中可以使用 Spring 中的<message>标签实现国际化消息的显示。首先需要导入 Spring 标签库，并且指定标签前缀为 spring，其中 message 标签中的 code 属性指定资源文件中的键，通过<a>标签传递 locale 参数，选择语言环境，提交请求后首先会被拦截器拦截，拦截器根据 locale 请求参数改变语言环境，请求会被控制器方法正常接收。i18nConreoller.java 如下所示，由于语言环境发生改变，方法 test 需要返回 i18n.jsp 页面显示。实现登录页面国际化结果如图 15-7 和图 15-8 所示。

```
@RequestMapping("i18n")
    public String test(Local local){
    //切换语言环境页面
    return "i18n";
}
```

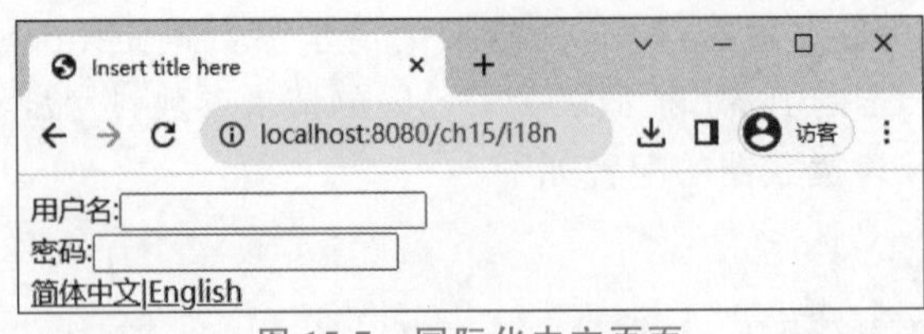

图 15-7 国际化中文页面

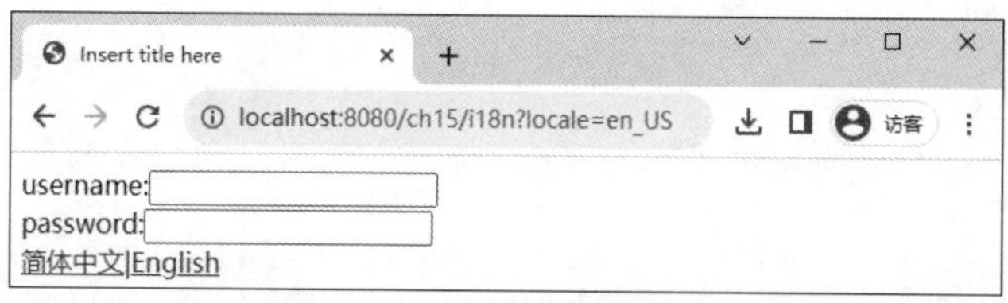

图 15-8　国际化英文页面

15.4 综合案例

本章案例主要实现上传、下载图片的功能。

15.4.1 案例设计

本章案例包括 Web 表现层。该层主要包括 Spring MVC 中的 Controller 类和 JSP 页面。Controller 类 FileController.java 负责处理用户上传、下载的请求，并将请求的处理结果返回给 JSP 页面进行展示。JSP 页面 upload.jsp 是上传页面，download.jsp 是下载页面。本章案例使用的主要文件如表 15-4 所示。

表 15-4　本章案例使用的主要文件

文　　件	所在包/路径	功　　能
web.xml	/WEB-INF/	Web 配置文件
Spring MVC.xml	src/	Spring MVC 配置文件
FileController.java	com.imut.controller	文件控制器
upload.jsp	/WEB-INF/pages/	上传页面
download.jsp	/WEB-INF/pages/	下载页面

15.4.2 案例演示

在浏览器地址栏中输入 http://localhost:8080/ch15_Spring MVC/login.jsp，登录系统，单击“上传下载”超链接，显示上传文件页面，如图 15-9 所示。单击“选择文件”，选择要上传的文件，单击“上传”按钮，显示上传成功页面，如图 15-10 所示。在该页面单击“下载”超链接，如图 15-11 所示，图中左下角显示文件下载成功。

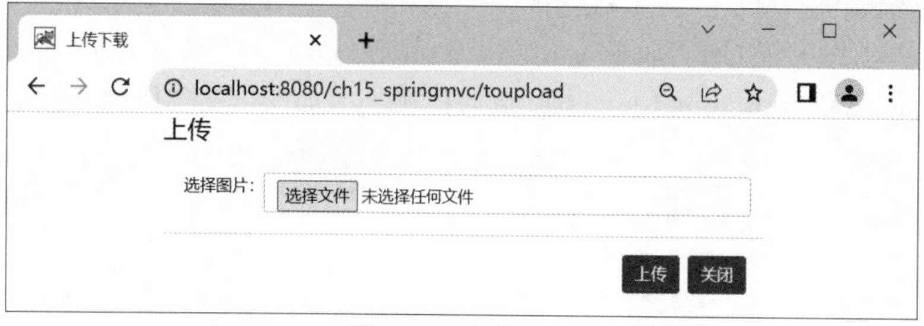

图 15-9　上传页面

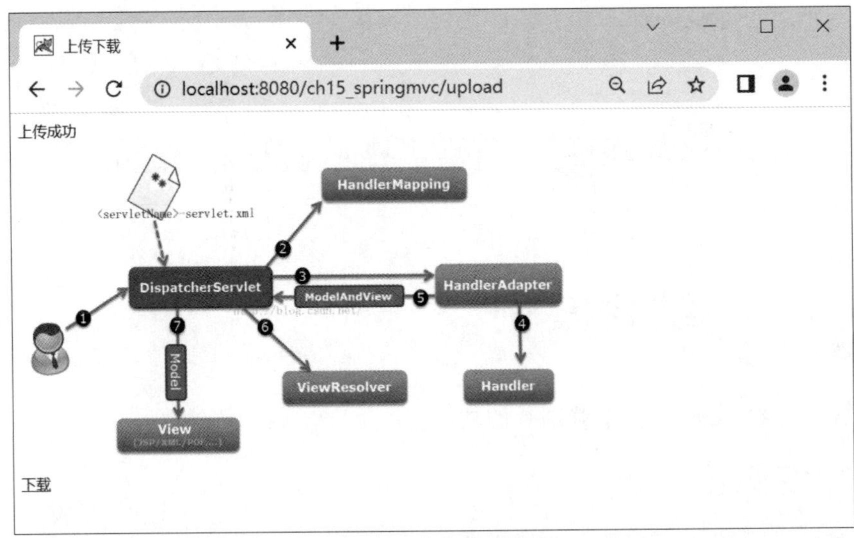

图 15-10　上传成功页面

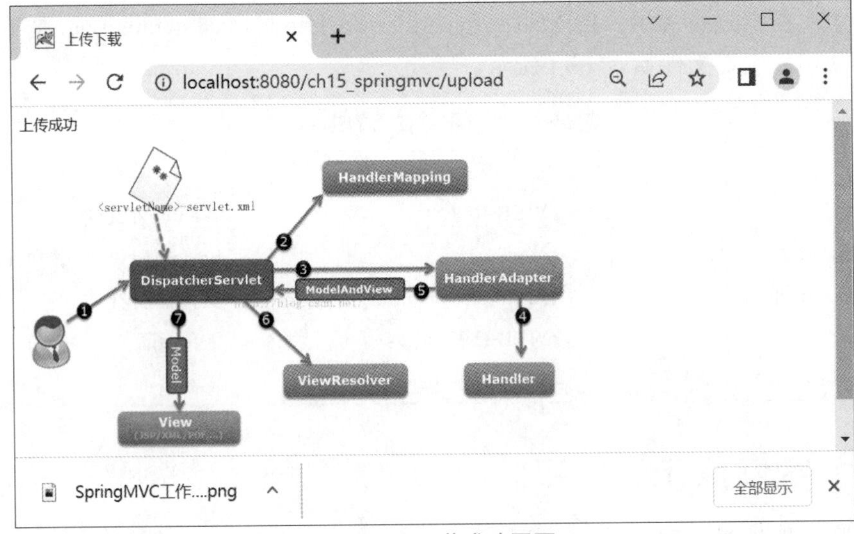

图 15-11　下载成功页面

15.4.3　代码实现

1. 创建项目，添加依赖

复制 Web 动态工程 ch14_Spring MVC 为 ch15_Spring MVC。在 pom.xml 文件中增加 commons-io 包和 commons-fileupload 包的依赖。增加的依赖代码如下所示。

```
<dependency>
    <groupId>commons-io</groupId>
    <artifactId>commons-io</artifactId>
    <version>2.13.0</version>
</dependency>
```

```
<dependency>
    <groupId>commons-fileupload</groupId>
    <artifactId>commons-fileupload</artifactId>
    <version>1.3.3</version>
</dependency>
```

2. 编写配置文件

web.xml 配置文件内容不改变。修改 Spring MVC.xml 配置文件，增加文件解析器配置。

配置文件 Spring MVC.xml 增加的代码如下。

```
<bean id="multipartResolver" class="org.springframework.web.multipart.commons.CommonsMultipartResolver">
    <property name="defaultEncoding" value="UTF-8"></property>
</bean>
```

3. 编写控制器类

编写控制器类 FileController.java。

程序 FileController.java 的代码如下。

```
@Controller
public class FileController {
    @RequestMapping("/toupload")
    public String toupload() {
        return "upload";
    }
    @RequestMapping("/upload")
    public ModelAndView doUpload(MultipartFile photo, HttpSession session) throws Exception {
        ModelAndView mv = new ModelAndView();
        String potoPath = session.getServletContext().getRealPath("\\images");
        String filename = photo.getOriginalFilename();
        File file=new File(potoPath,filename);
        photo.transferTo(file);
        mv.addObject("filename", filename);
        mv.setViewName("download");
        return mv;
    }
    @RequestMapping("/download")
    public ResponseEntity<byte[]> download(HttpServletRequest request,String filename) throws IOException {
        String path=request.getServletContext().getRealPath("/images/");
        filename=filename.replace("_", "%");
        filename=java.net.URLDecoder.decode(filename, "UTF-8");
        String downfilename="";
        if(request.getHeader("USER-AGENT").toLowerCase().indexOf("msie")>0) {
            filename=URLEncoder.encode(filename,"UTF-8");
```

```
            downfilename=filename.replaceAll("+", "%20");
        }else {
            downfilename=new String(filename.getBytes("UTF-8"),"ISO-8859-1");
        }
        File file=new File(path,filename);
        HttpHeaders headers = new HttpHeaders();
        headers.setContentDispositionFormData("attachment", downfilename);
        //application/octet-stream二进制流数据(最常见的文件下载)
        headers.setContentType(MediaType.APPLICATION_OCTET_STREAM);
        //文件下载的Http中的状态最好使用HttpStatus.OK
HttpStatus statusCode = HttpStatus.OK;
        ResponseEntity<byte[]> entity = new ResponseEntity<byte[]>(FileUtils.
readFileToByteArray(file), headers, statusCode);
        return entity;
    }
}
```

4. 编写 JSP 页面文件

修改 index.jsp 文件，增加"上传下载"超链接。编写 upload.jsp 文件，编写 download.jsp 文件。由于篇幅原因，JSP 文件只放<body></body>之间的代码。

程序 index.jsp 的代码如下。

```
<body  style="width: 300px;margin:0px auto;">
<div>
    <ul class="nav nav-pills nav-stacked">
        <li class="active"><a href="listStudent">学生管理</a></li>
        <li class="active"><a href="listCourse">课程管理</a></li>
        <li class="active"><a href="listCcourse">选课管理</a></li>
        <li class="active"><a href="toupload">上传下载</a></li>
    </ul>
</div>
</body>
```

程序 upload.jsp 的代码如下。

```
<body style="width: 600px;margin:0px auto;">
    <div class="panel panel-info" style="margin:0px auto;">
        <div class="panel-heading"><h4>上传</h4></div>
        <div class="panel-body">
            <form action="upload" method="post" enctype="multipart/form-data">
                <div class="modal-body">
                    <div class="input-group" style="margin: 5px;">
                        <span class="input-group-addon">选择图片:</span>
            <input type="file"  class="form-control" name="photo" accept=
"image/*" />
                    </div>
                </div>
                <div class="modal-footer">
```

```
                    <button class="btn btn-primary" type="submit">上传</button>
                     <button type="button" class="btn btn-primary" onclick="
javascript:history.go(-1)">关闭</button>
                </div>
            </form>
        </div>
    </div>
</body>
```

程序 download.jsp 的代码如下。

```
<body>
    上传成功<br>
    <img src="images/${filename }" width="600px"/><br>
    <a href="download? filename=${filename}">下载</a>
</body>
```

15.5 习题

1. 选择题

(1) 下面关于 MultipartFile 接口的说法,错误的是(   )。

A. getOriginalFilename()用于获取上传文件的初始化名

B. getSize()用于获取上传文件的大小,单位是 KB

C. getInputStream()用于读取文件内容,返回一个 InputStream 流

D. transferTo(File file)用于将上传文件保存到目标目录下

(2) 下面属于 CommonsMultipartResolver 属性的是(   )。

A. getContentType　　　　B. getInputStream

C. isEmpty　　　　D. defaultEncoding

(3) 下面关于文件下载方法内容描述,错误的是(   )。

A. 响应头信息中的 MediaType 代表的是 Interner Media Type(即互联网媒体类型),也叫作 MIME 类型

B. MediaType. APPLICATION _ OCTET _ STREAM 的值为 application/octet-stream,即表示以二进制流的形式下载数据

C. HttpStatus 类型代表的是 Http 中的状态

D. HttpStatus.OK 表示 500,即服务器已成功处理了请求

(4) 以下有关 CommonsMultipartResolver 类的属性及说法,错误的是(   )。

A. maxUploadSize:上传文件最大长度(以字节为单位)

B. maxInMemorySize:缓存中的最大尺寸

C. defaultEncoding:默认编码格式,默认为 UTF-8

D. resolveLazily:推迟文件解析

2. 填空题

(1) ResponseEntity 对象的作用类似于@ResponseBody 注解，它用于直接返回________。

(2) 使用 Servlet API 中提供的 URLEncoder 类中的________方法将中文转为 UTF-8 编码。

(3) HttpStatus 类型代表的是 Http 中的状态，示例中的 HttpStatus.OK 表示________，即服务器已成功处理了请求。

(4) Spring MVC 提供了一个________类型的对象，使用它可以很方便地定义返回的 HttpHeaders 对象和 HttpStatus 对象。

(5) 在 Spring MVC 中，实际上是使用 MultipartResolver 接口的实现类________来完成文件上传工作的。

(6) 实现文件下载时，下载页面链接的________属性要指定后台文件下载的方法以及文件名。

(7) 进行文件下载时，需要使用 HttpHeaders 的________方法定义以流的形式下载返回文件数据。

3. 判断题

(1) 上传文件时，必须保证所上传的文件不重名，为此可以通过 UUID 等方式对上传的文件名称进行重命名。 (   )

(2) 通过 maxUploadSize 属性可以对上传文件缓存中的最大尺寸进行设置。 (   )

(3) Spring MVC 的文件上传是通过 MultipartResolver 对象实现的。 (   )

(4) 在进行文件下载时，必须向 lib 目录加入 commons-filedownload.jar。 (   )

(5) 在进行文件上传时，通常可以使用 GET 方式进行。 (   )

4. 简答题

(1) 请简述文件上传时表单需要满足的 3 个条件。

(2) 请简述如何解决中文文件名称下载时的乱码问题。

5. 程序设计题

(1) 编写程序，实现本地图片上传、下载功能。

(2) 编写程序，实现语言区域的选择，支持至少两种语言。

第4部分

MyBatis＋Spring＋Spring MVC 整合篇

第16章 MyBatis+Spring+Spring MVC框架整合

通过之前的学习，我们了解到轻量级 SSM 框架是由 Spring、Spring MVC 和 Mybatis 三个开源框架整合而成。SSM 框架的出现大大简化了 JavaEE 的开发，因此被广泛应用。其中 Spring 框架就像是整个项目中装配容器的工厂，作为后端组件的容器来实例化对象。Spring MVC 是 Spring 提供的一个基于 MVC 设计模式的轻量级 Web 开发框架，本质上相当于 Servlet。Mybatis 框架就是对 JDBC 的封装，它让数据库底层操作变得透明。

本章首先讲解 SSM 框架的整合，并通过学生选课案例进一步讲解 SSM 框架的整合和使用。

16.1 MyBatis+Spring+Spring MVC 整合

SSM 框架整合可以分为两部分，首先是 Spring 和 MyBatis 的整合，MyBatis 官方提供了与 Spring 整合的中间件——MyBatis-Spring，它实现了两个框架的无缝结合。其次是 Spring 和 Spring MVC 的整合，因为 Spring MVC 是隶属于 Spring 框架的一部分，所以两者整合自然且容易。除了实现业务功能框架整合，SSM 还需要考虑异常和日志等基本功能的整合。

16.1.1 整合思路

整合主要分为 4 个步骤。首先是框架环境的导入，使用 SSM 框架之前，需要分别引入 Spring 框架、Spring MVC 框架和 MyBatis 框架搭建环境的 JAR 包。除此之外，还需要引入框架间整合环境，即 MyBatis-Spring、数据访问相关环境和其他功能环境。其次是搭建 Spring 框架环境，主要是配置 web.xml 文件。然后进行 MyBatis 框架整合，在 Spring 的配置文件中完成 Spring 对 Dao 层(数据访问)的支持配置和 Service 层(事务)的支持配置。最后进行 Spring 和 Spring MVC 框架的整合,主要是对 web.xml 和 Spring MVC 的配置文件进行相关配置。

SSM 整合的实现方式可分为 2 种,即基于 xml 配置文件方式和基于注解方式,本章将以基于 xml 配置文件的方式对 SSM 框架进行整合。由于 Spring 和 Spring MVC 处于同一体系,所以 Spring 和 Spring MVC 也谈不上整合,只是搭建相关环境,因此 SSM 的整合关键就是 Spring 对 MyBatis 的整合。

16.1.2 基础环境

整合框架之前需要做一些基础工作,包括以下内容。

(1) 为使得项目符合 MVC 开发规范,需要在 src 目录 com.imut 包下新建 controller、service、service.impl、mapper、pojo 这几个包。

(2) 在 src 目录下新建数据库的配置文件 db.properties,文件内容包括用于配置连接数据库的参数信息,如数据库连接驱动、URL 和数据库登录所需的用户名、密码等信息。在 db_ssm 数据库创建的 person 表带有 id、name 和 age 3 个属性,输入 4 条假数据方便程序与数据库交互,person 表信息如图 16-1 所示。

(3) 除了数据库相关配置文件外,日志是开发、调试过程中确定问题来源的重要信息,需要新建日志配置文件 log4j.properties,配置包括日志的打印、日志的级别等,这些都是可以由开发者直接控制的。src 目录结构如图 16-2 所示。

ssm
db_ssm
表
person
视图
函数
事件
查询
报表

对象 | person @db_
开始事务

id	name	age
1	test1	10
2	test2	20
3	test3	30
4	test4	40

图 16-1 person 表信息

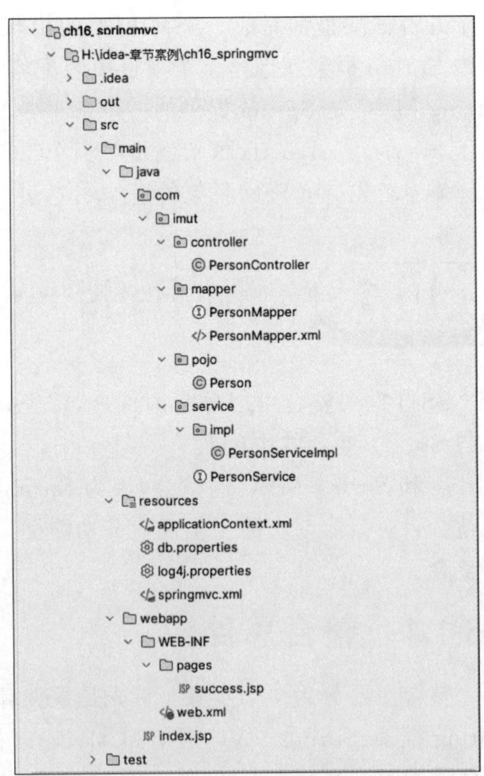

图 16-2 src 目录结构

（4）为了更加详细地讲解整个框架环境的依赖，下面将工程需要引入的 JAR 包以及 JAR 包间的作用以表格形式给出。有关 Spring、Spring MVC 以及数据访问所依赖的 JAR 包及其作用如表 16-1 所示。

表 16-1　Spring、Spring MVC 的相关依赖

依赖 JAR 包名称	作用
spring-core	Spring 的核心包，实现 IOC（控制反转）和 DI（依赖注入）
spring-beans	提供 BeanFactory，Spring 通过这个模块管理 Bean
spring-aop	提供 AOP
spring-aspects	提供与 AspectJ 框架的整合
spring-expression	提供 Spring 表达式语言
spring-context	在 IOC 基础上提供拓展服务，如国际化、事件、生命周期管理、邮件服务等其他服务的支持
spring-web	用于处理 Web 的基础模块
spring-webmvc	定义了中央控制器及相关类
spring-jdbc	对 JDBC 的简单封装
spring-tx	为 JDBC、Hibernate 等提供的一致的声明式和编程式事务管理

MyBatis 的依赖 JAR 包如表 16-2 所示。

表 16-2　MyBatis 相关依赖

依赖 JAR 包名称	作　用
mybatis	Mybatis 核心 JAR 包
mybatis-spring	Mybatis 和 Spring 整合所需

除了上述不同框架的依赖 JAR 包外，还有 Web 开发常用的工具包，相关依赖如表 16-3 所示。其中日志功能实现包括日志门面接口和日志实现，目前 Log4j＋CommonLogging 是比较流行的日志组合。

表 16-3　其他相关依赖

依赖 JAR 包名称	作　用
jstl	用于前端页面标签库 JSTL 环境配置
mysql-connector-java	提供 MySQL 数据库的连接驱动
log4j	日志实现，提供日志输出控制工具类的依赖
common-logging	日志门面，提供 log4j 等日志管理系统的接口

16.1.3　Spring 框架搭建

SSM 框架是以 Spring 为核心，协同其他两个框架调用有关资源，高效运作。所以环境

导入后，首先需要搭建 Spring 框架，在 web.xml 中配置并指定 Spring 的配置文件。Spring 框架使用 applicationContext.xml 配置文件，对非 web 端的组件扫描、数据源、MyBatis 会话工厂和事务等进行配置。

(1) 导入 Spring 框架依赖包后，在 web.xml 中通过 contextConfigLocation 来加载 Spring 容器，contextConfigLocation 参数定义了要装入的 Spring 配置文件。通过注册 ServletContext 监听器的实现类 ContextLoaderListener 实现容器创建和加载 ServletContext 作用域，配置如下。

```
<context-param>
    <param-name>contextConfigLocation</param-name>
    <param-value>classpath:applicationContext.xml</param-value>
</context-param>
<listener>
    <listener-class>org.springframework.web.context.ContextLoaderListener
</listener-class>
</listener>
```

(2) 在 src 目录下新建 Spring 的配置文件 applicationContext.xml，配置包扫描器，使用<context:component-scan>组件扫描标签指定路径下包含注解的类。除此之外，还需要引入 db.properties 配置文件对数据源 dataSource 进行配置，相关配置如下。

```
    <!—配置包扫描器 -->
    <context:component-scan base-package="com.imut"/>
    <!-- 引入外部文件 db.properties -->
<context:property-placeholder location="classpath:db.properties"/>
<!-- 配置数据源 -->
<bean id="dataSource" class="org.springframework.jdbc.datasource.
DriverManagerDataSource">
    <property name="driverClassName" value="${jdbc.driver}"/>
    <property name="url" value="${jdbc.url}"/>
    <property name="username" value="${jdbc.username}"/>
    <property name="password" value="${jdbc.password}"/>
    </bean>
```

16.1.4 Spring 整合 MyBatis

通过之前的学习，我们了解到这两个框架貌似没有实质联系，但从 Spring 框架角度来看，就是接管 MyBatis 框架中核心对象的创建任务。Spring 提供了很多集成类，唯独没有提供 MyBatis 的集成类，但通过 MyBatis 提供的集成库 MyBatis-Spring 可以实现和 Spring 框架的无缝衔接。

(1) Spring 对 Dao 层支持。

Spring 整合 MyBatis 主要是在 applicationContext.xml 中完成与数据访问相关的配置。例如数据源、会话工厂、事务处理等。在与 Spring 框架整合后，MyBatis 全权交给 Spring 容器和事务管理器来管理。通过配置 SqlSessionFactory 的 Bean 并注入 DataSource，SqlSessionFactoryBean 就可以完全取代 MyBatis 的全局配置文件。映射文件和 MyBatis 参数设置通过 SqlSessionFactoryBean

方式进行配置。由于映射文件和类在同一文件夹下，所以可以省略指定 xml 映射文件位置的属性配置。通过配置 MapperScannerConfigurer 可以实现自动扫描 mapper 接口，并自动代理生成映射器。Spring 对 Dao 层的支持具体配置如下。

```
<!--配置 sqlSessionFactory-->
<bean id="sqlSession" class="org.mybatis.spring.SqlSessionFactoryBean">
    <property name="dataSource" ref="dataSource"></property>
    <property name="mapperLocations" value="classpath*:com/imut/mapper/*.
xml"></property>
</bean>
<!--配置 mapper 扫描器-->
<bean class="org.mybatis.spring.mapper.MapperScannerConfigurer">
    <property name="basePackage" value="com.imut.mapper"/>
</bean>
```

（2）Spring 对 Service 层支持。

由于 Service 层主要功能是处理业务逻辑，包括控制事物、处理事务和调用 Dao 等。所以下面主要配置 Spring 与 MyBatis 事务的整合。需要在 Service 层通过 Spring 进行连接的绑定。Spring 提供的事务管理器 DataSourceTransactionManager，并不直接产生连接对象，而是通过管理事务，以确保 Dao 层和 Service 层使用的是同一个连接对象。所以和 SqlSessionFactoryBean 引入的数据源需要一致。具体配置如下。

```
<!--注册事务管理器-->
<bean id="transactionManager" class="org.springframework.jdbc.datasource.
DataSourceTransactionManager">
    <property name="dataSource" ref="dataSource"/>
</bean>
```

16.1.5 Spring 整合 Spring MVC

对 Spring MVC 框架进行整合，在 web.xml 中配置中央控制器，并指定 Spring MVC 的配置文件进行其他配置。Spring MVC 框架使用 Spring MVC.xml 配置文件实现 Web 端的组件扫描、视图解析器和其他配置。

（1）Spring MVC 框架环境的搭建也是通过 web.xml 配置文件完成的。主要对 Spring MVC 的中央控制器 DispatcherServlet 和字符集编码过滤器的注册任务进行配置。配置 DispatcherServlet 时需要注意，在 Spring MVC 的配置中，<param-name>需要和 Spring 配置中的<param-name>名称相同，且文件名要与 Spring 配置文件名称格式相同。Spring 容器初始化时，Spring MVC 配置文件中注册的 Bean 也会被创建，这样加载 Spring 配置文件时也会把 Spring MVC 的配置文件加载进来。通过注册 Spring MVC 的字符集过滤器 filter 来解决请求参数时的中文乱码问题，相关配置如下。

```
<servlet>
    <servlet-name>DispatcherServlet</servlet-name>
    <servlet-class>org.springframework.web.servlet.DispatcherServlet
</servlet-class>
```

```
    <init-param>
        <param-name>contextConfigLocation</param-name>
        <param-value>classpath:Spring MVC.xml</param-value>
    </init-param>
    <load-on-startup>1</load-on-startup>
</servlet>
<servlet-mapping>
    <servlet-name>DispatcherServlet</servlet-name>
    <url-pattern>/</url-pattern>
</servlet-mapping>
<!--编码过滤器 -->
<filter>
    <filter-name>CharacterEncodingFilter</filter-name>
    <filter-class>org.springframework.web.filter.CharacterEncodingFilter
</filter-class>
    <init-param>
        <param-name>encoding</param-name>
        <param-value>UTF-8</param-value>
    </init-param>
    <init-param>
        <param-name>forceEncoding</param-name>
        <param-value>true</param-value>
    </init-param>
</filter>
<filter-mapping>
    <filter-name>CharacterEncodingFilter</filter-name>
    <url-pattern>/*</url-pattern>
</filter-mapping>
```

(2) 在 src 目录下新建 Spring MVC 的配置文件 Spring MVC.xml,它主要配置注解驱动、控制器、视图解析器和静态资源拦截器。对于控制器的配置,同样使用 Spring 配置文件中<context:component-scan>标签路径扫描的方式。通过 Spring MVC 注解驱动配置,使得访问路径与方法的匹配可以通过注解配置,开启的注解包括@RequestMapping、@RequestParam 等。配置 InternalResourceViewResolver 视图解析器来处理 Controller 返回的 ModelAndVeiw 类型数据。最后是静态资源放行配置处理,其中 js、css 等静态资源不需要进行请求转发,直接放行即可。

相关配置如下。

```
<!--注解驱动-->
<mvc:annotation-driven/>
<!--配置包扫描器,扫描 controller-->
<context:component-scan base-package="com.imut.controller"/>
<!--配置视图解析器-->
<bean class="org.springframework.web.servlet.view.InternalResourceViewResolver">
<!--配置视图前缀-->
<property name="prefix" value="/WEB-INF/pages/"></property>
<!--配置视图后缀-->
```

```
<property name="suffix" value=".jsp"></property>
</bean>
<!--配置静态资源的访问-->
<mvc:default-servlet-handler/>
<mvc:resources location="/js/"  mapping="/js/**" />
<mvc:resources location="/css/"  mapping="/css/**" />
```

至此,所有的环境搭建和框架整合步骤已经完成,16.2 节将通过设计测试用例来验证 SSM 框架是否可行。

16.2 用例测试

本节将通过一个 Person 信息查询例子验证 SSM 框架搭建是否成功。测试之前,各层之间的依赖关系如图 16-3 所示。Controller 层负责具体的业务模块流程的管制,主要调用 Service 层里面的接口控制具体的业务流程。Service 层是建立在 Mapper 层之上的,而 Service 层又是在 Controller 层之下的,因而 Service 层应该既调用 Mapper 的接口,又要提供接口给 Controller 层的类来进行调用,它刚好处于一个中间层的位置。每个模型都有一个 Service 接口,每个接口分别封装各自的业务处理方法。Mapper 主要负责与数据库交互,通过创建 Mapper 接口在配置文件中定义该接口的实现类,就能实现在模块中调用 Mapper 接口进行数据业务的处理,而不用关注此接口的具体实现类是哪个。Pojo 主要用于定义与数据库对象相应的属性,提供 get/set 方法、tostring 方法以及有参无参构造方法。用例创建步骤如下。

图 16-3 获取 Person 信息页面

(1) 在 com.imut.pojo 包下创建 Person.java 实体类,包括 id、name 和 age 3 个属性并分别生成 getters 和 setters 方法,部分代码如下所示。

```
public class Person {
    private int id;
    private String name;
    private String age;
    public int getId() {
        return id;
    }
    public void setId(int id) {
        this.id = id;
    }
```

```
    ...
}
```

(2) 在 com.imut.mapper 包下创建 PersonMapper.java 接口和 PersonMapper.xml，注意两个名字必须相同。PersonMapper.java 接口中定义一个获取 Person 对象的接口 getPerson()。PersonMapper.xml 文件中主要配置 SQL 语句，用来和数据库交互，这里只写查询全部 Person 信息的方法。部分代码如下所示。

```
PersonMapper.java:
    public interface PersonMapper {
        public List<Person> getPerson();
    }
PersonMapper.xml:
    <mapper namespace="com.imut.mapper.PersonMapper">
        <select id="getPerson" resultType="com.imut.pojo.Person">
            select * from person
        </select>
    </mapper>
```

(3) 在 com.imut.service 下创建 PersonService.java 接口，接口中同样定义一个获取 Person 对象的接口 getPerson()。除此之外，Service 层还需要一个实现类来实现这个接口，在包 com.imut.service.impl 中创建 PersonServiceImpl.java 类，并实现 PersonService.java 接口中的 getPerson()方法，结果返回 Person 列表对象。其中 PersonServiceImpl.java 实现类需要@Service 注解标注，作用相当于在 xml 配置文件中配置一个 id 为 service 的 Bean。如果该类被@Service 注解标注将自动注册到 Spring 容器，不需要在配置文件中定义 Bean 类似于@Controller 注解。部分代码如下所示。

```
PersonService.java:
    public interface PersonService {
        public List<Person> getPerson();
    }
PersonServiceImpl.java:
    @Service
    public class PersonServiceImpl implements PersonService{
    @Autowired
    private PersonMapper personMapperImpl;
    @Override
    public List<Person> getPerson() {
        //TODO Auto-generated method stub
        return personMapperImpl.getPerson();
    }
}
```

(4) 在 com.imut.controller 包下新建 personController.java，创建 getPerson()方法返回 ModelAndView 对象，把从数据库查询出来的 Person 信息通过 ModelAndView 对象传至前端。部分代码如下所示。

```
@Resource
private PersonService personServiceImpl;
@RequestMapping("test")
public ModelAndView getPerson() {
    List<Person> list = new ArrayList<Person>();
    list = personServiceImpl.getPerson();
    System.out.println(list);
    ModelAndView mv = new ModelAndView();
    mv.addObject("list", list);
    mv.setViewName("success");
    return mv;
}
```

(5) 在 WebContent 下创建主页面 index.jsp,通过超链接请求控制器方法获取 Person 信息。在 WEB-INF 下创建 pages 文件夹,用于存放 jsp 页面,创建 success.jsp 页面,使用核心标签<c:forEach>循环遍历 Person 对象,并在页面输出。部分代码如下。

```
index.jsp:
    <a href="test">获取 Person 信息</a>
success.jsp:
    <table>
        <thead>
            <tr>
                <th>编号</th>
                <th>姓名</th>
                <th>年龄</th>
            </tr>
        </thead>
        <tbody>
            <c:forEach items="${list}" var="list">
                <tr>
                    <td>${list.id}</td>
                    <td>${list.name}</td>
                    <td>${list.age}</td>
                </tr>
        </c:forEach>
    </tbody>
</table>
```

最后检查代码,启动服务器在浏览器地址栏输入 localhost:8080//ch16/index.jsp,成功获取数据库中 Person 的信息,运行结果如图 16-4 和图 16-5 所示。到目前为止,已经使用 SSM 框架完成了数据库的查询功能,程序正常运行表示框架搭建成功。

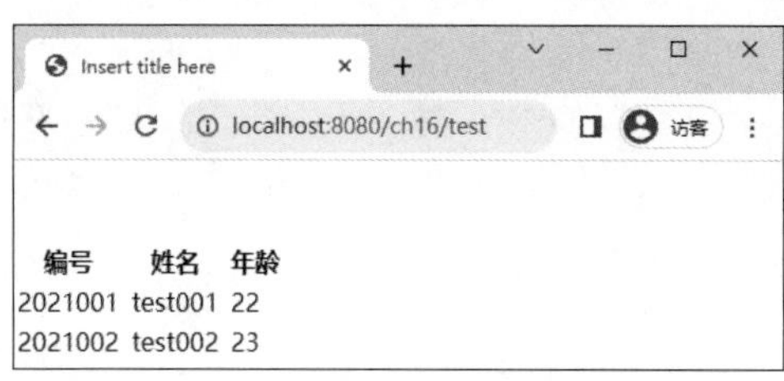

图 16-4 获取 Person 信息页面

图 16-5　Person 信息的输出页面

16.3　综合案例

本章案例实现用户登录、学生管理、课程管理、选课管理模块的所有功能。案例整合了Spring、Spring MVC 以及 MyBatis 框架完成整个系统的功能。

16.3.1　案例设计

本章案例划分为持久层、数据访问层、业务逻辑层和 Web 表现层。其中，持久层由若干实体类组成。数据访问层由接口和 MyBatis 映射文件组成，接口的名称统一以 Mapper 结尾，且 MyBatis 映射文件名称与接口名称相同。业务逻辑层由 Service 接口和实现类组成，接口名称统一使用 Service 结尾，实现类名称统一在接口名称后面加 Impl。Web 表现层主要包括 Spring MVC 中的 Controller 类和 JSP 页面，Controller 类主要负责拦截用户的请求，并调用业务逻辑层中相应组件的方法来处理用户请求，然后将相应的结果返回给 JSP 页面。本章案例使用的公共文件如表 16-4 所示。

表 16-4　本章案例使用的公共文件

文　　件	所在包/路径	功　　能
web.xml	/WEB-INF/	Web 配置文件
log4j.properties	src/	日志输出配置文件
db.properties	src/	数据库连接的属性文件
applicationContext.xml	src/	Spring 的配置文件
Spring MVC.xml	src/	Spring MVC 配置文件
index.jsp	/	首页

在操作该系统前需要先登录。本模块使用的主要文件如表 16-5 所示。

表 16-5 “用户登录”模块使用的主要文件

文　　件	所在包/路径	功　　能
User.java	com.imut.pojo	封装用户信息的类
UserMapper.java	com.imut.mapper	定义操作用户表方法的接口
UserMapper.xml	com.imut.mapper	接口的映射文件
UserService.java	com.imut.service	操作用户信息的业务逻辑接口
UserServiceImpl.java	com.imut.service.impl	业务逻辑接口的实现类
UserController.java	com.imut.controller	操作用户信息的控制器类
LoginInterceptor.java	com.imut.interceptor	登录拦截器类
login.jsp	/	登录页面

“学生管理”模块包括列出所有学生信息、增加学生信息、修改学生信息以及删除学生信息的功能。本模块使用的主要文件如表 16-6 所示。

表 16-6 “学生管理”模块使用的主要文件

文　　件	所在包/路径	功　　能
Student.java	com.imut.pojo	封装学生信息的类
StudentMapper.java	com.imut.mapper	定义操作学生表方法的接口
StudentMapper.xml	com.imut.mapper	接口的映射文件
StudentService.java	com.imut.service	操作学生信息的业务逻辑接口
StudentServiceImpl.java	com.imut.service.impl	业务逻辑接口的实现类
StudentController.java	com.imut.controller	操作学生信息的控制器类
studentList.jsp	WEB-INF/pages	显示学生信息、添加学生信息页面
updateStudent.jsp	WEB-INF/pages	修改学生信息页面

“课程管理”模块包括中列出所有课程信息、增加课程信息、修改课程信息以及删除课程信息的功能。本模块使用的主要文件如表 16-7 所示。

表 16-7 “课程管理”模块使用的主要文件

文　　件	所在包/路径	功　　能
Course.java	com.imut.pojo	封装课程信息的类
CourseMapper.java	com.imut.mapper	定义操作课程表方法的接口
CourseMapper.xml	com.imut.mapper	接口的映射文件
CourseService.java	com.imut.service	操作课程信息的业务逻辑接口
CourseServiceImpl.java	com.imut.service.impl	业务逻辑接口的实现类
CourseController.java	com.imut.controller	操作课程信息的控制器类
courseList.jsp	WEB-INF/pages	显示课程信息、添加课程信息页面
updateCourse.jsp	WEB-INF/pages	修改课程信息页面

“选课管理”模块包括查看选课列表、选课、修改选课信息以及删除选课信息的功能。本模块使用的主要文件如表 16-8 所示。

表 16-8 “选课管理”模块使用的主要文件

文　　件	所在包/路径	功　　能
Ccourse.java	com.imut.pojo	封装选课信息的类
CcourseMapper.java	com.imut.mapper	定义操作选课表方法的接口
CcourseMapper.xml	com.imut.mapper	接口的映射文件
CcourseService.java	com.imut.service	操作选课信息的业务逻辑接口
CcourseServiceImpl.java	com.imut.service.impl	业务逻辑接口的实现类
CcourseController.java	com.imut.controller	操作选课信息的控制器类
ccourseList.jsp	WEB-INF/pages	显示选课信息、添加选课信息页面
updateCcourse.jsp	WEB-INF/pages	修改选课信息页面

16.3.2 案例演示

在浏览器地址栏中输入 http://localhost:8080/ch16_ssm/login.jsp，登录系统，单击“学生管理”超链接，进入学生列表页面，如图 16-6 所示。单击“添加学生”按钮，出现图 16-7 的页面，在该页面中输入学生信息，单击“提交”按钮，跳转到学生列表页面，如图 16-8 所示。在学生列表页面单击对应学生后面的“修改”按钮，显示修改页面，如图 16-9 所示。修改完学生信息后，单击“提交”按钮显示学生列表页面，如图 16-10 所示。在该页面单击对应学生后面的“删除”按钮，弹出确认删除提示框，如图 16-11 所示，单击“确认”按钮，将先删除该学生的选课信息，然后删除学生信息，删除完成后跳转到学生列表页面，如图 16-12 所示。

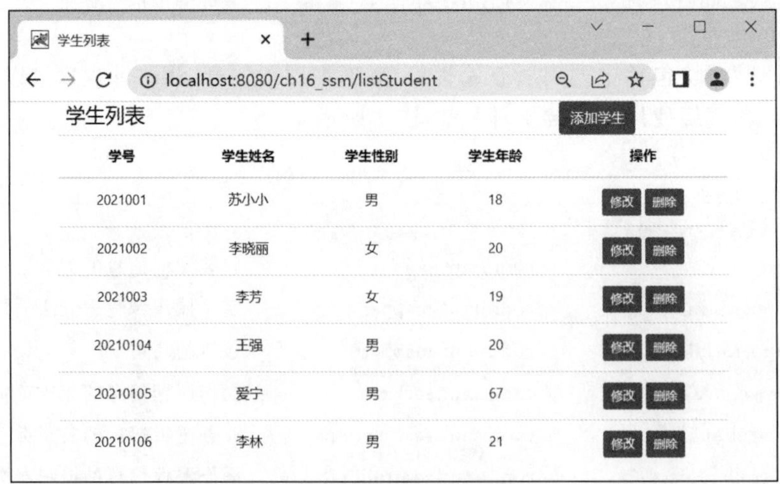

图 16-6 学生列表页面

图 16-7 添加学生信息页面

图 16-8 添加学生信息后的页面

图 16-9 修改学生页面

学号	学生姓名	学生性别	学生年龄	操作
2021001	苏小小	男	18	修改 删除
2021002	李晓丽	女	20	修改 删除
2021003	李芳	女	19	修改 删除
20210104	王强	男	20	修改 删除
20210105	爱宁	男	67	修改 删除
20210106	李林	男	21	修改 删除
20210107	王欢	男	19	修改 删除

图 16-10　修改学生信息后的页面

图 16-11　删除提示信息

学生列表

localhost:8080/ch16_ssm/listStudent

学生列表　添加学生

学号	学生姓名	学生性别	学生年龄	操作
2021001	苏小小	男	18	修改 删除
2021002	李晓丽	女	20	修改 删除
2021003	李芳	女	19	修改 删除
20210104	王强	男	20	修改 删除
20210105	爱宁	男	67	修改 删除
20210107	王欢	男	19	修改 删除

图 16-12　删除学生信息后的页面

单击“课程管理”超链接，进入课程列表页面，如图 16-13 所示。单击“添加课程”按钮，出现如图 16-14 所示的页面，在该页面中输入课程信息，单击“提交”按钮，跳转到课程列表页面，如图 16-15 所示。在课程列表页面单击对应课程后面的“修改”按钮，显示课程修改页面，如图 16-16 所示。修改完课程信息后，单击“提交”按钮显示课程列表页面。在该页面单击对应课程后面的“删除”按钮，弹出确认删除提示框，单击“确认”按钮，将先删除该课程被选修的信息，然后删除该课程信息，删除完成后跳转到课程列表页面，如图 16-17 和图 16-18 所示。

图 16-13　课程列表页面

图 16-14　添加课程信息页面

图 16-15　添加课程信息后的页面

图 16-16　修改课程信息页面

图 16-17　修改课程信息后的页面

课程列表

localhost:8080/ch16_ssm/listCourse

课程列表　添加课程

编号	课程编码	课程名称	操作
1	020625001	大学语文	修改 删除
2	020625002	高等数学	修改 删除
3	020625003	大学英语	修改 删除
5	020625005	程序设计基础	修改 删除
7	020625006	大学物理A	修改 删除

图 16-18　删除课程信息后的页面

单击“选课管理”超链接，进入选课列表页面，如图 16-19 所示。单击“添加选课”按钮，出现图 16-20 的页面，在该页面的下拉框中选择选课的学生和要选修的课程，单击“提交”按钮，跳转到选课列表页面，如图 16-21 所示。在选课列表页面单击对应课程后面的“修改”按钮，显示修改页面，如图 16-22 所示。在下拉框中选择要选修的课程后单击“提交”按钮，显示课程列表页面。在该页面单击对应选课后面的“删除”按钮，弹出确认删除提示框，单击“确认”按钮，删除该选课信息，删除完成后跳转到选课列表页面，如图 16-23 和图 16-24 所示。

图 16-19　选课列表页面

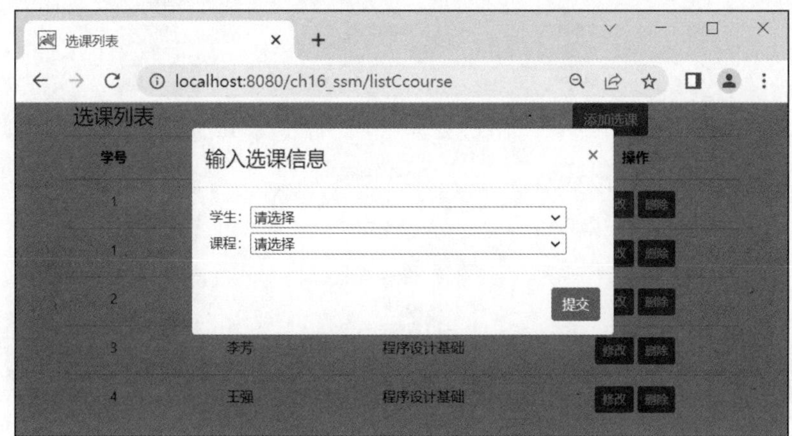

图 16-20　添加选课信息页面

图 16-21　添加选课信息后的选课列表页面

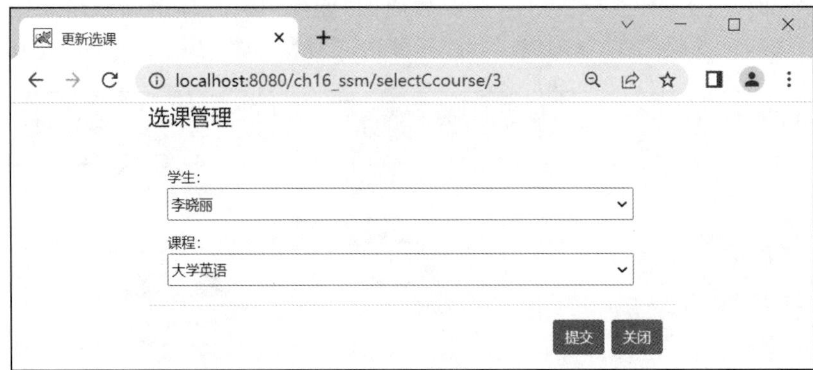

图 16-22　修改选课信息页面

图 16-23　修改选课信息后的页面

图 16-24　删除选课信息后的页面

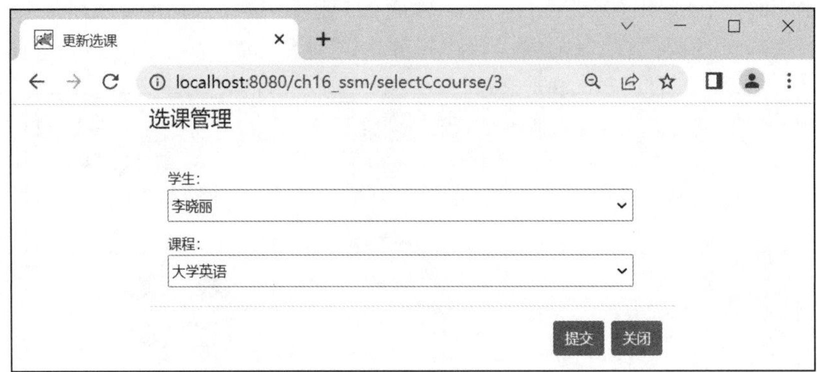

图 16-22　修改选课信息页面

选课列表　添加选课

localhost:8080/ch16_ssm/listCcourse

学号	学生姓名	课程名称	操作
1	苏小小	大学语文	修改 删除
1	苏小小	高等数学	修改 删除
1	苏小小	程序设计基础	修改 删除
2	李晓丽	大学语文	修改 删除
3	李芳	程序设计基础	修改 删除
4	王强	程序设计基础	修改 删除

图 16-23　修改选课信息后的页面

选课列表　添加选课

localhost:8080/ch16_ssm/listCcourse

学号	学生姓名	课程名称	操作
1	苏小小	大学语文	修改 删除
1	苏小小	高等数学	修改 删除
2	李晓丽	大学语文	修改 删除
3	李芳	程序设计基础	修改 删除
4	王强	程序设计基础	修改 删除

图 16-24　删除选课信息后的页面

16.3.3 代码实现

1. 创建项目，添加依赖

在 IDEA 中使用 Maven 创建 Web 动态工程 ch16_ssm，在 pom.xml 文件中增加以下包的依赖：MyBatis 的核心 JAR 包、日志包、MySQL 数据库的驱动 JAR 包、jstl 包、MyBatis 与 Spring 整合的包以及 Spring 的 10 个包。增加的依赖代码如下所示。

```
<dependencies>
    <dependency>
      <groupId>commons-logging</groupId>
      <artifactId>commons-logging</artifactId>
      <version>1.1</version>
    </dependency>
    <dependency>
      <groupId>javax.servlet</groupId>
      <artifactId>jstl</artifactId>
      <version>1.2</version>
      <scope>compile</scope>
    </dependency>
    <dependency>
      <groupId>log4j</groupId>
      <artifactId>log4j</artifactId>
      <version>1.2.17</version>
    </dependency>
    <dependency>
      <groupId>org.mybatis</groupId>
      <artifactId>mybatis</artifactId>
      <version>3.5.6</version>
    </dependency>
    <dependency>
      <groupId>org.mybatis</groupId>
      <artifactId>mybatis-spring</artifactId>
      <version>2.0.6</version>
    </dependency>
    <dependency>
      <groupId>mysql</groupId>
      <artifactId>mysql-connector-java</artifactId>
      <version>8.0.11</version>
    </dependency>
    <dependency>
      <groupId>org.springframework</groupId>
      <artifactId>spring-aop</artifactId>
      <version>5.2.5.RELEASE</version>
    </dependency>
    <dependency>
      <groupId>org.springframework</groupId>
      <artifactId>spring-aspects</artifactId>
      <version>5.2.5.RELEASE</version>
```

```
    </dependency>
    <dependency>
      <groupId>org.springframework</groupId>
      <artifactId>spring-beans</artifactId>
      <version>5.2.5.RELEASE</version>
    </dependency>
    <dependency>
      <groupId>org.springframework</groupId>
      <artifactId>spring-context</artifactId>
      <version>5.2.5.RELEASE</version>
    </dependency>
    <dependency>
      <groupId>org.springframework</groupId>
      <artifactId>spring-core</artifactId>
      <version>5.2.5.RELEASE</version>
    </dependency>
    <dependency>
      <groupId>org.springframework</groupId>
      <artifactId>spring-expression</artifactId>
      <version>5.2.5.RELEASE</version>
    </dependency>
    <dependency>
      <groupId>org.springframework</groupId>
      <artifactId>spring-jdbc</artifactId>
      <version>5.2.5.RELEASE</version>
    </dependency>
    <dependency>
      <groupId>org.springframework</groupId>
      <artifactId>spring-tx</artifactId>
      <version>5.2.5.RELEASE</version>
    </dependency>
    <dependency>
      <groupId>org.springframework</groupId>
      <artifactId>spring-web</artifactId>
      <version>5.2.5.RELEASE</version>
    </dependency>
    <dependency>
      <groupId>org.springframework</groupId>
      <artifactId>spring-webmvc</artifactId>
      <version>5.2.5.RELEASE</version>
    </dependency>
</dependencies>
```

2. 编写配置文件

在 WEB-INF 下编写 web.xml 配置文件。在/src/main/resources 下编写 db.properties、log4j.properties、applicationContext.xml 和 Spring MVC.xmlSpring MVC.xml 等配置文件。

配置文件 web.xml 的代码如下。

```
<?xml version="1.0" encoding="UTF-8"?>
<web-app xmlns:xsi="http://www.w3.org/2001/XMLSchema-instance" xmlns=
"http://java.sun.com/xml/ns/javaee" xsi:schemaLocation="http://java.sun.com/
xml/ns/javaee http://java.sun.com/xml/ns/javaee/web-app_3_0.xsd" id="WebApp_
ID" version="3.0">
    <display-name>ch16_ssm</display-name>
    <welcome-file-list>
        <welcome-file>index.jsp</welcome-file>
    </welcome-file-list>
    <context-param>
        <param-name>contextConfigLocation</param-name>
        <param-value>classpath:applicationContext.xml</param-value>
    </context-param>
    <listener>
        <listener-class>org.springframework.web.context.
ContextLoaderListener</listener-class>
    </listener>
    <servlet>
        <servlet-name>DispatcherServlet</servlet-name>
        <servlet-class>org.springframework.web.servlet.DispatcherServlet
</servlet-class>
        <init-param>
            <param-name>contextConfigLocation</param-name>
            <param-value>classpath:Spring MVC.xml</param-value>
        </init-param>
        <load-on-startup>1</load-on-startup>
    </servlet>
    <servlet-mapping>
        <servlet-name>DispatcherServlet</servlet-name>
        <url-pattern>/</url-pattern>
    </servlet-mapping>
    <filter>
        <filter-name>CharacterEncodingFilter</filter-name>
        <filter-class>org.springframework.web.filter.CharacterEncodingFilter
</filter-class>
        <init-param>
            <param-name>encoding</param-name>
            <param-value>UTF-8</param-value>
        </init-param>
        <init-param>
            <param-name>forceEncoding</param-name>
            <param-value>true</param-value>
        </init-param>
    </filter>
    <filter-mapping>
        <filter-name>CharacterEncodingFilter</filter-name>
        <url-pattern>/*</url-pattern>
    </filter-mapping>
</web-app>
```

配置文件 db.properties 的代码如下。

```
jdbc.driver=com.mysql.jdbc.Driver
jdbc.url=jdbc:mysql://localhost:3306/db_student? characterEncoding=
utf8&serverTimezone=GMT&useSSL=false
jdbc.username=root
jdbc.password=123456
```

配置文件 log4j.properties 的代码如下。

```
log4j.rootLogger=ERROR, stdout
log4j.logger.com.imut.mapper=DEBUG
log4j.appender.stdout=org.apache.log4j.ConsoleAppender
log4j.appender.stdout.layout=org.apache.log4j.PatternLayout
log4j.appender.stdout.layout.ConversionPattern=%5p [%t] - %m%n
```

配置文件 applicationContext.xml 的代码如下。

```
<?xml version="1.0" encoding="UTF-8"?>
<beans xmlns="http://www.springframework.org/schema/beans"
    xmlns:context="http://www.springframework.org/schema/context"
    xmlns:aop="http://www.springframework.org/schema/aop"
    xmlns:tx="http://www.springframework.org/schema/tx"
    xmlns:xsi="http://www.w3.org/2001/XMLSchema-instance"
    xsi:schemaLocation="
        http://www.springframework.org/schema/beans
        http://www.springframework.org/schema/beans/spring-beans.xsd
        http://www.springframework.org/schema/context
        http://www.springframework.org/schema/context/spring-context.xsd
        http://www.springframework.org/schema/aop
        http://www.springframework.org/schema/aop/spring-aop.xsd
        http://www.springframework.org/schema/tx
        http://www.springframework.org/schema/tx/spring-tx.xsd">
    <context:component-scan base-package="com.imut" >
        <context:exclude-filter type="annotation" expression="com.imut.
controller"/>
    </context:component-scan>
    <context:property-placeholder location="classpath:db.properties"/>
    <bean id="dataSource" class="org.springframework.jdbc.datasource.
DriverManagerDataSource">
        <property name="driverClassName" value="${jdbc.driver}"/>
        <property name="url" value="${jdbc.url}"/>
        <property name="username" value="${jdbc.username}"/>
        <property name="password" value="${jdbc.password}"/>
    </bean>
    <bean id="sqlSession" class="org.mybatis.spring.SqlSessionFactoryBean">
        <property name="dataSource" ref="dataSource"></property>
    </bean>
    <bean id="transactionManager" class="org.springframework.jdbc.
datasource.DataSourceTransactionManager">
```

```
        <property name="dataSource" ref="dataSource"/>
    </bean>
    <bean class="org.mybatis.spring.mapper.MapperScannerConfigurer">
        <property name="basePackage" value="com.imut.mapper"/>
    </bean>
    <tx:advice id="txadvice" transaction-manager="transactionManager">
        <tx:attributes>
            <tx:method name="add * " propagation="REQUIRED" rollback-for=
"Exception" />
            <tx:method name="del * " propagation="REQUIRED" rollback-for=
"Exception"/>
            <tx:method name="list * " propagation="REQUIRED" rollback-for=
"Exception" />
            <tx:method name="update * " propagation="REQUIRED" rollback-for=
"Exception"/>
            <tx:method name="selcet * " propagation="REQUIRED" rollback-for=
"Exception"/>
        </tx:attributes>
    </tx:advice>
</beans>
```

配置文件 Spring MVC.xmlSpring MVC.xml 的代码如下。

```
<?xml version="1.0" encoding="UTF-8"?>
<beans xmlns="http://www.springframework.org/schema/beans"
    xmlns:xsi="http://www.w3.org/2001/XMLSchema-instance"
    xmlns:aop="http://www.springframework.org/schema/aop"
    xmlns:tx="http://www.springframework.org/schema/tx"
    xmlns:context="http://www.springframework.org/schema/context"
    xmlns:mvc="http://www.springframework.org/schema/mvc"
    xsi:schemaLocation="
        http://www.springframework.org/schema/beans
        http://www.springframework.org/schema/beans/spring-beans.xsd
        http://www.springframework.org/schema/context
        http://www.springframework.org/schema/context/spring-context.xsd
        http://www.springframework.org/schema/tx
        http://www.springframework.org/schema/tx/spring-tx.xsd
        http://www.springframework.org/schema/aop
        http://www.springframework.org/schema/aop/spring-aop.xsd
        http://www.springframework.org/schema/mvc
        http://www.springframework.org/schema/mvc/spring-mvc.xsd">
    <mvc:annotation-driven/>
    <context:component-scan base-package="com.imut.controller"/>
    <bean class="org.springframework.web.servlet.view.
InternalResourceViewResolver">
        <property name="prefix" value="/WEB-INF/pages/"></property>
        <property name="suffix" value=".jsp"></property>
    </bean>
    <mvc:resources location="/js/" mapping="/js/**"></mvc:resources>
```

```
    <mvc:interceptors>
        <mvc:interceptor>
            <mvc:mapping path="/**"/>
            <mvc:exclude-mapping path="/login"/>
            <bean class="com.imut.interceptor.LoginInterceptor"/>
        </mvc:interceptor>
    </mvc:interceptors>
</beans>
```

3. 编写实体类

在/src/main/java 目录下新建包 com.imut.pojo，在该包下创建实体类 User.java、Student.java、Course.java 以及 Ccourse.java。

程序 User.java 代码与第 14 章综合案例中的 User.java 内容相同，程序 Student.java、Course.java 以及 Ccourse.java 代码与第 3 章综合案例中对应文件的代码相同，这里不再列出。

4. 编写数据持久层接口及映射文件

在/src/main/java 下新建包 com.imut.mapper，在该包下新建接口文件 UserMapper.java、StudentMapper.java、CourseMapper.java 和 CcourseMapper.java。新建对应的映射文件 UserMapper.xml、StudentMapper.xml、CourseMapper.xml 以及 CcourseMapper.xml。

程序 StudentMapper.java 和 StudentMapper.xml 代码与第 3 章综合案例对应文件的代码相同，这里不再列出。

程序 UserMapper.java 的代码如下。

```
public interface UserMapper {
    public User getUserByNameAndPwd(@Param("username")String username,@Param
("password") String password);
}
```

程序 CourseMapper.java 的代码如下。

```
public interface CourseMapper {
    public Course getCourseById(int cid);
    public List<Course> listCourses();
    public List<Ccourse> getCcourseById(int cid);
    public List<Course> listSelect(int sid);
    public void addCourse(Course course);
    public void updateCourse(Course course);
    public void deleteCourse(int cid);
}
```

程序 CcourseMapper.java 的代码如下。

```
public interface CcourseMapper {
    public Ccourse getCcourseById(int id);
    public List<Ccourse> listCcourses();
```

```
    public void addCcourse(Ccourse ccourse);
    public void updateCcourse(Ccourse ccourse);
    public void deleteCcourse(int id);
}
```

程序 UserMapper.xml 的代码如下。

```
<mapper namespace="com.imut.mapper.UserMapper">
    <select id="getUserByNameAndPwd" resultType="com.imut.pojo.User">
        select * from user where username=#{username} and password=#{password}
    </select>
</mapper>
```

程序 CourseMapper.xml 的代码如下。

```
<mapper namespace="com.imut.mapper.CourseMapper">
    <select id="getCourseById" resultType="com.imut.pojo.Course">
        select * from course where cid = #{cid}
    </select>
    <select id="listCourses" resultType="com.imut.pojo.Course">
        select * from course
    </select>
    <select id="getCcourseById" resultType="com.imut.pojo.Ccourse">
         select cc.id, c.cid, c.cname, s.sid, s.sname from ccourse cc, course c,
student s
            where c.cid=cc.cid and cc.sid=s.sid and c.cid=#{cid}
    </select>
    <select id="listSelect" resultType="com.imut.pojo.Course">
        select course.* FROM course where cid not in
            (select c.cid from course c,ccourse cc,student s
                where cc.cid=c.cid and cc.sid=s.sid and s.sid=#{sid})
    </select>
    <insert id="addCourse">
        insert into course (cno,cname)
            values(#{cno},#{cname});
    </insert>
    <update id="updateCourse">
        update course
        <set>
            <if test="cno!=null and cno!=''">cno=#{cno},</if>
            <if test="cname!=null and cname!=''">cname=#{cname}</if>
        </set>
        where cid=#{cid}
    </update>
    <delete id="deleteCourse">
        delete from course where cid=#{cid}
    </delete>
</mapper>
```

程序 CcourseMapper.xml 的代码如下。

```
<mapper namespace="com.imut.mapper.CcourseMapper">
    <select id="getCcourseById" resultType="com.imut.pojo.Ccourse">
        select * from ccourse where id = #{id}
    </select>
    <select id="listCcourses" resultType="com.imut.pojo.Ccourse">
        select cc.*,s.sname,c.cname from ccourse cc,student s,course c
            where cc.sid=s.sid and cc.cid=c.cid order by s.sid
    </select>
    <insert id="addCcourse">
        insert into ccourse ( cid,sid ) values (#{cid},#{sid})
    </insert>
    <update id="updateCcourse">
        update ccourse set sid=#{sid},cid=#{cid}
            where id=#{id}
    </update>
    <delete id="deleteCcourse">
        delete from ccourse where id=#{id}
    </delete>
</mapper>
```

5. 编写 Service 层接口及实现类

在/src/main/java 目录下新建包 com.imut.service，在该包下创建接口 UserService.java、StudentService.java、CourseService.java 和 CcourseService.java。新建包 com.imut.service.impl，在该包下新建接口实现类 UserServiceImpl.java、StudentServiceImpl.java、CourseServiceImpl.java 和 CcourseServiceImpl.java。

程序 UserService.java 的代码如下。

```
public interface UserService {
    public User login(String username,String password);
}
```

程序 StudentService.java 的代码如下。

```
public interface StudentService {
    public Student getStudentById(int sid);
    public List<Student> listStudents();
    public List<Ccourse> getCcoursesById(int sid);
    public void addStudent(Student student);
    public void updateStudent(Student student);
    public void deleteStudent(int sid);
}
```

程序 CourseService.java 的代码如下。

```
public interface CourseService {
    public Course getCourseById(int cid);
    public List<Course> listCourses();
    public List<Ccourse> getCcourseById(int cid);
```

```
    public List<Course> listSelect(int sid);
    public void addCourse(Course course);
    public void updateCourse(Course course);
    public void deleteCourse(int cid);
}
```

程序 CcourseService.java 的代码如下。

```
public interface CcourseService {
    public Ccourse getCcourseById(int id);
    public List<Ccourse> listCcourses();
    public void addCcourse(Ccourse ccourse);
    public void updateCcourse(Ccourse ccourse);
    public void deleteCcourse(int id);
}
```

程序 UserServiceImpl.java 的代码如下。

```
@Service
public class UserServiceImpl implements UserService {
    @Autowired
    private UserMapper userMapper;
    @Override
    public User login(String username,String password) {
        return userMapper.getUserByNameAndPwd(username,password);
    }
}
```

程序 StudentServiceImpl.java 的代码如下。

```
@Service
public class StudentServiceImpl implements StudentService {
    @Autowired
    StudentMapper studentMapper;
    @Autowired
    CcourseMapper ccourseMapper;
    @Override
    public Student getStudentById(int sid) {
        return studentMapper.getStudentById(sid);
    }
    @Override
    public List<Student> listStudents() {
        return studentMapper.listStudents();
    }
    @Override
    public List<Ccourse> getCcoursesById(int sid) {
        return studentMapper.getCcoursesById(sid);
    }
    @Override
```

```
    public void addStudent(Student student) {
        studentMapper.addStudent(student);
    }
    @Override
    public void updateStudent(Student student) {
        studentMapper.updateStudent(student);
    }
    @Override
    public void deleteStudent(int sid) {
        List<Ccourse> list=studentMapper.getCcoursesById(sid);
        for(Ccourse cc:list) {
            ccourseMapper.deleteCcourse(cc.getId());
        }
        studentMapper.deleteStudent(sid);
    }
}
```

程序 CourseServiceImpl.java 的代码如下。

```
@Service
public class CourseServiceImpl implements CourseService {
    @Autowired
    CourseMapper courseMapper;
    @Autowired
    CcourseMapper ccourseMapper;
    @Override
    public Course getCourseById(int cid) {
        return courseMapper.getCourseById(cid);
    }
    @Override
    public List<Course> listCourses() {
        return courseMapper.listCourses();
    }
    @Override
    public List<Ccourse> getCcourseById(int cid) {
        return courseMapper.getCcourseById(cid);
    }
    @Override
    public List<Course> listSelect(int sid) {
        return courseMapper.listSelect(sid);
    }
    @Override
    public void addCourse(Course course) {
        courseMapper.addCourse(course);
    }
    @Override
    public void updateCourse(Course course) {
        courseMapper.updateCourse(course);
    }
```

```
    @Override
    public void deleteCourse(int cid) {
        List<Ccourse> list=courseMapper.getCcourseById(cid);
        for(Ccourse cc:list) {
            ccourseMapper.deleteCcourse(cc.getId());
        }
        courseMapper.deleteCourse(cid);
    }
}
```

程序 CcourseServiceImpl.java 的代码如下。

```
@Service
public class CcourseServiceImpl implements CcourseService {
    @Autowired
    private CcourseMapper ccourseMapper;
    @Autowired
    private StudentMapper studentMapper;
    @Autowired
    private CourseMapper courseMapper;
    @Override
    public Ccourse getCcourseById(int id) {
        Ccourse ccourse=ccourseMapper.getCcourseById(id);
        Student student=studentMapper.getStudentById(ccourse.getSid());
        Course course=courseMapper.getCourseById(ccourse.getCid());
        ccourse.setSname(student.getSname());
        ccourse.setCname(course.getCname());
        return ccourse;
    }
    @Override
    public List<Ccourse> listCcourses() {
        List <Ccourse> list=ccourseMapper.listCcourses();
        return list;
    }
    @Override
    public void addCcourse(Ccourse ccourse) {
        ccourseMapper.addCcourse(ccourse);
    }
    @Override
    public void updateCcourse(Ccourse ccourse) {
        ccourseMapper.updateCcourse(ccourse);
    }
    @Override
    public void deleteCcourse(int id) {
        ccourseMapper.deleteCcourse(id);
    }
}
```

6. 编写控制器类

在/src/main/java 下新建包 com.imut.controller，在该包下新建控制器类 UserController.java、StudentController.java、CourseController.java 和 CcourseController.java。

程序 UserController.java 的代码如下。

```
@Controller
public class UserController {
    @Autowired
    private UserService userService;
    @RequestMapping("/login")
    public String login(String username, String password,HttpSession session) {
        User user=userService.login(username,password);
        if(user!=null) {
            session.setAttribute("user", user);
            return "redirect:index.jsp";
        }else {
            return "redirect:login.jsp";
        }
    }
}
```

程序 StudentController.java 的代码如下。

```
@Controller
public class StudentController {
    @Autowired
    private StudentService studentService;
    @RequestMapping("/listStudent")
    public ModelAndView findAllStudent() {
        ModelAndView mv=new ModelAndView();
        List<Student> list=studentService.listStudents();
        mv.addObject("list", list);
        mv.setViewName("studentList");
        //跳转到:WEB-INF/pages/studentList.jsp 页面
        return mv;
    }
    @RequestMapping("/addStudent")
    public String addStudent(Student student) {
        studentService.addStudent(student);
        return "redirect:/listStudent";
    }
    @RequestMapping("/selectStudent/{sid}")
    public ModelAndView selectStudentById(@PathVariable int sid) {
        Student student=studentService.getStudentById(sid);
        ModelAndView mv=new ModelAndView();
        mv.addObject("student", student);
        mv.setViewName("updateStudent");
        return mv;
```

```
    }
    @RequestMapping("/selectStudent/updateStudent")
    public String updateStudent(Student student) {
        studentService.updateStudent(student);
        return "redirect:/listStudent";
    }
    @RequestMapping("/deleteStudent/{sid}")
    public String deleteStudent(@PathVariable int sid) {
        studentService.deleteStudent(sid);
        return "redirect:/listStudent";
    }
}
```

程序 CourseController.java 的代码如下。

```
@Controller
public class CourseController {
    @Autowired
    private CourseService courseService;
    @RequestMapping("/listCourse")
    public ModelAndView  findAllCourse() {
        ModelAndView mv=new ModelAndView();
        List<Course> list=courseService.listCourses();
        mv.addObject("list", list);
        mv.setViewName("courseList");
        return mv;
    }
    @RequestMapping("/addCourse")
    public String addCourse(Course course) {
        courseService.addCourse(course);
        return "redirect:/listCourse";
    }
    @RequestMapping("/selectCourse/{cid}")
    public ModelAndView selectCourseById(@PathVariable int cid) {
        Course course=courseService.getCourseById(cid);
        ModelAndView mv=new ModelAndView();
        mv.addObject("course", course);
        mv.setViewName("updateCourse");
        return mv;
    }
    @RequestMapping("/selectCourse/updateCourse")
    public String updateCourse(Course course) {
        courseService.updateCourse(course);
        return "redirect:/listCourse";
    }
    @RequestMapping("/deleteCourse/{cid}")
    public String deleteCourse(@PathVariable int cid) {
        courseService.deleteCourse(cid);
        return "redirect:/listCourse";
    }
}
```

程序 CcourseController.java 的代码如下。

```
@Controller
public class CcourseController {
    @Autowired
    private StudentService studentService;
    @Autowired
    private CourseService courseService;
    @Autowired
    private CcourseService ccourseService;
    @RequestMapping("/listCcourse")
    public ModelAndView listCcourse() {
        List<Ccourse> list=ccourseService.listCcourses();
        List<Student> studengList=studentService.listStudents();
        List<Course> courseList=courseService.listCourses();
        ModelAndView mv=new ModelAndView();
        mv.addObject("studentList", studengList);
        mv.addObject("courseList", courseList);
        mv.addObject("list", list);
        mv.setViewName("ccourseList");
        return mv;
    }
    @RequestMapping("/addCcourse")
    public String addCcourse(Ccourse ccourse) {
        ccourseService.addCcourse(ccourse);
        return "redirect:/listCcourse";
    }
    @RequestMapping("/selectCcourse/{id}")
    public ModelAndView selectCcourseById(@PathVariable int id) {
        ModelAndView mv=new ModelAndView();
        Ccourse ccourse=ccourseService.getCcourseById(id);
        List<Course> courseList=courseService.listSelect(ccourse.getSid());
        Course course=courseService.getCourseById(ccourse.getCid());
        courseList.add(course);
        mv.addObject("courseList", courseList);
        mv.addObject("ccourse", ccourse);
        mv.setViewName("updateCcourse");
        return mv;
    }
    @RequestMapping("/selectCcourse/updateCcourse")
    public String updateCcourse(Ccourse ccourse) {
        ccourseService.updateCcourse(ccourse);
        return "redirect:/listCcourse";
    }
    @RequestMapping("/deleteCcourse/{id}")
    public String deleteCcourse(@PathVariable int id) {
        ccourseService.deleteCcourse(id);
        return "redirect:/listCcourse";
    }
}
```

7. 编写 JSP 页面文件

在根目录 webapp 下编写 login.jsp 文件和 index.jsp 文件,这两个文件内容与第 14 章综合案例对应文件的内容相同,不再列出。在 WEB-INF 下新建 pages 文件夹,在该文件夹下新建 studentList.jsp、updateStudent.jsp、courseList.jsp、updateCourse.jsp、ccourseList.jsp、updateCcourse.jsp 文件。由于篇幅原因,JSP 文件只放<body></body>之间的代码。

程序 studentList.jsp 代码与第 13 章综合案例的 studentList.jsp 代码内容相同,这里不再列出。

程序 updateStudent.jsp 的代码如下。

```
<body style="width: 600px;margin:0px auto;">
    <div class="panel panel-info" style="margin:0px auto;">
        <div class="panel-heading"><h4>学生管理</h4></div>
        <div class="panel-body">
            <form action="updateStudent" method="post">
                <div class="modal-body">
                    <div class="input-group" style="margin: 5px;display: none;" >
                        <span class="input-group-addon">编号:</span>
                         <input type="text" class="form-control" name="sid"
disabled="disabled" value="${student.sid}">
                         <input type="hidden" class="form-control" name="sid"
value="${student.sid}">
                    </div>
                    <div class="input-group" style="margin: 5px">
                        <span class="input-group-addon">学生学号:</span>
                        <input type="text" class="form-control" name="cno"
value="${student.sno}">
                    </div>
                    <div class="input-group" style="margin: 5px">
                        <span class="input-group-addon">学生姓名:</span>
                        <input type="text" class="form-control" name="cname"
value="${student.sname}">
                    </div>
                    <div class="input-group" style="margin: 5px">
                        <span class="input-group-addon">学生性别:</span>
                        <input type="text" class="form-control" name="sgender"
value="${student.sgender}">
                    </div>
                    <div class="input-group" style="margin: 5px">
                        <span class="input-group-addon">学生年龄:</span>
                         <input type="text" class="form-control" name="sage"
value="${student.sage}">
                    </div>
                </div>
                <div class="modal-footer">
                    <button class="btn btn-primary" type="submit">提交</button>
                    <button type="button" class="btn btn-primary" onclick=
"javascript:history.go(-1)">关闭</button>
```

```
            </div>
        </form>
    </div>
  </div>
</body>
```

程序 courseList.jsp 的代码如下。

```
<body style="width: 800px;margin:0px auto;">
    <div style="margin-left:10px;width:150px;float: left;">
        <h4>课程列表</h4></div>
    <div style="width:200px;float:right;">
        <button type="button" class="btn btn-primary" data-toggle="modal" data
-target="#myModal">添加课程</button>
    </div>
    <div class="modal fade" id="myModal" tabindex="-1" role="dialog" aria-
labelledby="myModalLabel">
        <div class="modal-dialog">
            <div class="modal-content">
                <div class="modal-header">
                    <h4 class="modal-title">输入课程信息</h4>
                    <button data-dismiss="modal" class="close" type="button">
                         <span aria-hidden="true">×</span><span class="sr-
only">Close</span>
                    </button>
                </div>
                <form action="addCourse" method="post">
                    <div class="modal-body">
                        <div class="input-group" style="margin: 5px">
                            <span class="input-group-addon">课程编码:</span>
                            <input type="text" class="form-control" name="cno">
                        </div>
                        <div class="input-group" style="margin: 5px">
                            <span class="input-group-addon">课程名称:</span>
                            <input type="text" class="form-control" name="cname">
                        </div>
                    </div>
                    <div class="modal-footer">
                        <button class="btn btn-primary" type="submit">提交
</button>
                    </div>
                </form>
            </div>
        </div>
    </div>
    <table id="mytb" class="table table-hover table-condensed">
        <thead>
            <tr>
                <th>编号</th>
```

```
                <th>课程编码</th>
                <th>课程名称</th>
                <th>操作</th>
            </tr>
        </thead>
        <tbody>
            <c:forEach items="${list}" var="course">
                <tr>
                    <td>${course.cid }</td>
                    <td>${course.cno }</td>
                    <td>${course.cname }</td>
                    <td>
                        <a href="selectCourse/${course.cid}">
                            <button type="button" class="btn btn-primary btn-
sm">修改</button>
                        </a>
                        <a href="deleteCourse/${course.cid}">
                            <button type="button" class="btn btn-primary btn-
sm" onclick="return confirm('确认要删除吗?')">删除 </button>
                        </a>
                    </td>
                </tr>
            </c:forEach>
        </tbody>
    </table>
</body>
```

程序 updateCourse.jsp 的代码如下。

```
<body style="width: 600px;margin:0px auto;">
    <div class="panel panel-info" style="margin:0px auto;">
        <div class="panel-heading"><h4>课程管理</h4></div>
        <div class="panel-body">
            <form action="updateCourse" method="post">
                <div class="modal-body">
                    <div class="input-group" style="margin: 5px;">
                        <span class="input-group-addon">课程序号:</span>
                          <input type="text" class="form-control" name="cid"
disabled="disabled" value="${course.cid}">
                          <input type="hidden" class="form-control" name="cid"
value="${course.cid}">
                    </div>
                    <div class="input-group" style="margin: 5px">
                        <span class="input-group-addon">课程编号:</span>
                          <input type="text" class="form-control" name="cno"
value="${course.cno}">
                    </div>
                    <div class="input-group" style="margin: 5px">
                        <span class="input-group-addon">课程名称:</span>
```

```
                <input type="text" class="form-control" name="cname"
value="${course.cname}">
              </div>
            </div>
            <div class="modal-footer">
              <button class="btn btn-primary" type="submit">提交</button>
              <button type="button" class="btn btn-primary" onclick=
"javascript:history.go(-1)">关闭</button>
            </div>
          </form>
        </div>
    </div>
</body>
```

程序 ccourseList.jsp 的代码如下：

```
<body style="width: 800px;margin:0px auto;">
    <div style="margin-left:10px;width:150px;float: left;">
        <h4>选课列表</h4></div>
    <div style="width:200px;float:right;">
        <button type="button" class="btn btn-primary" data-toggle="modal" data
-target="#myModal">添加选课</button>
    </div>
    <div class="modal fade" id="myModal" tabindex="-1" role="dialog" aria-
labelledby="myModalLabel">
        <div class="modal-dialog">
            <div class="modal-content">
                <div class="modal-header">
                    <h4 class="modal-title">输入选课信息</h4>
                    <button data-dismiss="modal" class="close" type="button">
                        <span aria-hidden="true">×</span><span class="sr-
only">Close</span>
                    </button>
                </div>
                <form action="addCcourse" method="post">
                    <div class="modal-body">
                        <div class="input-group" style="margin: 5px">
                            <span class="input-group-btn">学生:</span>
                            <select name="sid" style="width: 80%;">
                                <option selected>请选择</option>
                                <c:forEach items="${studentList}" var="student" >
                                    <option value="${student.sid }">${student.
sname }</option>
                                </c:forEach>
                            </select>
                        </div>
                        <div class="input-group" style="margin: 5px">
                            <span class="input-group-btn">课程:</span>
                            <select name="cid" style="width: 80%;">
```

```
                        <option selected>请选择</option>
                        <c:forEach items="${courseList}" var="course" >
                            <option value="${course.cid }">${course.
cname }</option>
                        </c:forEach>
                    </select>
                </div>
            </div>
            <div class="modal-footer">
                <button class="btn btn-primary" type="submit">提交
</button>
            </div>
        </form>
    </div>
</div>
</div>
<table id="mytb" class="table table-hover table-condensed" >
    <thead>
        <tr>
            <th>学号</th>
            <th>学生姓名</th>
            <th>课程名称</th>
            <th>操作</th>
        </tr>
    </thead>
    <tbody>
        <c:forEach items="${list}" var="ccourse">
            <tr>
                <td>${ccourse.sid }</td>
                <td>${ccourse.sname }</td>
                <td>${ccourse.cname }</td>
                <td>
                    <a href="selectCcourse/${ccourse.id }">
                        <button type="button" class="btn btn-primary btn-
sm">修改</button>
                    </a>
                    <a href="deleteCcourse/${ccourse.id }">
                        <button type="button" class="btn btn-primary btn-
sm" onclick="return confirm('确认要删除吗?')">删除 </button>
                    </a>
                </td>
            </tr>
        </c:forEach>
    </tbody>
</table>
</body>
```

程序 updateCcourse.jsp 的代码如下：

```
<body style="width: 600px;margin:0px auto;">
```

```
    <div class="panel panel-info" style="margin:0px auto;">
        <div class="panel-heading"><h4>选课管理</h4></div>
        <div class="panel-body">
            <form action="updateCcourse" method="post">
                <div class="modal-body">
                    <div class="input-group" style="margin: 5px">
                        <input type="hidden" class="form-control" name="id"
value="${ccourse.id}">
                    </div>
                    <div class="input-group" style="margin: 5px">
                        <span class="input-group-addon" style="margin-top:
7px;">学生:</span>
                        <select name="sid" style="width: 93.4%;height: 36px;">
                            <option value="${ccourse.sid }">${ccourse.sname }
</option>
                        </select>
                    </div>
                    <div class="input-group" style="margin: 5px">
                        <span class="input-group-addon" style="margin-top:
7px;">课程:</span>
                        <select name="cid" style="width: 93.4%;height: 36px;">
                            <c:forEach items="${courseList}" var="course" >
                                <option value="${course.cid}" <c:if test=
"${course.cid == ccourse.cid}">selected</c:if>>${course.cname }</option>
                            </c:forEach>
                        </select>
                    </div>
                </div>
                <div class="modal-footer">
                    <button class="btn btn-primary" type="submit">提交</button>
                    <button type="button" class="btn btn-primary" onclick=
"javascript:history.go(-1)">关闭</button>
                </div>
            </form>
        </div>
    </div>
</body>
```

16.4 习题

1. 选择题

(1) 在下列选项中，不需要配置在 web.xml 中的是(　　)。

A. Spring 的监听器　　B. 编码过滤器

C. 视图解析器　　D. 前端控制器

(2) 在下列选项中，属于 Spring MVC 所必需的 Jar 包的是(　　)。

A. spring-web-4.3.6.RELEASE.jar

B. spring-webmvc-portlet-4.3.6.RELEASE.jar

C. spring-webmvc-4.3.6.RELEASE-javadoc.jar

D. spring-websocket-4.3.6.RELEASE.jar

(3) 在 MyBatis+Spring 的项目中,以下有关事务的相关说法正确的是(　　)。

A. 在 MyBatiS+Spring 的项目中,事务是由 MyBatis 来管理的

B. 在项目中,数据访问层既是处理业务的地方,又是管理数据库事务的地方

C. 进行注解开发时,需要在配置文件中配置事务管理器,并开启事务注解

D. 进行注解开发时,需要使用@Transactional 注解来标识表现层中的类

(4) 以下不属于 MapperScannerConfigurer 类,在 Spring 配置文件中使用时需要配置的属性是(　　)。

A. basePackage　　　　B. annotationClass

C. sqlSessionFactoryBeanName　　　　D. mapperInterface

(5) Spring 与 MyBatis 整合,下列说法错误的是(　　)。

A. MyBatis-Spring 的 jar 包由 Spring 提供

B. 可以不用给出 Dao 的实现类,而由映射接口实现

C. MyBatis 的配置文件可以写在 MyBatis 本身的配置文件中,也可以在 Spring 中指定

D. Spring 和 yBatis 集成后同样可以使用 XL 配置声明式事务

(6) 下面选项中,不属于整合 SSM 框架所编写的配置文件的是(　　)。

A. db.properties　　　　B. applicationContext.xml

C. mybatis-config.xml　　　　D. struts.xml

2. 填空题

(1) SSM 框架整合主要是________的整合,以及________的整合。

(2) 为了避免 Spring 配置文件中的信息过于臃肿,通常会将 Spring 配置文件中的信息按照________分散在多个配置文件中。

(3) @Transactional 注解主要是针对数据的________、________、________进行事务管理。

(4) ________是 MyBatis-Spring 团队提供的一个用于根据 Mapper 接口生成 Mapper 对象的类。

(5) 进行 Spring 与 MyBatis 整合时,Spring 框架需要准备的 JAR 包共 10 个,其中包括 4 个核心模块 JAR,AOP 开发使用的 JAR,________和事务的 JAR。

(6) 在 MyBatis+Spring 的项目中,事务由________来管理。

(7) Spring 与 MyBatis 框架整合时,可以通过 Spring ________,然后调用实例对象中的查询方法来执行 MyBatis 映射文件中的________,如果能够正确查询出数据库中的数据,就可以认为 Spring 与 MyBatis 框架整合成功。

3. 判断题

(1) 在实际的项目开发中,Spring 与 MyBatis 都是整合在一起使用的。　　(　　)

（2）@Autowired 注解需要标注在 Service 层的实现类上，这样才能实现依赖注入。（　）

（3）@Transactional 注解主要是针对数据的增加、修改、删除和查询进行事务管理。（　）

（4）Spring 与 Spring MVC、Spring MVC 与 MyBatis 需要相互整合。（　）

（5）在 Spring MVC 的配置文件中，视图解析器是必须配置的。（　）

4. 简答题

（1）请简述 SSM 框架整合思路。

（2）Spring 框架与 MyBatis 框架在整合时需要配置的主要步骤是什么？